LEARNING RESOURCES CTR/NEW ENGLAND TECH.
3 0147 0000 8346 2

Behavioral Pediatrics
Research and Practice

Behavioral Pediatrics

Research and Practice

Edited by Dennis C. Russo

Children's Hospital Medical Center
Harvard Medical School
Boston, Massachusetts

and James W. Varni

Orthopaedic Hospital
University of Southern California School of Medicine
Los Angeles, California

PLENUM PRESS • NEW YORK AND LONDON

Library of Congress Cataloging in Publication Data

Main entry under title:

Behavioral pediatrics.

Includes bibliographical references and indexes.
1. Pediatrics—Psychosomatic aspects. 2. Pediatrics—Psychological aspects. I. Russo, Dennis C., 1950- . II. Varni, James W., 1948- . [DNLM: 1. Pediatrics. 2. Child behavior. WS 100 B419]
RJ47.5.B435 618.92'9 82-3799
ISBN 0-306-40961-5 AACR2

A Division of Plenum Publishing Corporation
233 Spring Street, New York, N.Y. 10013

Printed in the United States of America

To our parents

Charles and Helen Russo

James and Theresa Varni

Contributors

Janice E. Abernathy, Platte Medical Clinic, Platte City, Missouri

Bruce L. Bird, University of New Orleans; Associated Catholic Charities of New Orleans, New Orleans, Louisiana

Michael F. Cataldo, John F. Kennedy Institute; Johns Hopkins University School of Medicine, Baltimore, Maryland

Hyman Chai, National Asthma Center, Denver, Colorado

Edward R. Christophersen, Department of Pediatrics, University of Kansas Medical Center, Kansas City, Kansas; Bureau of Child Research, University of Kansas, Lawrence, Kansas

Thomas J. Coates, Johns Hopkins University School of Medicine, Baltimore, Maryland

Thomas L. Creer, Ohio University, Athens, Ohio

Jerry Dash, Childrens Hospital of Los Angeles; University of Southern California School of Medicine, Los Angeles, California

Shirley Graves, University of Florida School of Medicine, Gainesville, Florida

Harvey E. Jacobs, Rehabilitation Medical Center, Los Angeles, California

William R. Jankel, John F. Kennedy Institute; Johns Hopkins University School of Medicine, Baltimore, Maryland

Ernest R. Katz, Childrens Hospital of Los Angeles; University of Southern California School of Medicine, Los Angeles, California

Jonathan Kellerman, Childrens Hospital of Los Angeles; University of Southern California School of Medicine, Los Angeles, California

Bruce J. Masek, Children's Hospital Medical Center; Harvard Medical School, Boston, Massachusetts

Barbara G. Melamed, University of Florida, Gainesville, Florida

Lynn Parker, Children's Hospital of New Orleans, New Orleans, Louisiana

Charles M. Renne, National Asthma Center, Denver, Colorado

Rochelle L. Robbins, Cornell University Medical Center, New York, New York

Mark C. Rogers, Johns Hopkins University School of Medicine, Baltimore, Maryland

Dennis C. Russo, Children's Hospital Medical Center; Harvard Medical School, Boston, Massachusetts

James W. Varni, Orthopaedic Hospital; University of Southern California School of Medicine, Los Angeles, California

William Whitehead, Gerontology Research Center; Johns Hopkins University School of Medicine, Baltimore, Maryland

Preface

As in any field of clinical or scientific endeavor, a cataloguing of the techniques and findings of behavioral pediatrics must follow shortly after the first few years of systematic work. This book has been designed to serve this initial summary function for the field. It represents a first attempt to bring together in one place the definition and scope of behavioral pediatrics and to outline current research and treatment approaches to various organic disorders, clinical settings, and problem areas in which a sufficient body of knowledge has accrued for authoritative statement.

As the first text in a rapidly expanding area, our decisions regarding the topics to be covered and the contributors to this volume were guided by our desire to represent what we would consider to exemplify the field in its early development: pragmatic and thorough study of significant problems from a base of sound scientific inquiry. Each of the topics addressed in the present volume develops, in a preliminary fashion, an epistemology for current practice and future study. All of the contributing authors have been involved in the development of their specialty areas through their research and, importantly, through their clinical work in the hospitals affiliated with the medical schools in which they hold appointments.

To some, the text may appear rather capricious in its selection of topics for coverage. For example, although chapters on neurological and cardiovascular disorders are included, equally important bodily systems such as the endocrine and immune systems are not. The decision to include a topic in this work was guided less by the importance of the particular bodily system or problem than by the fact that an established body of knowledge in behavioral pediatrics was currently available. Future editions of the text will probably highlight other areas of work as these receive sufficient study.

In addition to its utility as a compendium of existent knowledge, the present work has been designed as a working definition of the scope and focus of behavioral pediatrics. As will be outlined in the first chapter, historical roots, current economic and governmental priorities, and theoretical perspective suggest a number of directions for possible development. We have attempted to distill from these a working definition and procedural guidelines for practice and research.

We have designed the text to provide background on both the medical and the psychological aspects of the topics covered based on our belief that unilateral intervention is not only limiting but perhaps dangerous. At best, unilateral intervention does not consider all of the variables available to collaborative endeavor. It is not surprising, therefore, that the present text includes medical as well as psychological contributors in both the present chapters and in the published work of the authors.

We wish to thank all of the authors for their patience with our editing, correspondence, and the quirks of protocol. Each has been more than patient and has produced a statement of value in defining an area of expertise for the emerging field.

In this work we have received much assistance from our colleagues, in both the initial formulation of this text and in our subsequent deliberation over detail. We are particularly appreciative of the editorial efforts of Tori Bognanni in the development of the text and in the preparation of this manuscript. Lastly, we wish to thank our respective medical faculties for their assistance and support and, most importantly, for their substantial aid in shaping the philosophies of this text.

DENNIS C. RUSSO
JAMES W. VARNI

Contents

Introduction

During the past five years, there has been a veritable explosion of interest in the clinical and scientific applications of behavioral science in medicine. A number of factors are presently aggregated in medicine, psychology, and government which support these efforts and suggest that studies of behavior/disease interactions and their clinical modification accompany shifts in the focus of medical education and treatment as well as national health policy. There is increasing evidence that the epidemiology of disease in North America and Europe has changed, challenging health care specialists to deal with problems inherent in chronic disease. Chronicity calls forward the development of methodologies for dealing with the patient as well as with the disease. The necessity for assisting patients to cope with potentially lifelong conditions in which stress, altered social environment, and frequent and protracted hospitalization influence successful management has prompted the revision of existing health care practices.

In light of the increasing costs of health care and the focus of such care on prevention and symptom management, a number of specific events foreshadow potential alterations in health care models: (1) the establishment of a Society of Behavioral Medicine and the Academy of Behavioral Medicine Research; (2) the creation of a new division (Division 38) of the American Psychological Association on Health Psychology; (3) the redefinition of the field of psychosomatic medicine toward a biopsychosocial model; (4) the development of a subspecialty in behavioral pediatrics by the American Board of Pediatrics; (5) the creation of a behavioral medicine study section within the National Institutes of

Health and the development of specific Behavioral Medicine Branches within several of the Institutes (NHLBI, NCI); and (6) the creation of new journals such as the *Journal of Behavioral Medicine, Behavioral Medicine Abstracts,* and the *Journal of Pediatric Psychology* as specific repositories for basic and applied research studies of behavioral and medical interactions. Supplementing these national developments, the increasing number of new appointments of behavioral psychologists to the faculties of American medical schools, the creation of multidisciplinary departments for the study and practice of behavioral medicine at many of these sites, and the rapid publication of a number of books outlining applications and findings strongly indicate the potential and future of behavioral medicine within medical practice in the 1980s.

In this regard, therefore, the present volume begins by tracing the development of behavioral pediatrics from roots in medicine, behavioral medicine, and health psychology. A working definition which represents both history and present effort is proposed to support continuing work. Although actions and works will produce a lasting definition, a preliminary effort at assessing the scope of the field appears appropriate at this time.

In the second section of this text, four specific chronic disease areas are reviewed. These areas were selected because significant developmental research has already been undertaken which will assist in the empirical direction of future efforts.

Pain and the stresses of hospitalization are highlighted in the third section of the book. These problems pervade much of current pediatric practice and deserve consideration apart from diseases or conditions. The fourth section evaluates the different concerns posed by the environments of pediatric medicine. The inpatient and ambulatory settings each require differing models of care and pose unique problems in patient management. Finally, the last section addresses two critical issues: the prevention of disease and adherence to prescribed treatments, areas of major potential in behavioral pediatrics.

This triad—the disease, the environment, and the patient—should form the basis for all intervention in behavioral pediatrics. To consider only one factor without its complements is to care inadequately for the child. The development of behavioral pediatrics is as much tied to this fact as to its methodology, techniques, or research. In today's medicine, a primary goal must be increased quality of life, adaptation, and appropriate learning for the pediatric patient.

DENNIS C. RUSSO
JAMES W. VARNI

I
INTRODUCTION TO BEHAVIORAL PEDIATRICS

1
Behavioral Pediatrics

Dennis C. Russo and James W. Varni

During the past five years there has been an accelerating interest in the clinical and scientific application of behavioral science to medicine. Although historically these two areas have held little affinity for one another, they have become increasingly merged in an effort to improve health care. A number of writers (Cataldo, Russo, Bird, & Varni, 1980; Ferguson & Taylor, 1980; Matarazzo, 1980; Pomerleau, 1979; Schwartz & Weiss, 1978; Varni & Russo, 1980) have attempted to provide explanation for this tremendous diversion of talent and energy to a previously little-studied area. As this chapter progresses, we shall review these attempts to place into perspective the behavioral medicine movement, their roots in previous theory, and the history of attempts to contend with the purported synergistic effects of psychological factors and disease.

Although efforts have been quite visible and have produced a proliferation of books, journals, conferences, and the like, little of the force of the behavioral medicine movement has yet been felt by pediatrics. This is quite ironic for, as we shall see, there exists within pediatrics the potential for great receptivity to behavioral medicine techniques. The existent literature has focused largely on the diseases and health risks of adults. In most texts, pediatric applications are quite limited. Although they are still in their early stages, these efforts in behavioral pediatrics

DENNIS C. RUSSO • Children's Hospital Medical Center; Harvard Medical School, Boston, Massachusetts 02115. JAMES W. VARNI • Orthopaedic Hospital; University of Southern California School of Medicine, Los Angeles, California 90007.

hold considerable promise. This book represents the first cohesive effort to bring together existing knowledge regarding pediatric effort in behavioral medicine and to define, from a broad perspective, the extent of the ongoing work in the field.

As a precursor to the data-based and topical chapters to follow, this introduction shall therefore focus on a systematic assessment of the current interest in behavioral pediatrics, both historically and demographically; evaluate the similarities and differences between behavioral pediatrics and other areas such as ambulatory, community, developmental, or well-child approaches in pediatrics; and assess the methodological base for clinical inquiry and research.

HISTORICAL AND DEMOGRAPHIC FACTORS

The history of attempts to evaluate the interrelationships between the mind and the body and their relative contributions to health and well-being, as well as to disease, extends back to the earliest recorded works. In medicine, such controversy has been the topic of debate since before the time of Galen and Descartes. For centuries noted thinkers, and more recently experimental scientists, have attempted to understand the similarities and differences, the cross-connections, and the unique contributions of the physical, mental, and spiritual dimensions. A fundamental dualism of mind and body has never been experimentally determined, however. Although different theories have been proposed to account for observed phenomena, their frequent unavailability for experimental testing has left many questions unanswered. Of considerable concern, such theoretical speculation has for years been the basis of clinical practice! Such deductive theorizing has been quite unsatisfying in nonmedical applications (Russo & Newsom, 1982), and its history in solving critical problems in disease and patient management has been equally unsatisfactory (Agras, 1975; Pomerleau, 1979).

The basis for the theoretical speculation is quite clear: there are myriad anecdotal examples of the behavior of individuals or groups apparently affecting health or illness. The observation of events of this nature has occurred with surprising frequency and yet has produced little in the way of improved management for at-risk or diseased populations. Freud (1905) chose to address this issue phenomenologically and presents an early observation of the basic problem:

> Persistent affective states of a . . . "depressive" nature (as they are called), such as sorrow, worry, or grief, reduce the state of nourishment of the whole body, cause the hair to turn white, the fat to disappear and the walls of the

> blood-vessels to undergo morbid changes. The major affects evidently have a large bearing on the capacity to resist infectious illness; a good example of this is to be seen in the medical observation that there is a far greater liability to contract such diseases as typhus and dysentery in defeated armies than in victorious ones. The affects, moreover—this applies almost exclusively to depressive affects—are often sufficient in themselves to bring about both diseases of the nervous system accompanied by manifest anatomical changes and also diseases of other organs . . . . There can be no doubt that the duration of life can be appreciably shortened by depressed affects.

The clarity of this observation does not change the relative empirical inaccessibility of depressive affects or other theoretical constructs to determine experimentally potential cause–effect relationships needed for clinical practice.

At present, little empirical data exist on mind–body interactions, although increasing scientific activity has occurred in recent years (Eisenberg, 1979; Levine, Gordon, & Fields, 1978; Snyder, 1977). Suffice it to say that we have yet to come much further than Galen or Descartes, or Freud, in analyzing a theoretical dualism or in applying derived hypotheses to clinical problems with much predictive value.

Why, then, the flourish and renewed interest which has accompanied the merging of medical and behavioral science methods? What are the factors which make behavioral medicine, or pediatric approaches, more likely to be successful than their predecessors? To understand more completely the differences and potentials which exist, it is necessary to review the factors supporting these endeavors and to evaluate the other previous attempts to access similar issues.

As we shall hope to show, what holds the clearest promise and what sets behavioral medicine apart from its predecessors is that, for the first time, significant *empirical methodology* is now available to study health-related behaviors and to complement the existing array of biomedical assessment and treatment techniques.

FACTORS SUPPORTING BEHAVIORAL MEDICINE

Simultaneous with methodological acumen have come a number of other developments in medicine and government which further suggest the timeliness of a systematic integration between the behavioral and medical sciences. The parallel developments in several fields have begun to merge, contiguous with social and epidemiological issues, to support the current efforts to formalize a new field of inquiry.

Primary among the issues is the simple fact that the epidemiology of disease has changed radically over the past half century (Cataldo *et al.*, 1980; Dubos, 1965; Schneiderman, Weiss, & Engel, 1979). Advances in science, sanitation, technology, medicine, and pharmacology have reduced or eliminated much of the threat of acute or infectious diseases such as cholera, typhus, and smallpox. These diseases, through wave after wave of epidemic, were the major killers of adults and children prior to the twentieth century. This success at reducing the significance of bacterial and infectious diseases has simultaneously produced improved actuarial longevity and increased awareness of heart disease, cancer, and other forms of chronic illness.

Much of the focus of current medical treatment has, therefore, shifted as well. Treatment of these disorders is often aimed at symptom reduction and the arrest of pathology, since the pathogenesis of many of these chronic disorders is not yet well understood or, in many cases, may not yet be therapeutically accessible. As Meenan has said (Meenan, Yelin, Nevitt, & Epstein, 1981):

> Barring significant basic research breakthroughs, we are probably at a point of diminishing returns in the treatment of most chronic diseases, since we have reached the stage where additional medical or surgical therapy is apt to produce progressively smaller improvements in individual health status.

Further, Culliton (1978) has remarked that despite all of the advances in medical technology and the costliness of care, little or no improvement in average longevity has been derived in the past decade. Although the maximum life span has not increased, the average age of initial infirmity may be raised through changes in lifestyle resulting in a more rectangular morbidity curve and extended vigor farther into a fixed life span (Fries, 1980). Such observations clearly call for the input of the behavioral sciences to medicine (Breslow, 1978; Fabrega & Van Egeren, 1976; Hamburg & Brown, 1978) in prevention (Masek, Epstein, & Russo, 1981), chronic care (Cataldo *et al.*, 1980; Varni & Russo, 1980), and rehabilitation (Cataldo & Russo, 1979; Ince, 1976).

As reviewed by a number of authors (Blanchard, 1977; Cataldo *et al.*, 1980; Matarazzo, 1980; Pomerleau, 1979; Pomerleau, Bass, & Crown, 1975; Russo, Bird, & Masek, 1980; Schwartz & Weiss, 1978; Varni & Russo, 1980), a host of other factors are presently aggregated, in addition to altered epidemiology, improved behavioral rigor, and the decreased benefit of new medical technologies, which are also responsible for the development of behavioral medicine. Factors such as high cost (Culliton, 1978), changes in the conduct of medical education (Engel, 1977), dissatisfaction with traditional psychiatric and psychosomatic explanations about the patient with disease, and altered governmental funding priorities have combined with new technologies both in behavioral and biochemical areas to produce the establishment of interdisciplinary professional groups on a national and international level.

While the meteoric growth of the field thus far and its political and academic roots are fairly clear, the scope and definition of what constitutes behavioral medicine, both clinically and scientifically, are not: a clear definition of behavioral medicine is needed as a precursor to any attempt to define or discuss behavioral pediatrics.

THE DEFINITION AND SCOPE OF BEHAVIORAL MEDICINE

The starting points of behavioral medicine are quite unclear. As in many other areas which have evolved, it appears difficult to identify a specific moment when behavioral medicine came to exist. The organized movement dates back to the late 1970s, although Blanchard (1977) refers to Birk's (1973) book *Behavioral Medicine: Biofeedback* as an early use of the term. Generally, significant organizing activity occurring in the years 1977 and 1978 marks the formal beginning of the behavioral medicine effort. During these years, the formation of the Society of Behavioral Medicine and the Academy of Behavioral Medicine Research grew out of the 1977 Yale Conference on Behavioral Medicine (Schwartz & Weiss, 1978). The first comprehensive texts (e.g., Williams & Gentry, 1977) were published and the formation of the *Journal of Behavioral Medicine* was in its developmental stages.

Within a relatively brief period of time a variety of centers and programs were created to house the new field. With enthusiasm and new ideas came the attempts to define the zeitgeist: definitions still actively discussed several years later as data are collected, clinics developed, and patients treated.

These varying ideas regarding the development of behavioral medicine reflect differences inherent in the philosophies of behavioral practitioners in general. The primacy of behavioral techniques, the proper area of work and study, the scope of interdisciplinary collaboration, and the relative focus on data and assessment are current areas in which behavioral medicine's role is less than clearly defined. Over the past several years distinct trends can be seen in both practice and definition.

Following Miller's studies (1975) on the self-regulation of physiological functions and Birk's (1973) equation of behavioral medicine and biofeedback, the field, at its birth, might well have been limited to physiological conditioning or biofeedback techniques. As a number of behavioral medicine centers developed at American medical schools such as Pennsylvania, Stanford, and Johns Hopkins, it became apparent that such a definition was unrepresentative of practice.

The Yale Conference on Behavioral Medicine (Schwartz & Weiss, 1978), in bringing together a larger, more varied group of scientists and practitioners, provided a more balanced and general definition of the field:

> Behavioral Medicine is the field concerned with the development of behavioral science knowledge and techniques relevant to understanding physical health and illness and the application of this knowledge and these techniques to prevention, diagnosis, treatment, and rehabilitation. Psychosis, neurosis, and substance abuse are included only insofar as they contribute to physical disorders as an end point.

In this definition both the broad base of inquiry and the interdisciplinary nature of behavioral medicine are defined. The context is also set: clinical and research efforts within the mainstream of medical efforts.

Similarly, the Academy of Behavioral Medicine Research, a more research-oriented group, chose a definition slightly briefer but along the same lines as that of the Yale Conference:

> Behavioral Medicine is the interdisciplinary field concerned with the development and integration of behavioral and biomedical science knowledge and techniques relevant to the understanding of health and illness and the application of this knowledge and these techniques to prevention, diagnosis, treatment, and rehabilitation.

Both definitions presuppose that behavioral medicine researchers and clinicians irrespective of their field of training bring a science base and set of methods for integration into the developing field. Rather than being solely technique-oriented, the field, as embodied in these definitions, provides outlets for theoretical and basic research as well. What is also inherent in these definitions is their insistence on intellectual integration and practical collaboration between previously uninvolved fields.

Not all attempts at defining behavioral medicine are so broadly based. Pomerleau (1979) proposed a more strictly behaviorally-centered view of behavioral medicine as

> (a) the clinical use of techniques derived from the experimental analysis of behavior—behavior therapy and behavior modification—for the prevention, management or treatment of physical disease or physiological dysfunction; and (b) the conduct of research contributing to the functional analysis and understanding of behavior associated with medical disorders and problems in health care.

Matarazzo (1980), however, took exception to such a view. He suggested that behavioral medicine as an endeavor be separated from work in the areas of behavioral health and health psychology. In this scheme, Matarazzo proposed that the term *behavioral medicine* represent the interdisciplinary field concerned with health and disease or related dysfunction such as obesity and hypertension; the term *behavioral health* define an interdisciplinary subspecialty within behavioral medicine concerned with maintenance of health and prevention of disease in healthy individuals; while the term *health psychology* be conceptualized as a discipline-specific field comprising psychology's contribution to behavioral medicine and health.

Each of these definitions brings out the particular biases and concerns of the authors or group producing the definition. What appears abundantly clear is that, in the long run, what will best define behavioral medicine is its practice, methods, and accomplishments. The initial guidelines have focused on general function and overall mission. In the present text, our definition will be more carefully rooted in the procedures and science base for behavioral pediatrics. As we shall see, in both nontraditional medicine and pediatrics trends and approaches coexist which tend to overlap with such broadly based explanatory definitions.

Psychological Medicine

During the twentieth century, medical investigations on the somatic aspects of health and disease have flourished. Advances in science and access to new technologies for assessing disease and biochemical parameters have resulted in steady progress in understanding, curing, or managing disease. On the other hand, similar investigative progress has not been evident for the psychological aspects of health and disease.

Psychological medicine has gone by many names: psychosomatic medicine, liaison psychiatry, medical psychology, preventive medicine, health education, health enhancement, holistic medicine, and others. Each of the approaches has been claimed to be better than its predecessors and more comprehensive than its contemporaries, as we shall suggest may be the case for behavioral medicine and behavioral pediatrics. What has been a uniting thread running through all of the techniques is that there have been *few scientifically verifiable successes* among the approaches: a phenomenon we hope will not be replicated by behavioral medicine and behavioral pediatrics. A number of reasons may be responsible if behavioral medicine and behavioral pediatrics are to be successful, but at the present time two appear to be especially salient: (1) the availability and rigorous application of empirical assessment tools and (2) a basis in process and outcome data analysis rather than theory.

Much of psychological medicine, particularly psychosomatic medicine, is based upon conceptual and theoretical speculation regarding aberrant psychological function of the patient with disease or physical symptoms. For example, psychosomatic medicine has traditionally been less concerned with factors such as noncompliance and health care behaviors, than with psychodynamic etiology or contributions thereof to physical diseases such as asthma or rheumatoid arthritis. (Alexander, 1950). Such psychodynamic interpretations have been particularly prevalent when the medical explanations for the etiology of the disease process are lacking. This "mental disease" approach has been overlaid on physical illness, with its focus on a psychodynamic etiology tending to minimize the active learning process of the patient with disease. This model has been rigorously held to despite the fact that the more recent psychological literature stresses brief supportive therapy as opposed to long-term in-depth analysis. A reconceptualization is needed to account for the currently available data on the relationships between health behaviors and disease manifestations and systematically to include the patient with physical disease as an active participant in his or her own treatment.

An Alternative Model

We should like to propose that the child with disease be viewed on an NPAS model: *normal person—abnormal situation.* Such an approach assumes that many individuals who are normally able to cope at least reasonably well with daily events find themselves with reoccurring symptoms of a disease requiring hospitalization, surgery, and other ongoing treatments. The families of the children themselves often fit such a model as well. The normal process of parenting is made more difficult or impossible by the chronic disease of the child.

The NPAS model adds logic and suggests predictability of the problems of children with chronic illness and their families. It need not apply to every case. It is clear that in some portion of the patient population other issues of a psychological nature may be present. The purpose of work within a behavioral framework is not to ignore these factors, nor to assume them prepotent. Certainly, a number of children who develop chronic disease had adjustment difficulties before their illness and in a certain proportion of families marital difficulties, financial issues, and other factors existed which may work to subvert treatment. Nevertheless, it is our contention that in the main, children with chronic diseases are a psychologically normal population.

A number of recent studies have demonstrated that for children with life-threatening or chronic diseases, the overall pattern is one of psychological normalcy rather than deviancy (Bedell, Giordani, Amour, Tavormina, & Boll, 1977; Kellerman, Zeltzer, Ellenberg, Dash, & Rigler, 1980; Tavormina, Kastner, Slater, & Watt, 1976; Zeltzer, Kellerman, Ellenberg, Dash, & Rigler, 1980). Rather than finding psychopathology, these studies found that children with life-threatening or chronic diseases are faced with acute or chronic stress situations which bring about major changes in daily living, requiring normal adaptive or coping mechanisms. When these results are combined with work by Fordyce (1976) and others looking at the ecology of pain and illness behaviors as well as an examination of the learning characteristics of the hospital environment (cf. Chapter 8), then it becomes increasingly apparent that childhood chronic disease and its treatment may provide numerous opportunities for the learning of aberrant behavior (Varni, Bessman, Russo, & Cataldo, 1980).

The medical environment and its practices are themselves, therefore, a target for study and modification. Hospitals were created for acutely ill individuals. Fifty years ago, the patient population of a typical hospital was quite different from what it is today. Care practices have changed as surgeons, physicians, nurses, and others have attempted to keep pace with rapidly changing patient needs. It is questionable as to whether, despite good intentions, the environment of care has advanced as rapidly as medical science.

It appears clear, therefore, that conceptualizations of maladaption which place responsibility largely on the patient's weaknesses and which do not consider the learning characteristics of the disease, the home, or the hospital environment are inadequate to explain, predict, or manage the problems of chronicity faced by patients. What behavioral medicine and behavioral pediatrics offer lies in their focus on learning and teaching, their emphasis on skills training rather than the etiology of skills deficits, and their empirical assessment of treatment process and outcome.

BEHAVIORAL PEDIATRICS

As we have previously reviewed, considerable disagreement exists as to the proper focus, definition, and conduct of behavioral medicine. Similar difficulties arise when one attempts to review nonbehavioral efforts in psychological medicine. It would be unreasonable to assume that by focusing specifically on behavioral pediatrics any of these ambiguities would be resolved. It is the case, in fact, that most speculation in psychiatry and psychological thought has followed the pattern of adult formulation preceding work on the child (Russo & Newsom, 1982). This has been the case for behavioral efforts in medicine as well. For this reason, and several others, the following sections trace issues specific to pediatric practice, through its history and current status, to arrive at a working definition and description of the field.

The History of Pediatrics

Since 1900, several waves of revolution and refinement have influenced the practice of pediatric medicine (Kenney & Clemmens, 1975). Although these revolutions cannot be wholly subsumed in the pediatric sphere, children may have benefited more than adults from many of the developments.

In the late 1800s and the early years of the twentieth century, a most important revolution occurred in the philosophy of medicine. The method by which medicine was taught changed from an apprenticeship to a science-based curriculum model (Engel, 1977). This method was to be a primary factor responsible for the meteoric rise of scientific medicine during the past 80 years. At Johns Hopkins, with other United States medical schools following, the turn of the century brought forth this basic science core curriculum. Early training in chemistry, biology, and anatomy preceded involved clinical work. This focus on methods of science as opposed to the mechanics of practice served as the major foundation block for contemporary medicine. This science base has been followed developmentally by a number of important phases in pediatric medicine.

The 1900s brought forward an era of *descriptive pediatrics.* During this time, improved observational techniques and methods of quantification resulted in a more precise description and classification of diseases and a charting of the natural course of pediatric illnesses (Kenney & Clemmens, 1975).

Improved classification led to improved treatment during the 1920s in what one pediatrician (Levine, 1960) has termed the "golden age" of *curative pediatrics.* Improved methods for the evaluation of etiology of illness, new laboratory methods for evaluating biotic diseases, and the development of antibiotic agents served greatly to reduce mortality from infectious diseases of childhood. Simultaneous improvements in public health and sanitation methods further assisted medicine in reducing infant and child mortality. Although these methods and discoveries were common to all medical practice, they had their greatest impact in pediatrics because of their precision in reducing the vulnerabilities of the developing and susceptible child.

In the 1950s, many childhood illnesses (polio, diphtheria, etc.), which had previously not been well managed, came under control. Viral illness became preventable in this period of *prophylactic pediatrics* through immunization with weakened or altered strains of the disease prior to actual exposure.

With the coming of the 1960s, it was apparent that the pediatric population had changed radically. As with adults (Dubos, 1965), childhood illness epidemiology had moved from acute, infectious diseases in which medical treatment was of singular importance to diseases, disorders, and problems in which somatic variables represented only one point of management (Cataldo *et al.*, 1980; Varni & Russo, 1980).

Simultaneous with these medical advances there occurred a social revolution similarly unparalleled. The dissolution of the nuclear family, increases in mobility, media and technological improvements, stratification by age and socioeconomic status, and suburban expansion created a new social ecology (Risley, Clark, & Cataldo, 1976). From these developments arose new problems, with pediatrics increasingly becoming the initial intervention profession for these diverse problems.

Although acute and infectious disease still existed, its importance in the minds of the parents of the 1960s was greatly diminished. Children simply did not die anymore of smallpox, measles, or tetanus, and polio and whooping cough were almost nonexistent due to aggressive immunization programs. What did concern parents were the immediate difficulties of raising children in the new social ecology; the problems of developmental disabilities, birth defects, and retardation became central and more realistic due to advances in prenatal care, nutrition, and neonatology; and the terror of life-threatening diseases such as cancer became more evident, given an apparent hopeless defense.

External social and governmental pressures and strong internal direction from pediatrics itself spearheaded many of these advances. A bimodal focus was seen within pediatric medicine, with large amounts of pediatric talent diverted into well-child, ambulatory care and with another major segment of activity in chronic disease management. Although the two paths were quite different, they did share several common factors. First, proper pediatric practice now required multiple and periodic visits. Second, rather than being solely problem-based, intervention was also time-based. Patients would be seen routinely to evaluate and prevent problems through follow-up check-ups. Third, in light of the first two factors, it was likely that problems of behavior, education, and socialization would be more reflected in the interplay between parent and pediatrician. Lastly, since the timing, frequency, and type of visit or hospitalization was at least in part tied to the expression of symptoms, repeated exposure to the doctor, the hospital, its staff, and the associated fears or worries of the child and parents could also serve as a basis for learning about the interrelation between symptoms and socioenvironmental influences.

What clearly emerged was a pediatrics in which the developing, learning, and adapting child must play a prominent role. The focus on disease management rather than on cure required modification to account for all of the factors which made up the patient with the disease. Although acute surgeries for tonsillitis or appendicitis and hospitalization for acute disease still existed, repeated care for chronic disease or preventive care for ambulatory problems had become the norm. In both environments, problems ranging from simple behavior management (e.g., toileting, feeding, tantrums) to skills deficits (e.g., social skills, coping skills, activities of daily living) to maladaptive behavior symptom expression (e.g., "operant" pain behaviors) could be seen with increasing frequency. Hence, pediatrics in the 1970s reached outward, even before adult disciplines, to other helping professions such as education, psychology, and social work for input to the new practice of pediatrics. It is within this context that the various approaches must be considered in arriving at a definition of behavioral pediatrics.

Current Approaches to Behavioral Pediatrics

In our attempt to define for this volume the nature and scope of behavioral pediatrics, several major obstacles must be hurdled, not the least of which is the fact that behavioral pediatrics is not a new term. Varni and Dietrich (1981) traced the roots of this movement in pediatric medicine, showing that within pediatrics, the historical focus of behavioral pediatrics has been on the biosocial development and learning difficulties of children and adolescents (Friedman, 1970, 1975; Richmond, 1975), with a primary emphasis on the identification of behavioral and psychosocial adjustment problems and developmental deficits or disabilities as seen by the practicing pediatrician (Brown, 1970; McClelland, Staples, Weisberg, & Bergen, 1980; Yancy, 1975). However, as a result of recent research and clinical practice, behavioral pediatrics has been undergoing a major reconceptualization and subsequently emerging as a more comprehensive and multidisciplinary endeavor (Varni & Dietrich, 1981).

Although Christophersen (Christophersen & Gleeson, 1980; Christophersen & Rapoff, 1980) has provided clear arguments as to why behavioral pediatric practice requires strong preparation in behavioral assessment and treatment methods, considerable latitude exists in defining the field. Although Green (1980) proposed several potential alternative terms to *behavioral* pediatrics such as *biosocial, developmental,* or *psychosocial* pediatrics, one might also add to this list such terms as *ambulatory* or *community* to cover this common approach. However, imprecise use of terms does little to create a specific role for behavioral pediatrics. It is our contention that the field deserves independent definition and is not subsumed within existing pediatric approaches. Drawing from the emerging literature, we suggest that behavioral pediatrics occupies a unique position in the array of pediatric subspecialties. It is clearly different from other areas of pediatrics and offers a richness of treatment and research application to both acute and chronic problems.

Varni and Dietrich (1981) carefully delineate these issues in pointing the way toward their operational definition of behavioral pediatrics. They further distinguish behavioral pediatrics as different from behavioral medicine in its focus on children, development, and parent–child interactions. Unlike pediatric psychology, behavioral pediatrics is interdisciplinary; additionally, it is distinguished from child behavior therapy by its primary focus on medical problems and medical settings. These differences will most likely become clearer as the field develops further within multidisciplinary medical environments.

In the interest of establishing a broad basis for behavioral pediatrics, we should like to propose that the field draw from the existent definitions of behavioral medicine and from the many areas of ongoing work as reported in this text. In this way, several subspecialties relating to acute and chronic diseases, ambulatory care, pediatric environments, or therapeutic adherence may be defined. The unique character of each would lie in the specific collaboration of biobehavioral and medical methods.

A definition of behavioral pediatrics should, therefore, describe the field as being:

1. Interdisciplinary in nature
2. Concerned with the management of disease and related symptoms *and* child management
3. Firmly rooted in the empirical methodologies of its contributing sciences
4. Concerned with long-term care, acute intervention, and prevention
5. Directed toward treatment which is data-based, concerned with both process and outcome measures, and occurring only upon collaborative decision
6. Interested in both ambulatory settings and inpatient care environments
7. Concerned with disease mechanisms and biochemical, physiological, and behavioral interrelationships

A definition derived from these parameters should provide a basis for the development of the field of behavioral pediatrics which best combines the talents of the various specialties and is likely to produce the most beneficial research and application. It does not limit the field to behavioral management of ambulatory problems nor does it prevent it. As a working definition, it should provide the necessary flexibility for initial work, yet safeguard the patient through multidisciplinary input. Taking into consideration these various factors, Varni and Dietrich (1981) proposed the following operational definition:

> Behavioral pediatrics represents the interdisciplinary integration between biobehavioral science and pediatric medicine, with emphasis on multidimensional and comprehensive diagnosis, prevention, treatment, and rehabilitation of physical disease and disabilities in children and adolescents.

To this definition we shall further add specific efforts to train practitioners in one field using the techniques of another (cf. Chapter 9), evaluation of the setting or context of therapeutic services (cf. Chapter 8), and general procedures to reduce stress or improve the quality or impact of care (cf. Chapters 6 and 7).

The above definition stresses the factors previously reviewed. It should serve simultaneously to integrate the contributions to this text and to define behavioral pediatrics in a broad context.

The Techniques of Behavioral Pediatrics

Much will be written about the "proper" set of techniques for behavioral pediatrics. Although these techniques will be a synthesis of existing methods (e.g., EEG and behavioral observations in seizure management), a host of articles will appear to discuss the merits of one approach versus another. If the flow of talent continues to be in the movement of behaviorists *to* medicine, it is likely, therefore, that the majority of such discussion articles will focus on particular behavioral approaches. Irrespective of what logically appears to be of value, what will influence the field are data, of the highest order available, demonstrating the efficacy of behavioral techniques to produce or maintain health care behaviors. By focusing on methodology rather than on technique, behavioral pediatrics may provide a new science base for the study of behavior and disease. It is worth considering that revolutions in science occur both as the result of seminal theorizing and as a result of improvements in assessment. What is unique to behavioral approaches is their science base, a radical departure from previous psychological medicine. The various subdisciplines within behavioral science offer a richness of diagnostic and treatment options. Therefore, the areas of potential techniques will be discussed only briefly in deference to what we would consider to be the issue of primary importance: assessment methodology.

Table I presents a listing of some of the behavioral techniques currently available for research and practice in behavioral pediatrics. We

Table I. Some Behavioral Techniques Used in Behavioral Pediatrics

Operant and Social Learning Procedures
1. Behavioral contracts to modify maladaptive behavior and promote maintenance of healthy behavior
2. Direct reinforcement
   A. Social
   B. Token (stars, charts, lotteries, etc.)
   C. Primary
3. Response cost/time out
4. Aversive procedures
5. Modeling or observational learning

Cognitive and Behavioral Self-Regulation Procedures
1. Relaxation training (autogenic, deep muscle, meditative)
2. Assertion training
3. Social skills training
4. Cognitive restructuring
5. Covert conditioning
6. Behavioral rehearsal
7. Systematic desensitization

Biofeedback and Physiological Self-Regulation Procedures
1. Direct specific training (e.g., sphincteric control)
2. Nonspecific reduction in sympathetic arousal (e.g., EMG/thermal conditioning for relaxation)

have not attempted to be exhaustive, nor have we attempted to delineate theoretical ideas which support the techniques pigeonholed in the table. It is apparent that three broad areas of inquiry subsume the specific techniques:

1. Operant and social learning procedures
2. Cognitive and behavioral self-regulation procedures
3. Biofeedback and physiological self-regulation procedures

In practice, the distinction between these areas is often blurred and some formulations clearly cross boundaries. As a partial catalogue of available behavioral techniques for use in behavioral pediatrics, Table I represents only a brief outline without regard to specific disease entities or problems. To have attempted a more detailed analysis of techniques and applications would have been a disservice. No specific catalogue of applications exists at this time in the development of the field, and it is often the case that multiple behavioral/medical techniques are used simultaneously. Since any of several approaches might be effective, it is suggested that behavioral pediatrics research and practice be guided by assessment and empirical process–outcome data rather than theory or technique.

Assessment Issues in Behavioral Pediatrics

In an article written previously by several of the contributors to this text (Russo, Bird, & Masek, 1980), the argument was put forth that methodology be the basis for behavioral medicine. Such an approach will be of no less importance in the development of behavioral pediatrics. The simultaneous assessment of treatment programs in terms of both behavioral and medical process–outcome measures is essential to an understanding of treatment effects and long-term acceptance of the field. Several issues must be stressed here which are picked up at numerous points throughout the text.

Process and Outcome Measures

Although medicine has been concerned with both process and outcome assessment measures, a number of specific methodological problems have been identified in this area (Dinsdale, Mossman, Gullickson, & Anderson, 1970; Greenfield, Solomon, Brook, & Davies-Avery, 1978; Nobrega, Morrow, Smoldt, & Offord, 1977). Behavioral methods may be of assistance in developing assessment strategies to rectify these flaws.

Williamson (1971) has parcelled assessment issues into four specific categories: (1) diagnostic outcomes representing data which delineate the prognosis and need for treatment; (2) therapeutic outcomes representing health status after a long-term intervention; (3) diagnostic process, which further details the identification of specific needs and therapy; and, finally, (4) therapeutic process, which includes the progress of the intervention on health status and such potentially contributory factors as noncompliance to treatment regimens. Each of these areas may most effectively be addressed by simultaneous assessment of both organismic and environmental variables. Some measures may most appropriately be those regarding disease status, some focusing on function and some evaluating both parameters in the same assessments. For example, the distinction between the process and outcome measurement sets may be a function of different assessment-targeted parameters (e.g., joint deterioration versus activities of daily living) or, rather, a matter of temporal ordering (e.g., measurements taken during the intervention versus pre- and post-assessments).

Multidimensional Assessment

A primary axiom in both research and practice in behavioral pediatrics must be that data be collected on a multidimensional level (e.g., biochemical, behavioral, self-report). Since the goals, assessment methods, and time base of medical and behavioral methods are often dissimilar, the design of process–outcome measures requires familiarity with issues in both spheres. As detailed by Varni and Dietrich (1981), the empirical development of these multidimensional and comprehensive process and outcome measurement instruments should be one of the first tasks of behavioral pediatrics. The demonstration of the effects of the intervention will not only aid in verifying treatment efficacy, but may contribute to knowledge pertinent to the mechanisms of behavioral-medical interrelationships.

As Fries has pointed out (Fries, Spitz, Kraines, & Holman, 1980), disability may be the most important outcome category in chronic disorders. Activities of daily living (ADL) scales are most often used to assess disability associated with chronic disease (Sheikh, Smith, Meade, Goldenberg, Brennan, & Kinsella, 1979). In the case of pediatric chronic illness, Pless and Pinkerton (1975) have emphasized the importance of assessing the functional impact of handicap as well as the extent to which the disease is treated successfully in a purely medical sense, with school attendance suggested as an important ADL outcome measure for children with chronic disease (Parcel, Gilman, Nader, & Bunce, 1979). Thus, for a comprehensive assessment of the health–illness status of a patient, measurement must be made in terms of departures from normal role behavioral functioning as well as in reference to health in a physical or medical sense (Reynolds, Rushing, & Miles, 1974).

Reliability and Validity as Simultaneous Considerations

Behavioral science has historically been more concerned with the reliability of measurement variables and the internal validity of independent variables and has until recently placed less emphasis on external validity. In contrast, medical science has traditionally emphasized external validity as the paramount concern (Harvey, Johns, Owens, & Ross, 1976). These differing conceptual bases compel simultaneous consideration. Validity is a matter of degree and not a unitary concept, requiring empirical testing and revision, with the optimum process of validation consisting of the simultaneous measurement of content, criterion (predictive and concurrent), and construct validity (Isaac & Michael, 1971). A merger of epistemologies may produce improvements in both fields with respect to the attribution of causality (Cataldo, Russo, & Freeman, 1980) or the improved reliability of measurement (Richens, 1975). Clearly, only with reliable and valid comprehensive and multidimensional outcome and process measurement instruments will behavioral pediatrics be seen as a significant contribution to the total care of children and adolescents (Varni & Dietrich, 1981).

Standards for Data

Much argument has occurred internally within the behavioral sphere about the sufficiency of various types and levels of data required to support particular theoretical positions or procedures. In traditional psychological thought such a discrepancy does not exist; we are taught that interval data are preferable to ordinal data, which in turn are preferable to nominal data (Hayes, 1963). When validity of data, both within one's own referent and with respect to other systems of thought, becomes the standard, such speculation is moot. That is, whether self-report or rating-scale data are as appropriate for assessing cognitive function as observational data are for assessing behavioral change is important only when a single datum stands in isolation. For example, refinement in the assessment of brain peptides (Snyder, 1977) and in the conceptualization and assessment of pain (Fordyce, 1976; Varni, 1981) signals a merger in methodologies to answer such fundamental questions as: Do changes in levels of certain chemicals produce changes in behavior? Do changes in the psychological elaboration of pain alter its biochemistry? Whether the issue is pain, seizures, or dermatitis, increased proficiency at multidimensional assessment through collaborative research is required for validation and acceptance of approaches in behavioral pediatrics as well as within behavioral medicine in general.

Assessment in behavioral pediatrics, therefore, suggests the following guidelines:

1. The requirement of multidimensional assessment
2. Concern for simultaneous reliability and validity
3. Utilization of both single-subject and group-based methodology; within the same context these may have interactive benefits
4. The criteria of multiple replication–single case studies represent only a first step toward broader application

It is likely that initial efforts in methodology will direct future work. Standards such as those above allow for the test of the value of behavioral pediatric formulation and technique and are likely to generate their own collaborative future. Such methodology stems directly from the definition as previously stated and provides a philosophy and a method for the field.

CONCLUSION

By way of introduction to the remainder of the text, we should like to summarize the field of behavioral pediatrics by compartmentalization of some appropriate areas of study. Efforts fall along several axes: the disease, the environment, and the patient. It is obvious that all three must be considered simultaneously within the bimodal pediatric population. Such consideration will likely support an NPAS model of disease: a normal person in an abnormal situation. Within such a context, the role of behavioral pediatric practitioners becomes facilitative of the continuous process of adapting and coping, the reduction of symptoms, the prevention of further difficulties in the spheres of both behavior and disease, and the enhancement of the quality of life. Consideration of the disease, the environment, and the person also allows the phrasing of experimental questions regarding patient/caregiver interaction and patient/environment interaction which might profitably be experimentally investigated.

Throughout this introduction we have held to one primary belief: that the conceptualization and practice of a field lead to its ultimate success or failure. A clear course for behavioral pediatrics lies in its continued collaborative development and its concern for both the philosophies and methods of biobehavioral and medical approaches.

REFERENCES

Agras, W. S. Foreword. In R. C. Katz & S. Zlutnick (Eds.), *Behavior therapy and health care: Principles and applications.* New York: Pergamon Press, 1975, pp. xi–xiii.

Alexander, F. G. *Psychosomatic medicine: Its principles and applications.* New York: Norton, 1950.

Bedell, J. R. Giordani, B., Amour, J. L., Tavormina, J., & Boll, T. Life stress and the psychological and medical adjustment of chronically ill children. *Journal of Psychosomatic Research,* 1977, *21,* 237–242.

Birk, L. (Ed.). *Biofeedback: Behavioral medicine.* New York: Grune & Stratton, 1973.

Blanchard, E. B. Behavioral medicine: A perspective. In R. B. Williams & W. D. Gentry (Eds.), *Behavioral approaches to medical treatment.* Cambridge, Mass.: Ballinger, 1977.

Breslow, L. Risk factor intervention for health maintenance. *Science,* 1978, *200,* 908–912.

Brown, G. W. Developmental and behavioral pediatrics: A realistic challenge? *Journal of Developmental and Behavioral Pediatrics,* 1979, *1,* 1–8.

Cataldo, M. F. & Russo, D. C. Developmentally disabled in the community: Behavioral/medical considerations. In L. A. Hamerlynck (Ed.), *Behavioral systems for the develop-*

mentally disabled: II. Institutional, clinic, and community environments. New York: Bruner/Mazel, 1979, pp. 105–143.

Cataldo, M. F., Russo, D. C., Bird, B. L. & Varni, J. W. Assessment and management of chronic disorders. In J. M. Ferguson & C. B. Taylor (Eds.), *Comprehensive handbook of behavioral medicine. Vol. 3: Extended applications and issues*. New York: Spectrum, 1980, pp. 76–95.

Cataldo, M. F., Russo, D. C., & Freeman, J. M. Behavior modification of a 4½ year old child with myoclonic and grand mal seizures. *Journal of Autism and Developmental Disorders*, 1980, *9*, 413–427.

Christophersen, E. R., & Rapoff, M. A. Biosocial pediatrics. In J. M. Ferguson & C. B. Taylor (Eds.), *Comprehensive handbook of behavioral medicine. Vol. 3: Extended applications and issues*. New York: Spectrum, 1980.

Christophersen, E. R., & Gleeson, S. Research in behavioral pediatrics. *Behavior Therapist*, 1980, *3*, 13–16.

Culliton, B. J. Health care economics: The high cost of getting well. *Science*, 1978, *200*, 883–885.

Dinsdale, S. M. Mossman, P. L., Gullickson, G., & Anderson, T. P. The problem-oriented medical record in rehabilitation. *Archives of Physical Medicine and Rehabilitation*, 1970, *51*, 488–492.

Dubos, R. *Man adapting*. New Haven, Conn.: Yale University Press, 1965.

Eisenberg, L. Is health a state of mind? *New England Journal of Medicine*, 1979, *301*, 1282–1283.

Engel, G. L. The care of the patient: Art or science? *The Johns Hopkins Medical Journal*, 1977, *140*, 222–232.

Fabrega, H., & Van Egeren, L. A behavioral framework for the study of human disease. *Annals of Internal Medicine*, 1976, *84*, 200–208.

Ferguson, J. M. & Taylor, C. B. (Eds.). *Comprehensive handbook of behavioral medicine* (Vols. 1–3). New York: Spectrum, 1980.

Fordyce, W. E. *Behavioral methods for chronic pain and illness*. St. Louis: C. V. Mosby, 1976.

Freud, S. Psychological [Mental] treatment (1905). In *The complete psychological works of Sigmund Freud*, Vol. 7. London: Hogarth Press, 1953.

Friedman, S. B. The challenge in behavioral pediatrics. *Journal of Pediatrics*, 1970, *77*, 172–173.

Friedman, S. B. Forward to symposium on behavioral pediatrics. *Pediatric Clinics of North America*, 1975, *22*, 515–516.

Fries, J. F. Aging, natural death, and the compression of morbidity. *New England Journal of Medicine*, 1980, *303*, 130–135.

Fries, J. F., Spitz, P., Kraines, R. G., & Holman, H. R. Measurement of patient outcome in arthritis. *Arthritis and Rheumatism*, 1980, *23*, 137–145.

Green, M. The pediatric model of care. *Behavior Therapist*, 1980, *3*, 11–13.

Greenfield, S., Solomon, N. E., Brook, R. H., & Davies-Avery, A. Development of outcome criteria and standards to assess the quality of care for patients with osteoarthrosis. *Journal of Chronic Diseases*, 1978, *31*, 375–388.

Hamburg, D. A., & Brown, S. S., The science base and social context of health maintenance: An overview. *Science*, 1978, *200*, 847–849.

Harvey, A. M., Johns, R. J, Owens, A. H., & Ross, R. S. (Eds.). *The principles and practice of medicine* (19th ed.). New York: Prentice Hall, 1976.

Hayes, W. L. *Statistics*. New York: Holt, Rinehart & Winston, 1963.

Ince, L. P. *Behavior modification in rehabilitative medicine*. Springfield, Ill.: Charles C Thomas, 1976.

Isaac, S., & Michael, W. B. *Handbook in research and evaluation.* San Diego: Edits, 1971.

Kellerman, J., Zeltzer, L., Ellenberg, L., Dash, J., & Rigler, D. Psychological effects of illness in adolescence: I. Anxiety, self-esteem and the perception of control. *Journal of Pediatrics,* 1980, *97,* 126–131.

Kenney, T. J., & Clemmens, R.L. *Behavioral pediatrics and child development.* Baltimore: Williams & Wilkens, 1975.

Levine, S. Z. Pediatric education at the crossroads: Presidential address to American Pediatric Society. *American Journal of Diseases of Children,* 1960, *100,* 650–657.

Levine, J. D., Gordon, N. C., & Fields, H. L. The mechanisms of placebo analgesia. *Lancet,* 1978, *8091*(2), 654–657.

Masek, B. J. Epstein, L. H., & Russo, D. C. Behavioral perspectives in preventive medicine, In S. M. Turner, K. S. Calhoun, & H. E. Adams (Eds.), *Handbook of clinical behavior therapy.* New York: Wiley, 1981, pp. 475–499.

Matarazzo, J. D. Behavioral health and behavioral medicine: Frontiers for a new health psychology. *American Psychologist,* 1980, *35,* 807–817.

McClelland, C. Q., Staples, W. I. Weisberg, I., & Bergen, M. E. The practitioner's role in behavioral pediatrics. *Journal of Pediatrics,* 1973, *82,* 325–331.

Meenan, R. F., Yelin, E. H., Nevitt, M., & Epstein, W. V. The impact of chronic disease: A sociomedical profile of rheumatoid arthritis. *Arthritis and Rheumatism,* 1981, *24,* 544–549.

Miller, N. E. Application of learning and biofeedback to psychiatry and medicine. In A. M. Friedman, H. I. Kaplans, & B. J. Sadock (Eds.), *Comprehensive textbook of psychiatry.* Baltimore: Williams & Wilkens, 1975.

Nobrega, F. T., Morrow, G. W., Smoldt, R. K., & Offord, K. P. Quality assessment in hypertension: Analysis of process and outcome methods. *New England Journal of Medicine,* 1977, *296,* 145–148.

Parcel, G. S., Gilman, S. C., Nader, P. R., & Bunce, H. A comparison of absentee rates of elementary school children with asthma and nonasthmatic schoolmates. *Pediatrics,* 1979, *64,* 878–888.

Pless, I. B., & Pinkerton, A. *Chronic childhood disorders: Promoting patterns of adjustment.* London: Henry Kimpton, 1975.

Pomerleau, O. F. Behavioral medicine: The contribution of the experimental analysis of behavior to medical care. *American Psychologist,* 1979, *34,* 654–663.

Pomerleau, O. F., Bass, F., & Crown, V. The role of behavior modification in preventive medicine. *New England Journal of Medicine,* 1975, *292,* 1277–1282.

Prazar, G., & Charney, E. Behavioral pediatrics in office practice. *Pediatric Annals,* 1980, *9,* 12–22.

Reynolds, W. J., Rushing, W. A., & Miles, D. L. The validation of a function status index. *Journal of Health and Social Behavior,* 1974, *15,* 271–288.

Richens, A. Quality control of drug estimations. *Acta Neurologica Scandanavica,* 1975, *60* (Supplement), 81–84.

Richmond, J. B. An idea whose time has arrived. *Pediatric Clinics of North America,* 1975, *22,* 517–523.

Risley, T. R., Clark, H. B., & Cataldo, M. F. Behavioral technology for the normal middle-class family. In E. J. Mash, L. A. Hamerlynck, & L. C. Handy (Eds.), *Behavior modification and families.* New York: Bruner/Mazel, 1976.

Russo, D. C., Bird, B. L., & Masek, B. J. Assessment issues in behavioral medicine. *Behavioral Assessment,* 1980, *2,* 1–10.

Russo, D. C., & Newsom, C. D. Psychotic disorders of childhood. In J. Lachenmeyer & M. Gibbs (Eds.), *Psychology of the abnormal child.* New York: Gardner, 1982.

Schneiderman, N., Weiss, T., & Engel, B. T. Modification of psychosomatic behaviors. In R. S. Davidson (Ed.), *Modification of pathological behavior.* New York: Gardner, 1979.

Schwartz, G. E., & Weiss, S. M. Yale conference on behavioral medicine: A proposed definition and statement of goals. *Journal of Behavioral Medicine,* 1978, *1,* 3–12.

Sheikh, K., Smith, D. S., Meade, T. W., Goldenberg, E., Brennan, P. J., & Kinsella, G. Repeatability and validity of modified activities of daily living (ADL) index in studies of chronic disability. *International Journal of Rehabilitation Medicine,* 1979, *1,* 51–58.

Snyder, S. Opiate receptors in the brain. *New England Journal of Medicine,* 1977, *296,* 266–271.

Tavormina, J. B., Kastner, L. S., Slater, P. M., & Watt, S. L. Chronically ill children: A psychologically and emotionally deviant population? *Journal of Abnormal Child Psychology,* 1976, *4,* 99–110.

Varni, J. W. Self-regulation techniques in the management of chronic arthritic pain in hemophilia. *Behavior Therapy,* 1981, *12,* 185–194.

Varni, J. W., Bessman, C. A., Russo, D. C., & Cataldo, M. F. Behavioral management of chronic pain in children. *Archives of Physical Medicine and Rehabilitation,* 1980, *61,* 375–379.

Varni, J. W., and Dietrich, S. L. Behavioral pediatrics: Towards a reconceptualization. *Behavioral Medicine Update,* 1981, *3,* 5–7.

Varni, J. W., & Russo, D. C. Behavioral medicine in health care: Hemophilia as an exemplary model. In M. Jospe, J. Nieberding, & B. Cohen (Eds.), *Psychological factors in health care.* Lexington, Mass.: Lexington Books, 1980.

Williams, R. B., & Gentry, W. D. (Eds.). *Behavioral approaches to medical treatment.* Cambridge, Mass.: Ballinger, 1977.

Williamson, J. W. Evaluating quality of patient care: A strategy relating outcome and process assessment. *Journal of the American Medical Association,* 1971, *218,* 564–569.

Yancy, W. S. Behavioral pediatrics and the practicing pediatrician. *Pediatric Clinics of North America,* 1975, *22,* 685–694.

Zeltzer, L., Kellerman, J., Ellenberg, L., Dash, J., & Rigler, D. Psychological effects of illness in adolescences: II. Impact of illness in adolescents: Crucial issues and coping styles. *Journal of Pediatrics,* 1980, *97,* 132–138.

II

BIOBEHAVIORAL FACTORS AND CHRONIC DISORDERS

2

The Application of Behavioral Techniques to Childhood Asthma

Thomas L. Creer, Charles M. Renne, and Hyman Chai

INTRODUCTION

The major purpose of this chapter is to outline a behaviorally-oriented assessment and treatment approach for pediatric asthma. The discourse will delineate problem areas associated, directly and indirectly, with asthma management and treatment and deal with issues related to behavior analysis and intervention for these problems.

The earlier section of the chapter provides some fundamental information about asthma, its causes, process, and impact. Various measurement strategies used in clinical and research work with the illness are described and critically appraised. The aim of this earlier segment is to place the reader in a better position to understand and evaluate both the problems and potential solutions as expounded in the following sections of the chapter.

THOMAS L. CREER • Ohio University, Athens, Ohio 45701. CHARLES M. RENNE • National Asthma Center, Denver, Colorado 80204. HYMAN CHAI • National Asthma Center, Denver, Colorado 80204. Preparation of this chapter was supported in part by Contract No. N01–HR–2972 and Grant No. HL–22021 from the Division of Lung Diseases of the National Institute of Heart, Lung, and Blood and by Grant No. 1 MH–32101 from the National Institute of Mental Health.

Definition of Asthma

No definition of asthma is acceptable to everyone. Several attempts have been made to arrive at one, but all have resulted in failure (e.g., Porter & Birch, 1971). For the purposes of this discussion, we will rely on a definition provided by Scadding (1966). He defined asthma as an intermittent, variable, and reversible airway obstruction. Chai (1975) has pointed out that each word in this definition is crucial: *Intermittent* refers to the fact that attacks generally occur on an aperiodic basis. Thus, although a patient may go for prolonged periods without suffering an asthma episode, he may experience several attacks in a period of a week or month. As we will note in a later section, the intermittent character of asthma can hamper both the prediction and prevention of asthma attacks. *Variable* signifies that asthma attacks vary in severity from mild episodes, characterized by a sensation of tightness in the chest or a slight wheeze, to status asthmaticus (Chai & Newcomb, 1973). Uncertainty about how severe an attack may become can influence both the type of treatment sought and its availability. *Reversible* implies that airways may revert to a normal condition either spontaneously or following treatment. Theoretically, the airways should be totally reversible to their normal state. It is becoming increasingly apparent in practice, however, that complete reversibility of lung function depends a great deal on the severity of the disease. Hence, the more severe the asthma, the less surely total reversibility can be anticipated despite maximal therapeutic efforts. Yet, some reversibility is characteristic of all asthma and is a component in any definition of the disorder, if the condition is to be differentiated from other reactive airway diseases.

EPIDEMIOLOGY

There is no single estimate of the incidence of asthma in the United States. Considerable data have been generated through various methods including (1) cases identified by practitioners, (2) cases identified by hospital-based physicians, and (3) cases identified by studies of population groups in which the definition of who suffers asthma has been left either to the patients themselves or to the parents of patients. On the basis of a review of available statistics, the American Lung Association (1975) estimated that between 5% and 15% of children $11\frac{1}{2}$ years of age and younger suffer asthma. Other reviews of available data (e.g., Davis, 1972) suggest that 7% of the total population of the U.S. has suffered asthma at one time or another.

Asthma is most frequent among boys younger than 14 years of age (American Lung Association, 1975). Below the age of 5 years, males are affected by asthma approximately 1.5 to 2.0 times as often as females; between 5 and 9 years of age, the incidence in males and females is approximately the same (Speizer, 1976). Thereafter, particularly in youngsters over 15 years of age, asthma is more frequent in girls (American Lung Association, 1975).

Considerable confusion surrounds previous discussions regarding the prognosis of asthma (Creer, 1979). Various studies, for example, have reported rather discrepant results regarding the percent of childhood asthma that persists beyond puberty. The figures reported vary from 26% (Rackemann & Edwards, 1952) to 78% (Johnstone, 1968). Such differences probably are traceable to a failure to develop a uniform concept of prognosis, as well as to differences in definitions of asthma, ages of patients, treatment methods, length of observation periods, and experimental designs employed in the studies (Creer, 1979). There are some conclusions that have stood up under these investigative discrepancies. In general, better prognosis is associated with an earlier age of onset of asthma, except in the case of infants for whom disease onset occurred under 2 years of age (Slavin, 1977). Speizer (1976), for example, concluded that patients in whom onset occurred in childhood have a 60% chance of total remission with no sequelae.

The remission rates for more severe cases of childhood asthma are low (Slavin, 1977). Aas (1963) reported that in a sample of children with asthma classified as mild, 73% were later considered cured and 82% termed well or improved. In youngsters with severe asthma, however, only 30% were later regarded as cured, while another 30% were regarded as improved.

Asthma-related mortality is low in comparison to other events that can lead to death (e.g., heart disease, cancer, automobile accidents, etc.). Nevertheless, it is estimated that asthma claims between 2,000 and 3,000 victims each year in the United States alone (Segal, 1976). The number of asthma-related deaths is probably underestimated; there is a distinct possibility that asthma is a contributing cause not reflected in official mortality figures (Gordis, 1973).

It is in its morbidity that the impact of asthma is most clearly seen (Creer, 1979), particularly with respect to the role it plays in restricting activity and in generating monetary costs.

1. *Activity Restriction.* Asthma can be especially restrictive with children (Creer, 1979). It is a leading contributor to school absenteeism among youngsters and in one survey accounted for nearly one fourth of days lost from school because of chronic illness (Schiffer & Hunt,

1963). More recent studies (e.g., Bharani & Hyde, 1976) reaffirm the fact that asthma remains a leading contributor to school absenteeism.

The pattern of days missed from school also may be of major significance. Douglas and Ross (1965) suggested that frequent brief absences from school were more harmful to the academic progress of youngsters than were occasional long absences. Unfortunately, the former pattern appears to be typical for children afflicted with asthma (Creer & Yoches, 1971).

In addition to the consequence of falling behind due to absenteeism, children with asthma often find themselves excluded from many of the games and activities enjoyed by their peers and in general ostracized from the mainstream of the community (Creer, 1977; Creer & Christian, 1976).

2. *Costs of Asthma.* Asthma is expensive. The disease accounts for one third as many visits to a physician's office as does essential hypertension, the leading medical diagnosis, and about half as many office visits as does diabetes (Davis, 1972). It also produces 75% of all admissions of patients with respiratory disorders to hospitals (American Lung Association, 1975). In terms of actual dollars spent for treatment, 1975 industry estimates reveal that $224.2 million was spent for bronchodilators, $24.7 million for corticosteriods, and $43 million for over-the-counter asthma remedies (Creer, 1977). That same year, direct costs for physicians' services, hospital care, and drugs were estimated to be $850 million, and costs associated with morbidity were $440 million. The total costs of asthma, exclusive of mortality, were estimated to be $1.3 billion for 1972 (Cooper, 1976).

Childhood asthma had undermined the financial stability of many families (Creer, 1979). Vance and Taylor (1971) indicated that anywhere from 2% to 30% of a family's income was spent in managing asthma in one child. In over half of the cases investigated, 18% of the families' income was spent on asthma. These expenses do not represent the total financial impact of the affliction. Other costs, such as travel expenses required in driving a child to a physician or hospital, depreciation of vehicles, food for special diets, lost work time, and required home renovations in the interest of achieving a contaminant-free environment, must be included as well.

PATHOGENESIS OF ASTHMA

Traditionally, the pathogenesis of asthma has been discussed according to some clinical classification scheme, particularly that proposed by Rackemann (1928), who divided asthma into extrinsic, intrinsic, and mixed types. The fact that many patients fell into the latter category prompted more elaborate descriptive classifications based on factors provoking episodes or responses to different treatments. Recently, Reed and Townley (1978) proposed that asthma be considered as a disorder with three components:

1. Stimuli that provoke airway obstruction
2. Persons on whom the stimuli act, with the physiological and biochemical steps that link the stimulus to the response
3. Physiological and pathological responses that constitute obstruction

Such a multifactoral approach to the classification of asthmatic patients, note Reed and Townley, is more accurate because causes and mechanisms of the disorder are not presumed. Their classification scheme also permits a clearer presentation of variables involved in the pathogenesis of asthma and hence provides a firm basis for planning optimum treatment for the individual patient.

Stimuli That Provoke Asthma

There are a variety of stimuli that provoke airway obstruction in asthmatic patients. Major precipitants include the following:

1. *Allergens.* To determine whether a specific substance is triggering a patient's asthma requires a number of tests and challenges (e.g., Chai, 1975). The essential requirement is to demonstrate that the presence of a foreign substance (i.e., the antigen) produces a specific antibody, mostly of the IgE or, in some cases, the IgG class, within the patient's blood. Without such a demonstration, it is unlikely that the specific substance is implicated in the patient's attacks. Ragweed-induced bronchospasm is a classical example of a true allergy. The presence of a positive skin test to ragweed—indicating the presence of the antibody—and a seasonal aggravation of asthma when ragweed counts are high help to confirm the diagnosis.

2. *Aspirin and Related Substances.* Aspirin and related substances provoke asthma in some patients. Again, however, it is important that reliable tests indicate that such stimuli do actually induce attacks.

3. *Infections.* Infections are major precipitants of asthma attacks in some patients. It appears that, particularly for children, such infections are usually viral in origin and rarely of a bacterial nature. Not all viruses induce attacks; some viruses, on the other hand, are frequently involved. The common cold and some influenza strains are the more common

viral precipitants of bronchial obstruction. Some fungal infections such as appergillosis induce asthma, while other fungal infections appear to have little role in triggering attacks.

4. *Emotional Factors.* The relationship between emotions and asthma is unclear. Consequently, any discussion of the role emotions may play in asthma is apt to be, in the words of Chai (1974), "very emotional!" There appears to be very little direct evidence that emotions *per se* trigger asthma attacks. There is some support, however, for the notion that emotions can produce responses which precipitate asthma attacks, either through mechanical means or motor activity (Creer, 1979). Responses such as shouting, laughing, screaming, coughing, crying, gasping, sneezing, and running often accompany emotional reactions and alone may produce bronchospasm. Many of these responses are parasympathetic stimulators and may cause bronchospasm by stimulating vagal fibers.

Emotions may also lead some patients to become agitated and to increase sharply their overall level of activity. This activity represents a type of exercise which can be a major precipitant of attacks in many patients.

5. *Exercise.* Exercise-induced asthma (EIA) affects well over 80% of asthmatic children (e.g., Ghory, 1975). The mechanisms involved in EIA are not totally understood, although airway cooling has recently been implicated as a causal element (Chen & Horton, 1977; McFadden & Ingram, 1979). The parasympathetic system is involved, essentially through the vagus nerve and its branches.

6. *Irritants.* All asthmatic children have airways that are subject to spasm when irritated by certain stimuli. Examples of such triggers are silica dusts, heavily chlorinated pools, perfumes, paint, and tobacco smoke. Cold air and changes in weather also precipitate asthma attacks in some youngsters.

Physiological Links between the Stimulus Precipitant and the Clinical Response

It is not our intention here to delineate each link between exposure to a stimulus trigger and asthma symptoms. Such a discussion is beyond the purview of the chapter and is redundant in face of more thorough discussions (e.g., Middleton, Reed, & Ellis, 1978; Weiss & Segal, 1976). However, a brief description of the anatomical and physiological aspects of the respiratory system will not only point to the complexity of the disorder but will help the reader to discern more clearly the problems associated with the assessment of asthma. One hopes that such a background will also provide some understanding of the logic underlying therapy, both medical and behavioral, for asthmatic patients.

The airways of the lung, basically the bronchi and bronchioles, consist of an initial large airway (the trachea) which divides at the carina into two branches. Each of the branches continues to divide in sequence until, when a person is about 20 years of age, some 300,000,000 bronchioles and alveoli have developed. Terminal bronchioles end in sacs (or alveoli) where gases in the air (principally oxygen and carbon dioxide) are exchanged. This occurs via capillaries surrounding the alveoli that receive oxygen from the alveoli while giving up carbon dioxide to be exhaled.

Each division of the bronchi is referred to as a generation. From the trachea down to the 12th or 13th generation, the bronchi are partially surrounded by cartilage that assist in keeping the airways open. Beyond the 12th or 13th generation, however, airways change their character in that the scaffolding provided by the cartilage disappears. The diameter of the bronchi at this point is about 2 mm. It has been customary to designate those airways greater than 2 mm as large airways; those airways smaller than 2 mm, on the other hand, are referred to as small airways. Large and small airways can also be differentiated by the fact that while abnormalities in large airways can generally be heard and detected, such changes in small airways can be clinically silent and difficult to detect.

Clinical Responses and Their Immediate Causes

Anatomically, airways have four elements that are commonly involved in asthma: (1) Smooth muscle surrounding the airway that, upon contraction, results in a narrowing of the airway; (2) an epithelial lining with cells that have whip-like flagella and carry any substance landing upon the lining toward the mouth; (3) mucus produced by an adjacent gland that entraps extraneous particles; and (4) a variety of cells which produce enzymes, act as scavengers, and provide other important physiological functions. These elements often occur in varying degrees from patient to patient.

It is the interaction between various stimuli and these anatomical responses which sometimes results in an asthma attack. Wheezing sounds are the most commonly observed responses associated with asthma attacks. Wheezing sounds are easily discernable with large airway constriction but may be detectable, if present, only through a stethoscope when the focal point of the illness resides in the small airways. Posture and breathing changes, basically by hunching the shoulders with extended inspiration and labored expiration, is usually evidence of a more severe attack. Such changes are secondary to the symptoms and serve more as adaptive responses that permit greater air exchange. Reports of feelings of tightness in the chest, particularly associated with small airways obstruction, is common. Chest pain may also be reported as accompanying tightness, particularly with exacerbation of the illness. Generally, the symptoms are due to one of three factors: (1) bronchoconstriction due to contraction or spasms in the smooth muscle surrounding the airways, (2) edema or a swelling of the epithelial lining of the bronchial tubes, and (3) mucus and mucus plugs that clog the narrow airways. The consequence is a breakdown in the exchange between oxygen and carbon dioxide, primarily characterized by the trapping of stale air in the alveoli. These factors, and hence the symptoms of attacks, vary from person to person and from attack to attack in the same person (Snider, 1976). If smooth muscle spasm of large airways is unaccompanied by secretory plugging of small airways, there may be a dramatic remission of the attack; if there is widespread obstruction of small airways, on the other hand, the attack may be prolonged.

ASSESSMENT

Assessment of asthma medically and behaviorally entails a similar process: an attempt to identify relevant response units and their controlling variables. The medical investigator seeks to determine the source of the airway obstruction producing symptoms, as well as the factors linking environmental and physiological variables to such episodes. The purpose is to understand and manage the asthma attacks. Similarly, behavioral scientists attempt to delineate relevant asthma symptoms and to determine the relationship between these responses and their controlling environmental and physiological variables (such as current physiological states and past learning history). They also seek to understand and alter asthma symptoms as well as to ameliorate any harmful consequences of the disease.

Specific assessment procedures have been identified more or less with the medical or behavioral science approaches to the asthma problem. One or more of the following procedures are typically emphasized in medical investigations of asthma.

1. *Clinical Examination.* Chai (1975) has enumerated steps that are followed in conducting a clinical examination, including obtaining a detailed history, performing skin tests, and carrying out bronchial challenges to different stimuli. Such an examination may not only help to

confirm or refute a diagnosis of asthma but also may yield information as to important characteristics or patterns of the patient's asthma and the stimuli that precipitate attacks.

2. *Physical Examination.* Observation and examination of patients during an asthma attack often provides valuable information as to the nature of the attacks, the effectiveness of different procedures in establishing control over episodes, and/or the precipitants of asthma. Although observing symptoms during attacks is necessary in order to arrive at a firm diagnosis, their presence alone is not sufficient for establishing the diagnosis. Gordis (1973), for example, pointed out that not all children who wheeze—a hallmark of asthma—actually suffer from asthma. In fact, Fry (1961) found that only 10% of a sample of 797 children who reported wheezing actually developed asthma when followed over a 10-year period.

3. *Pulmonary Function Measurements.* Pulmonary function measurement using spirometry, full-body plythesmography, or gas exchange assessment is a major component of a thorough examination for asthma. These measures essentially establish the condition. Pulmonary tests also permit an evaluation of the extent of central and peripheral airway involvement. This is important in that the two systems may not be involved concurrently in the disease process (Chai, 1975). For example, the focal point of asthma might be detected in the small airways, while symptoms or other clinical signs of abnormality may be absent in the large airways.

Besides their value in confirming a diagnosis, pulmonary physiology measurements are invaluable in providing an objective index of changes in the condition over time. For this purpose, however, such measurements must be obtained on a frequent basis (Chai, Purcell, Brady, & Falliers, 1968). Infrequent measurements are nearly meaningless because of the steady state variability which characterizes them (Creer, 1979). Figure 1 depicts this phenomenon. It shows peak expiratory flow rates (PEFR) obtained from a child who, both in the morning and evening, blew into a Wright Peak Flow Meter. This meter, which is an effort-dependent portable instrument, assesses the amount of air flow in the initial 0.1 second of a maximal expiratory blow. Peak flow measurements have been shown to be reliable indicants of a patient's pulmonary functioning even though effort-dependent and, therefore, a function of the motivation of the patient (Chai *et al.*, 1968; Wright & McKerrow, 1959). Figure 1 illustrates the considerable variation typically found between PEFR values from day to day and between values obtained on any given day. The Wright meter, and more recently developed, low-cost meters such as the Mini-Wright and Pulmonary Monitor meters, produce PEFRs that are highly controlled (Burns, 1979). Because of their low cost and

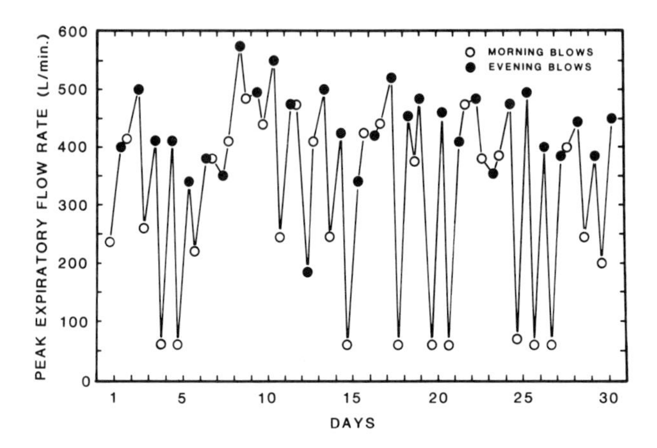

Figure 1. Peak expiratory flow rates obtained from an asthmatic child over a 30-day period (○ = morning blows; ● evening blows).

portability, increasing numbers of behavioral scientists are employing them in their clinical and research endeavors.

4. *Patient Reports.* To a considerable extent, effective medical treatment relies on the verbal input of patients and other family members. Chai and his colleagues (1968) suggested that, despite inherent flaws, information from patients represented the most important single diagnostic tool in assessing the precipitants of asthma attacks.

A drawback to patient report is that it is difficult to assess reliability and validity. As we will note later, studies on this topic have reported poor agreement between information obtained via verbal report and other indices of asthma, e.g., pulmonary function information.

5. *Medication Scores.* On the surface, it appears that medications taken to control asthma provide an objective and quantifiable way to assess the affliction. Not only can medications and dosage levels be specified, but medications can be ranked according to their potency. From such information, medication scores can be calculated across designated time spans (e.g., days) regardless of the number, types of drugs, or doses taken. Variations in medication scores over time have been used to provide "objective evidence" as to the course of asthma.

Creer (1979) has raised a major question about the reliability of medication scores. The question centers on the issue of compliance with medication regimens, particularly among children with asthma. For example, by analyzing serum and salivatory theophylline levels—theophylline-based drugs are widely prescribed for the management of asthma—Eney and Goldstein (1976) found therapeutic levels were reached in only 11% of their patient sample. Sublett, Pollard, Kadlec, and Karibo (1979) found even less compliance in the sample of asthmatic children they investigated in that only 1 of 50 youngsters attained the therapeutic criterion. The conclusion reached in both studies was that the subjects simply were not taking their prescribed medications. The lesson derived from these findings is that compliance must be assured if an investigator expects to place any reliance on medication scores as a dependent variable.

6. *Frequency and Duration of Hospital Admissions.* It has been suggested by Chai and his co-workers (1968) that the number of hospitalizations experienced by a patient over a period of time would be an excellent guide to the severity of asthma. Hospitalization at a high rate, for example, might imply that patients are suffering severe asthma. Indeed, several studies have found frequency and duration recordings of hospitalizations for asthma useful dependent measures (Creer, 1970; Creer, Weinberg, & Molk, 1974).

A problem in many attempts to assess frequency and duration of hospital admissions is that the measures are often influenced by variables independent of asthma, certainly of asthma severity *per se.* For example, two studies (Creer, 1970; Creer *et al.*, 1974) have demonstrated that removing reinforcement contingencies alone, staff attention or television, for example, within a hospital setting led to decreases in both the frequency and duration of hospitalizations of asthmatic children. The influence of such variables on these measures raises serious questions about their validity.

Assessment procedures most closely identified with behavioral scientists studying asthma are:

1. *Psychometric Testing.* Historically, psychometric testing has been the most prevalent form of measurement used in the assessment of asthmatic children. A primary aim in using tests was an attempt to isolate a personality pattern common to these youngsters. The results of this effort have been equivocal at best and have generally failed to demonstrate the existence of such patterns (Purcell & Weiss, 1970; Creer, 1978). An unfortunate outcome of this endeavor among behavioral scientists is the tendency to view psychometric testing as lacking any value as an assessment tool with asthmatic patients. This is lamentable because evaluation with such instruments can provide useful information, both for clinical and research purposes. Suess (1980), for example, conducted a series of studies to assess the effects of medications on the intellectual performance of children with asthma. He found that intellectual performance was influenced by medications taken to manage asthma and that this performance varied according to the elapsed time between

ingesting the medications and testing. Such findings have implications for designing treatment regimens which transcend the single-minded goal of merely controlling the patient's asthma.

2. *Interviews, Questionnaires, and Checklists.* Whether primarily self-report or information obtained from the verbal report of secondary sources, data gathered by these techniques can be valuable in behavioral assessment, particularly in suggesting potential target behaviors for study or change or in indicating changes in behavior that are correlated with intervention. The major weakness with these techniques is that they represent subjective reporting and as such are open to biases or errors which are difficult to detect or measure.

3. *Direct Observation.* The direct observation of a person responding in a particular situation is the backbone of behavioral assessment. It has been the cornerstone of asthma research and health-care services for asthmatic children at the National Asthma Center (Creer, 1979).

Measurement parameters such as those associated with the reliability and validity of the target behaviors, observer accuracy, and bias and observer reactivity must be given full consideration by any investigator using the direct observation method. Such parameters are as critical to this method as they are to other techniques. In addition, experience in using direct observation in work on asthma has led to the identification of other problems. First, there are inherent difficulties due to the intermittent nature of attacks and the low rate of occurrence of asthma-related target behaviors. The low rate of occurrence presents problems in assessing measurement parameters, for example, reliability. But because the intermittency also carries with it considerable difficulty in predicting individual attacks, observation of the event by a trained observer is virtually impossible over a reasonably short time span. The problem is compounded by the fact that asthma attacks and asthma-related behavior often occur according to different schedules (Creer, 1979). As a result, there is rarely point-by-point agreement between attacks and a given target behavior. The upshot is that the acquisition of direct observation data requires a lengthy period of time because of missed opportunity or periods of disease quiescence. Under such extended time conditions, there is also an increased likelihood that changes in the behaviors under study may occur independent of any action by an investigator, for instance, sudden changes in asthma produced by seasonally-related variables or medication changes.

The challenge to the scientist is to increase the predictability of an individual's asthma attacks and to observe a series of asthma episodes over a short period of time. Greater predictability would permit standby observers or perhaps allow automatic monitoring devices to be available for recording the episodes.

4. Psychophysiological Measurement. These measures have been used extensively in asthma research and clinical application. For example, Hahn (1966) and Hahn and Clark (1967) found differences in heart rate, respiration rate, and skin temperature between asthmatic and normal children in response to tone, shock and problem solving tests. Electromyograph (EMG) has been employed often in clinically oriented projects with biofeedback-assisted relaxation training (Davis, Saunders, Creer, & Chai, 1973; Kotses, Glaus, Crawford, Edwards, & Scherr, 1976; Miklich, Renne, Creer, Alexander, Chai, Davis, Hoffman, & Danker-Brown, 1977).

5. Epidemiological Information. Knowledge about the social and economic impact of asthma has generally been ignored as a dependent measure in asthma research, primarily because like most epidemiological data it is difficult to gather. This is deplorable because such information is relevant to the accurate assessment of the disease. It expands our understanding not only of the illness but also of its consequences on the lives of patients and their families.

One weakness of epidemiological information is that it is primarily correlational in nature; this, therefore, precludes the possibility of demonstrating functional relationships between specific treatment variables and given social or economic factors.

It bears repeating that while the foregoing assessment techniques have been grouped in relation to the progressions with which they have been most closely identified, increasing numbers of medical personnel and behavioral scientists often employ identical procedures in their respective research and clinical efforts.

The procedures described above can be grouped into one of three categories—overt motor, physiological, or cognitive-verbal. Each category in its own right provides relevant information about asthma. Although the categories may produce outcome measures that covary in the same direction with introduction of an intervention procedure, such covariation does not always occur. Studies measuring overt motor, physiological, and cognitive-verbal behavior have found only moderate correlation in the results produced by the three categories (e.g., Hartshorne & May, 1928; Chai *et al.*, 1968). In fact, often there is little or no correlation between measures from the three categories obtained on the same patients for diagnostic purposes. In a study by Rubinfeld and Pain (1976), for example, patients claimed they were not experiencing any form of respiratory distress while, at the same time, pulmonary function tests indicated a 50% impairment in their breathing ability. Alexander, Miklich, and Hershkoff (1972) found improvement in the pre- and post-flow rates of asthmatic children who had undergone relaxation, yet, no concurrent changes in either medication scores or hospitalizations for asthma were noted with these patients. Creer (1970) and Creer *et al.* (1974) reduced the frequency and duration of children's hospitalizations for asthma but found no concomitant alterations in medication requirements or the pulmonary physiology of these youngsters. Finally, Renne and Creer (1976) showed that by shaping children's medication-taking behavior on the intermittent positive pressure breathing apparatus—a machine that converts liquid bronchodilator medication into a nebulized form and delivers the inhalable substance under pressure to the patient's lungs—they could reduce the amount of symptomatic medication required by the youngsters. Again, however, there were no concurrent changes in other measures of asthma, including pulmonary function tests or the rate and duration of hospitalizations. This lack of correlation between the three categories and the capability to manipulate each category independently suggest that multiple investigative procedures be incorporated into any asthma-related study or clinical assessment program. As emphasized by Chai and his co-workers (1968), the accurate assessment of asthma requires the use of all possible measures in order to obtain a fuller understanding of asthma and its effects on the patient.

MANAGEMENT OF ASTHMA

Having presented a background on childhood asthma, we now turn to what should be the most significant contribution to this chapter—how behavioral principles and techniques have been applied to the management and prevention of asthma attacks. This approach represents a blending of medicine and behavioral technology, a marriage that, although sometimes experiencing the trials and tribulations that mark any serious attempt to be truly collaborative, has proved to be successful. The approach is dynamic in the sense that treatment constantly undergoes scrutiny and, where necessary, change. The approach is prevention-oriented but permits a quick response to problems as they occur. Rather than seeking that elusive "cure" for asthma, we have focused attention on the management of both asthma-related behaviors and the consequences of the affliction to patients and their families. This does not mean that asthma *per se* is ignored. It is obvious, for example, that by improving medication compliance or reducing panic during attacks the process of the disorder is also affected. Finally, the approach has from time to time involved considerable trial and error in the search to unravel and understand the mystery of childhood asthma. As will be told later, this has resulted in numerous excursions into barren culs-de-sac.

Sequence of Events

An asthma attack may begin with a sensation of tightness in the chest or shortness of breath. Coughing or wheezing may follow. The attack can further intensify, sometimes reaching the state of status asthmaticus. The aim of treatment is to prevent the attack or to abort it as rapidly as possible, hopefully in its mild stage. Most physicians and medical personnel follow a proven strategy in treating an attack. They usually begin by administering an inhaled medication to determine if control over the attack can be established early. If not, treatment proceeds to the next step, generally involving the administration of an additional stronger medication. Other steps may be taken sequentially until control is achieved. The point is that, before an attack occurs, most experienced, competent physicians know what sequence of steps is to be followed in the treatment of an asthma attack.

Patients can be taught to manage their attacks in a similar manner (Creer, 1980a, 1980b). In a self-management study sponsored by the Division of Lung Diseases of the National Institute of Heart, Lung and Blood, children and their parents were trained to follow a step-wise treatment regimen. The treatment plans were drawn up in conjunction with their physicians. The children and their parents were taught first one step and then the next until the condition was brought under control. Decision-making centered around when to take certain actions rather than on what action to take since the steps were laid out in a predetermined plan.

Steps within the asthma treatment program can be conceived of as links in a chain. Response links or steps typically included in a response chain for a given asthmatic child are shown in Figure 2. This figure graphically depicts the process of treatment that might be followed by children in controlling their asthma. It reflects both patient decisions and the behaviors occurring in response to these decisions. It will be shown that each of these links is significant to the management of an attack.

L_1 *Symptom Discrimination.* Generally children encounter few difficulties in discriminating the onset of an asthma attack. They correctly interpret physical sensations such as tightness in their chest or a shortness of breath. Parents, too, by observing the behaviors of their youngsters, are capable of discerning cues of an incipient attack such as coughing or wheezing. Recognition of these behaviors and sensations as asthmatic symptoms leads to the initial question that must be answered: Are these events indicants of an attack? It may be that they are not. In some instances, such behaviors or sensations may only be a warning

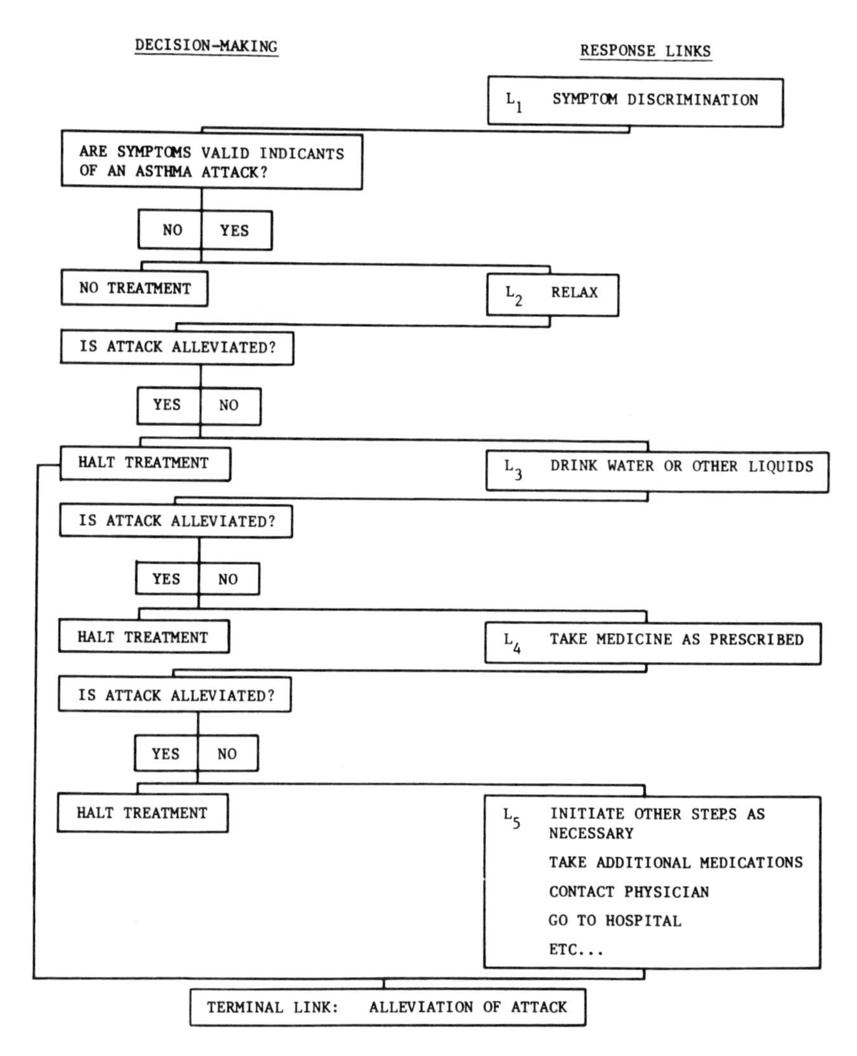

Figure 2. Patient behaviors followed to alleviate an asthma attack.

that an attack may occur. For example, slight wheezing sounds may portend an exercise-induced attack if the child does not stop engaging in some ongoing physical activity. Or, coughing may be a natural reflex to the inhalation of some foreign stimulus. Once the child stops exercising or the foreign substance is removed, the wheezing and coughing sounds may stop with no progression into an asthma attack.

Children and their parents decide at this point what their next step should be. If the coughing and wheezing sounds cease, then no treatment is necessary; if, on the other hand, the wheezing and coughing persist after the removal of the offending stimuli, they must proceed to the next link in the treatment chain.

L_2 *Relaxation.* Once the signs of asthma are recognized, the initial step of treatment is to induce relaxation. Relaxation is imperative for at least three reasons. First, relaxation permits the patient to establish control over the attack. In some cases, relaxation may keep the attack from becoming more severe. Second, there is some evidence that relaxation may actually improve the pulmonary functioning of some patients (e.g., Alexander *et al.* 1972), although no one has systematically demonstrated the effect while patients are having an attack. Finally, by systematically relaxing during an attack, patients are actually exercising some personal control over the affliction. In so doing, they become part of the solution to the problem of alleviating the attack rather than becoming a source of other problems that must be managed concurrent with asthma management.

Again, patients and their parents must judge whether the attack has been mollified. If the condition has improved, the next treatment step need not be taken. If the asthma has not been relieved, on the other hand, the next link in the treatment chain must be initiated.

L_3 *Drink Water and Other Liquids.* The next step often involves drinking water or other liquids. It is recommended that the patient sip a full glass approximately every 15 minutes. Warm liquids are preferred because they apparently promote bronchodilation; very cold liquids, on the other hand, may increase bronchoconstriction.

There are two major reasons why drinking liquids are beneficial (Creer, 1980c). First, liquids prevent dehydration and help to loosen mucous, making it easier to expectorate. Second, it is another way in which the child can become personally engaged in a self-management skill and directly contribute to controlling the disease.

If the attack improves with the ingestion of liquid, treatment might be halted at that point. If the attack continues, the patient and parents must decide whether to move to the next treatment link.

L_4 *Take Medicine as Prescribed.* In the event of an attack, most physicians prescribe medications which, regardless of precipitating stimuli, rapidly relieve the condition. These medications are in bronchodilator form and work by dilating the bronchi (Chai, 1980). Bronchodilators fall into two major groupings. First, there are the so-called sympathomimetics. These compounds, including epinephrine, isoproterenol, metaproterenol, and terbutaline, stimulate the sympathetic nervous system and produce bronchodilation. Second, there are purine derivatives called methylxanthines. These medications, the most common of which are theophylline and aminophylline, are of only moderate value in treating acute attacks. The value of these preparations lies in the fact that they are slowly absorbed and, therefore, are ideal for continuous therapy or for acute situations that require intravenous rather than bronchodilation treatment.

In the majority of cases, asthma control will be established by following the four steps outlined above. At this point, the focus of treatment for the patient shifts back to carrying out prescribed preventative measures that constitute the routine treatment program designed to prevent acute asthmatic episodes from occurring.

L_5 *Initiate Other Steps as Necessary.* If the attack persists, additional steps must be taken. This will quite likely include the consumption of more medicine ranging from additional doses of bronchodilator substances to other more potent medicines. It is possible, for example, that corticosteriods will be required. Physicians are reluctant to prescribe corticosteriods because of their potential side effects; however, in instances in which the patient is suffering severe asthma, particularly status asthmaticus, they can be lifesaving. In these instances, there is often no alternative but to administer corticosteriods.

Close monitoring of the attack is necessary to determine whether there is an improvement in the condition. An amelioration of the episode provides a signal to discontinue treatment of the attack while continuing to monitor the patient closely. If the episode persists or intensifies, the next step is usually to notify the physician or to journey to the emergency room of the local hospital. At this stage, parents must be prepared to act immediately according to whatever strategy they deem appropriate. It is at this point that it becomes imperative to know what to do and when to act.

The treatment steps as presented may convey the impression that treating an asthma attack is a simple matter. Nothing could be further from the truth. Asthma is a complex disorder. As we pointed out earlier, attacks are characterized both by the intermittent schedule on which they occur and by variability in severity from attack to attack. For this reason, some children may require emergency room treatment at a nearby hospital as a first step in treatment. For these patients, the attack can almost implode and immediately paralyze breathing. At other times, on the other hand, an attack may not even disrupt ongoing activities. Even with some discomfort, the patient and parents may elect not to initiate any treatment. It must be emphasized that, at all times, the patient must be an ally with the physician in the treatment decision-making process. Only such collaboration offers anything akin to a guarantee that asthma can be properly managed.

Too often, patients either lack knowledge as to what steps to take in managing an asthma attack or behave in ways that interfere with efforts to treat the attack. Behavioral techniques have played a significant role in helping to resolve such problems. Below are problems and how they have been resolved through behavioral technology.

Failure to Discriminate Attack Onset

Not all patients readily discriminate an incipient attack. For one thing, Creer (1980c) noted that they may attend to the wrong responses while ignoring those responses which are valid stimuli in heralding an attack. This was demonstrated by Chai and his co-workers (1968) and by Rubinfeld and Pain (1976), who pointed out the lack of agreement between subjective reports of symptoms as reported by patients and objective indexes of asthma. As Rubenfeld and Pain noted, there evolved a situation in which a patient would report that he was experiencing no impediment in breathing while at the same time pulmonary physiology tests indicated that he was exhaling less that 50% of his predicted value. The reverse can also be true: patients may claim to be experiencing respiratory distress which cannot be verified through a pulmonary physiology examination. Creer (1980c) suggested that somewhere in the history of such patients certain stimuli, either external or internal, became associated with the attacks. These stimulus cues may lead these patients to make an incorrect decision about their resporatory functioning. For example, they may associate a feeling of fear with an attack. Thereafter, a sensation of fear is incorrectly interpreted as portending asthma.

A second explanation as to why there is a discrepancy between subjective feelings and objective indices of asthma centers on the function of the context within which attacks occur. Hence, patients may claim to experience more attacks in particular settings, for instance, when the context is seen as aversive and an attack provides a means of escape, and fewer attacks in other settings, for instance, when the context is viewed as positive and escape would be a form of punishment. For example, many children have claimed to be experiencing asthma when faced with an examination at school. These same children may be wheezing and in obvious distress, however, and yet claim to be asthma-free in order to remain at a movie or party.

A final explanation for the failure to discriminate the symptoms of asthma that is valid for a minority of patients is that they lack hypoxia drive and therefore are incapable of receiving stimulus input signaling that an asthma attack is starting. Here it is a failure to receive the appropriate stimulation, rather than a failure to attend to the correct stimulus, which results in their not discriminating the onset of asthma.

Figure 3 depicts a strategy that can be used to improve symptom discrimination in patients with asthma. Initially, they may attend more closely to distal events. This means that there may be antecedent conditions which, while occurring at a time distal to an attack, may nevertheless be good predictors of the impending episode. Peak flow values are illustrative of a distal antecedent condition. By obtaining flow values in both the morning and evening, the physician can monitor the pulmonary physiology of an asthmatic patient over time. If the data are plotted as a function of time, as shown in Figure 1, it is possible to predict the likelihood of an attack's occurring. If children and parents are taught this method of pulmonary monitoring, they can take steps either to prevent the episode or to keep the attack mild. Regardless of what eventually occurs, however, the patient has some forewarning that an attack may be imminent and will be in a better position to take corrective action.

Taplin and Creer (1978) described how peak flow rates could be used in predicting more precisely whether asthma attacks would occur. Peak flow measurements were gathered both morning and evening in the homes of two asthmatic children. At the same time, their mothers accurately maintained daily records on the frequency and severity of asthmatic episodes. On the basis of the calculation of base rates of attacks and flow rates, two conditional probabilities were calculated for each youngster: first, the probability of asthma in a 12-hour period following a flow rate score equal to or less than the critical value and, second, the probability of an attack within a 12-hour period following a peak flow score greater than the critical value. Results of the exploratory study suggested that peak flow data collected from the patient increased the predictability of asthma attacks. Creer (1979) noted the implications of these findings, especially for patients diagnosed as suffering intrinsic

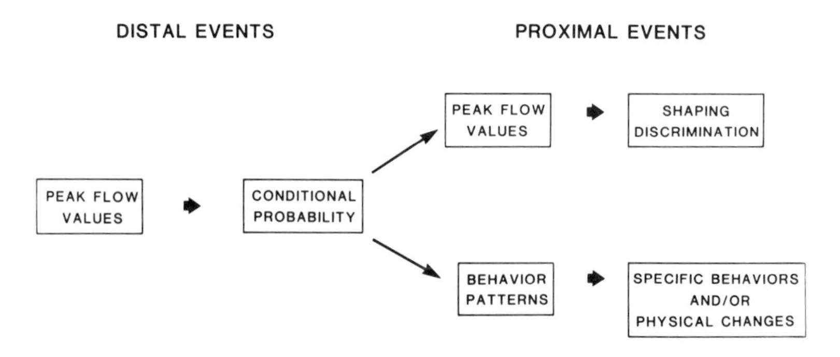

Figure 3. Distal and proximal conditions that may occur antecedent to an asthma attack.

asthma, in improving symptom discrimination and in managing the disorder.

Asthmatic children and their parents can also be taught how to discriminate symptoms as they occur. They may, as shown in Figure 3, better attend to those events that occur proximal to an attack. In one study, Renne, Nau, Dietiker, and Lyon (1976) taught youngsters to attune their asthmatic symptoms to an objective index of their condition. The training procedure used was not unlike classical psychophysical techniques in that the children correlated the magnitude of one response system—private, subjective feelings—to the magnitude of a second response system—public, objective data. Renne and his co-workers at the National Asthma Center instructed the subjects to report for symptomatic treatment of asthma at increasingly higher peak flow values. Reinforcement was added by giving the youngsters scrip exchangeable for gifts or the opportunity to chart their progress if they reported at or above the criteria. Peak flow measures were taken only after they reported for treatment. Results of this pilot study indicated that children were able to report for treatment at much higher levels of pulmonary functioning, some as high as 30% below their normal, asthma-free levels. As a consequence, children learned to discriminate asthmatic symptoms earlier on in an attack and to report to the center's hospital for treatment while the episodes were relatively mild.

Children and their parents involved in a self-management project at the center were taught to attend to any physical and behavioral clues which predicted their asthma attacks. This is another approach to monitoring proximal events. Some common behavioral patterns reported by these families as precursors of attacks are listed in Table I.

A number of specific behavior patterns or physical changes are even more surely predictive of asthma. Some commonly observed actions or physical changes are also noted in Table I. These range from verbal complaints, for instance, complaints of fatigue, to changes in breathing patterns. Facial changes such as a reddened face or the appearance of dark circles under the eyes, and motor patterns like repeated stroking of the chin or throat and hunching of the shoulders may also signal the onset of asthma. Parents on the self-management project were taught to monitor for any such repeated behavior/physical changes preceding asthma episodes. As a consequence of such careful monitoring, many were able to predict when an attack was imminent. They were then instructed to begin the treatment sequence by having the patient relax and drink liquids with the appearance of these cues. The more common behavioral patterns, however, are apt to be witnessed at times other than when a youngster is experiencing an asthmatic episode. For this

Table I. Behaviors and Behavioral Patterns That May Be Exhibited Antecedent to an Attack of Asthma

Behavioral patterns	
Aggressive	Moody
Quiet	Hyperactive
Frustrated	Obnoxious
Angry	
Specific behaviors and physical changes	
Verbal complaints	Breathing changes
Complaints of fatigue	Shallow breathing
Complaints of tightness	Breathing rapidly
Complaints of chest or neck changes	Breathing from neck up
	Prolonged coughing
Facial changes	Other changes
Red face	Postural changes
Perspiration	Stroking chin or throat
Dark circles under eyes	Hands on top of head
Eye color changes	Hunching of shoulders
Flaring nostrils	Indentation of neck
Swelling of face	Bluish fingernails
Pale face	Bad breath
Bluish lips	Voice changes
Bluish complexion	

reason, it is important that the children and their parents base the initiation of any treatment on the behavioral or physical change that most reliably predicts the onset of an asthma attack.

At an 18-month follow-up interview, parents reported that learning to monitor the physical and behavioral changes in their children had proved to be the most reliable means of predicting the onset of an attack. By initiating treatment at the first indication of a problem, many parents asserted that serious asthma attacks were aborted after completion of only one or two of the treatment steps described earlier. These children, too, began to monitor their illness more closely. Many learned to be aware that such symptoms as an itch in the throat or neck could portend the onset of an attack. Often, on their own, they would initiate treatment.

Failure to Take Medications Properly

An inhaled medication is often prescribed for use in the initial stage of treatment. This is because nebulized substances, frequently dispensed by inhalation therapy equipment, quickly reach the bronchial area to promote bronchodilation. A common problem associated with this form of treatment is that patients do not know how to use the equipment properly and, therefore, are unable to derive maximum benefit from the medication it provides. This point was brought home in a study conducted by Renne and Creer (1976). The aim of the study was to teach asthmatic youngsters how to use the intermittent positive pressure (IPPB) machine correctly. Basically, the IPPB converts liquid bronchodilator medication into a nebulized form and delivers the inhalable substance under pressure to a patient's airway where it can reduce bronchial obstruction and alleviate an asthma attack. Improper use of the IPPB, on the other hand, impedes the process, mainly because the medication fails to reach the site of obstruction in the lung. Under these conditions, the asthma may worsen and require additional medications and treatment.

The failure of the medication to alleviate the attack due to impeded transport is often interpreted as medication ineffectiveness and results in unnecessary changes in the treatment program. The first step to take when equipment-dispensed medication is ineffective is to observe the patients and determine whether they use the equipment correctly. Systematic observation of asthmatic children using the IPPB machine suggested that three separate behaviors had to occur concurrently for correct operation of the equipment: eye fixation, facial posturing, and diaphragmatic breathing.

Eye fixation was defined as looking directly at the dial on the front of the IPPB apparatus whenever the mouthpiece was in place. An analysis of the children's behavior indicated that eye fixation was critical to the youngsters' receiving immediate feedback on their performance and helping to reduce interference or distracting events. In general, eye fixation facilitated their attending more closely to the task of using the machine properly.

Facial posturing was defined as holding the mouthpiece unit at a 90° angle to the face, with the lips held motionless and firmly secured around the mouthpiece. No visible movement of the cheeks, for example, puffing, or nostrils flaring was allowed. The significance of the behavior was that the nebulized medication was less likely to be lost through either the corners of the mouth or the nostrils.

Diaphragmatic breathing was defined as a distension of the abdominal wall upon inspiration and contraction upon expiration. The abdominal wall had to move visibly outward on inspiration and inward on expiration. Diaphragmatic breathing insured that the vaporized air reached the deeper recesses of the lung in greater concentrations than was otherwise possible with intercostal directed breathing.

Renne and Creer employed a multiple baseline design (Baer, Wolf, & Risley, 1968) in which, following a period when baselines were obtained on all three behaviors—eye fixation, facial posturing, and diaphragmatic breathing—the responses were sequentially shaped. Scrip that could be exchanged for reinforcement—inexpensive gifts costing $2.00 or less at a nearby variety store—was provided following each correct response.

The results, depicted in Figure 4, suggested that the technique was highly effective in teaching youngsters with asthma to operate respi-

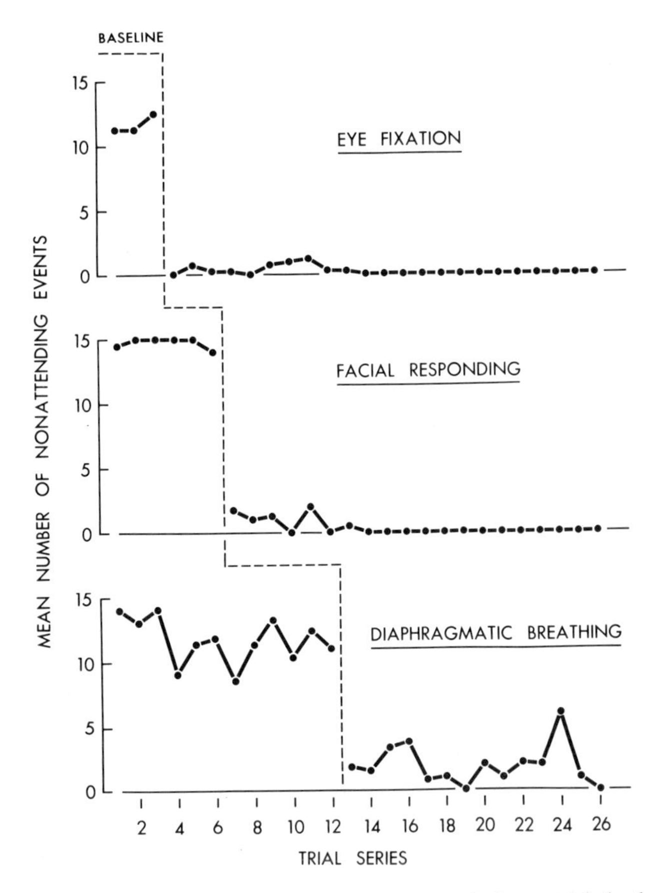

Figure 4. The mean number of inappropriate events recorded over trials for four subjects on three target responses: eye fixation, facial posturing, and diaphragmatic breathing. From "The Effects of Training on the Use of Inhalation Therapy by Children with Asthma" by C. M. Renne and T. L. Creer, *Journal of Applied Behavior Analysis*, 1976, *9*, 1–11. Copyright 1976 by the University of Kansas. Reprinted by permission.

ratory equipment properly; the multiple baseline procedure, on the other hand, permitted the training strategy to be convincingly assessed.

In addition to the demonstrated training success, it was found that the children required less symptomatic medication to alleviate their asthma attacks after they had been taught to use the IPPB machine correctly. A pre–post mean treatment effectiveness score was calculated with 41% effectiveness obtained prior to intervention and 82% effectiveness obtained after training on the apparatus for the same youngsters. Another benefit derived from the project was that the method of instructing children to use the IPPB machine appropriately was readily taught to nurses who, in turn, began routinely to instruct other children in its use. Their implementation of the training program as standard procedure improved the treatment effectiveness of the machine in general and, as a result, diminished some of the difficulties previously encountered in establishing control over the asthma attacks of some patients.

Panic

A problem commonly associated with asthma attacks has frequently been labeled as *panic* (Creer, 1971, 1974, 1979; Creer & Renne, 1968). In some patients panic exacerbates ongoing asthmatic episodes and interferes with treatment. Chai and Newcomb (1973) pointed out that children in panic during an attack are unable to cooperate with treatment and often exhaust themselves physically. In addition, panic can be contagious, often engulfing other family members or members of the treatment team (Creer, 1974). The upshot of such behavior is not only a child whose asthma rapidly increases in severity because of his or her own panic responses, but attending personnel who are handicapped by their own fears and anxieties.

In our approach to the problem of panic, the initial assumption was that the term identified a circumscribed pattern of behaviors (Creer and Renne, 1968). The basis for the premise was another assumption: Since physicians and nurses could independently compile identical lists of children observed to display panic, they must be referring to a unitary behavioral mosaic. Thus, the identification of specific behavioral components to panic became the aim.

The initial behavioral analysis involved having nurses and/or physicians complete a behavior checklist of commonly linked panic responses after each asthma attack for all children in the panic group. Eighteen patients fell into this category. Data were gathered on 16 subjects following 26 attacks. The results are shown in Table II, which lists 14 specific responses and the percentage of asthma attacks where the

Table II. Behaviors and Physical Changes Noted by Attending Medical Personnel[a]

Behavior and/or physical change	Percentage of hospitalizations when change was detected
1. Blue complexion	65
2. Tenseness	62
3. Overbreathing	58
4. Chattering speech	46
5. Fearful expression in eyes	46
6. Hyperactivity	35
7. Holding breath	35
8. Hyperventilation	27
9. Overexaggeration of condition	27
10. Screaming	27
11. Demanding	23
12. Impatience with treatment	19
13. Difficulty in taking medications	8
14. Disregard of explanations provided by medical staff	8

[a]Behaviors and physical changes noted by attending medical personnel during 26 hospital admissions of 16 children reported as panicking during attacks.

responses were observed. The patients turned blue in the face during 65% of the attacks. They were also seen as being tense and overbreathing in a high percentage of the episodes. Verbal behavior consisting of chattering, screaming, complaints about treatment, and failure to accept explanations of treatment were also prominent.

The initial efforts to define panic involved direct observation, interviews with children, nurses and physicians, and a television monitoring system set up in the emergency room of the center's hospital. From information thus gathered, it appeared that a synonym for panic might be "fearful agitation." Creer (1979) described children who displayed such panic in the following manner:

> They seemingly found it difficult to remain in bed, but were always trying to get up and cry out their anguished pleas for more and more attention. The youngsters would demand larger dosages of medications or the initiation of different types of treatment. They kept asking to see the attending physician in order to complain that the emergency respiratory equipment was not working rapidly enough, that other steps should be started, or, in many cases, that the physician was incompetent to treat them. Many youngsters appeared to be hyperventilating. In short, these children labeled as panickers galvanized the attention of everyone not only by their constant verbal bickering, but because their sudden movements could disrupt ongoing treatment. (p. 176)

Initially it was assumed that a single topographical pattern of behavior characterized by the above description accounted for all panic reactions. Later, however, it became apparent that some patients assigned by the staff to the panic group were going undetected by the 14-point behavior checklist. Direct observation of them during asthma attacks revealed why this was happening: They were undetected by this instrument because their responses constituted a second response pattern which was almost antithetical to what was being sampled on the checklist (Creer, 1974). A good description of this second type of response pattern, also termed panic, is as follows:

> The best adjective for patients displaying this pattern is 'frozen.' Wide-eyed and immobile in their beds, they stare at the ceiling, refusing to signal for help even when they require it. If not watched closely, they are apt to become cyanotic or, in some cases, unconscious, without making any signal that they are experiencing acute distress. In many respects, these patients frighten attending medical personnel most, simply because they do not make their presence known on busy pediatric wards. (Creer, 1979, p. 176)

The commonality shared by patients who exhibited either panic response pattern was that their behavior interacted with their asthma to intensify the attacks (Creer, 1979). The panic rendered their asthma more difficult to manage medically and in extreme cases threatened their very lives. Fortunately, both forms of panic readily lend themselves to alteration through application of a behavior therapy technique known as systematic desensitization with reciprocal inhibition (Wolpe, 1958).

The initial step is to teach a patient to perform a response that is incompatible with panic responses. Relaxing is one such response and is usually taught by the application of an abbreviated version of Jacobson's (1938) progressive relaxation method. Basically, this procedure entails alternate tensing and relaxing of successive muscle groups, on cue from the therapist, along with suggestions of calm and pleasant affect. In most instances, teaching children to discriminate muscle sensations so that they may determine whether they are tense or relaxed and to relax on self-cue can be achieved in two to four sessions. Three conclusions can be drawn from the work with children at the center (Creer, 1979). First, biofeedback was useful in teaching relaxation to some patients, but it was not a necessary condition for relaxation training to occur. Biofeedback was particularly useful with the younger children, possibly because it resembles something out of *Star Wars*. Second, children as young as 5 years can be taught to relax using the progressive relaxation method. Finally, many patients believe that relaxation, by itself without systematic desensitization, was of benefit in helping them to manage their asthma. More on this will be noted later.

The second step of systematic desensitization by reciprocal inhibition involves identifying the stimuli which provoke panic. Here situations are described in detail with an effort to ascertain the degree of panic they induce. Then the situations are arrayed in hierarchical order from least disturbing to most provocative. This can be best illustrated by presenting a hierarchy obtained from a child labeled as a panicker (Creer, 1979, pp. 178–179).

a. "I got asthma while I was asleep. It was really a bad attack."

b. "I guess that I got out of bed, but I don't remember much about it. I can remember going out into the hall, but that's it. My parents said that my head hit their door when I passed out."

c. "The next thing I remember, we were racing down the street in the car. I can remember seeing some of the lights flashing off the windows of the stores we passed; everything was kinda hazy."

d. "I remember coming to in the emergency room of the hospital. My dad was running around trying to get someone to help me. Then I passed out again."

e. "I remember coming to on the stretcher. The people in the emergency room were taking me down the hall as fast as they could."

f. "They gave me oxygen. It seemed to help a lot. I also got my first IV about then."

g. "It wasn't until I was in a room under an oxygen tent that I really started to see what was happening. My parents and the doctor were there."

h. "My mother was crying, everyone looked scared. It was then that I really got scared."

A common thread emerged in hearing children describe occasions on which they had been especially frightened or panicky: the importance of modeling processes in the acquisition of their panic. The majority of youngsters contended that they panicked because they had observed their parents, nurses or physicians display similar behaviors. In many ways, the children's descriptions of their parents' and other adults' reactions during asthma attacks corresponded almost word for word with the accounts of the staff in describing the children's panic responses. As Creer (1979) pointed out:

> The sight of a child gasping for breath or gradually becoming cyanotic is frightening to any witness, physicians and nurses included. When the child begins inexorably edging toward status asthmaticus, despite the best efforts of attending personnel, the fears of everyone around the youngster are compounded; their behavior, in turn, is imitated by the child. (p. 179)

The final steps of systematic desensitization treatment consists of a juxtapositioning of panic responses with relaxation as scenes from the hierarchies are imagined or visualized by the patient. The stimulus scenes in the hierarchies are presented so that the scenes producing the least amount of panic are presented first. Once the patient can remain relaxed while imagining the scene, the next scene in the hierarchy is presented. The procedure continues until all of the scenes can be visualized by the patient with no panic reaction to them.

Creer (1979) made two observations with respect to the alteration of panic through systematic desensitization by reciprocal inhibition: first, generalization of the treatment effect to real-life asthma episodes occurred. Reports from children, nurses, parents, and the medical staff indicated that the youngsters appeared to relax during subsequent attacks of asthma and that panic had abated in all the children treated in the program. Many of the patients claimed to have aborted incipient attacks by relaxing. While such reports are anecdotal, they do suggest that relaxation training can be useful in helping patients achieve a modicum of self-control over their affliction. Second, children received a considerable amount of social reinforcement when they attempted to relax in the face of an attack; this reinforcement served to strengthen and maintain the relaxation procedures they had learned.

One final point should be made regarding the construction and use of the panic behavior checklist. Initially, the 14-point checklist excluded several of the responses that later were important in defining panic. The exclusion was eventually corrected, but it does emphasize one common weakness found in constructing and using behavior checklists: too much reliance is often placed upon the use of a face validity approach while neglecting information from other sources, particularly direct behavior observation. Like the original 14-point checklist, such instruments probably suffer from a low-validity estimate and consequently may fail to provide the user with some critical information.

Misuse of Hospital Facilities

There are some children who appear actually to seek admission into a hospital and who seemingly make an effort to remain confined there even with remission of their illness. This pattern has been defined as malingering (Creer, 1970; Creer, *et al.*, 1974), and can be best illustrated by two case histories: Jack frequently complained of stomach pains upon hearing of his impending discharge from the hospital following the establishment of control over his asthma attack. His complaints invariably led to a series of examinations and tests, all of which postponed his release from the hospital. Another patient, Don, reported chest pains upon learning of his impending release. His complaints immediately mobilized the medical staff since, in the past, he had experienced two cardiac arrests associated with status asthmaticus. After several episodes during which the remarks of the boys could not be verified and after observing that the youngsters were spending an increasing amount of time in the hospital, the medical staff became convinced that the pair were malingering. They were reluctant to ignore the children's behavior, however, for fear that sooner or later the complaints might prove valid.

At this time, a behavior analysis was conducted. Both boys were performing poorly in school, and admission to the hospital provided a way to escape from what was seemingly an aversive situation. Also, when hospitalized, the boys had access to a gamut of potential reinforcers, ranging from viewing television and reading comic books to resting in bed and/or socializing with other youngsters admitted to the facility or with the hospital staff. After identifying the factors possibly reinforcing and thus maintaining malingering behavior, the next step was to determine whether planned manipulation of these variables would alter the behavioral pattern. This assessment of the intervention effect was made by employing a single-subject reversal or A-B-A-B design. This design is best illustrated by describing the procedure in a step-by-step manner.

1. *Baseline Condition (A).* Baseline data on the duration and frequency of hospitalizations were gathered for Jack and Don over a period of six weeks. As the graph shows in Figure 5, Don and Jack spent 67% and 55% of the time, respectively, in the hospital during the baseline

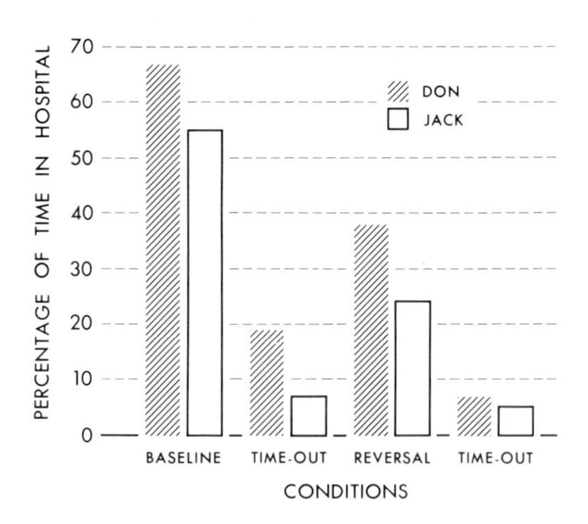

Figure 5. Percentage of time in hospital as a function of experimental conditions. From *Asthma Therapy: A Behavioral Health Care System for Respiratory Disorders* by T. L. Creer, New York: Springer, 1979, p. 189. Copyright 1979 by Springer Publishing Company. Reprinted by permission.

period. (Is there any wonder both boys performed poorly at school? They were rarely in attendance!)

2. *Time-Out Condition (B).* Intervention consisted of removing all available sources of positive reinforcement. To effect this condition the nursing staff adhered to the following instructions whenever Jack or Don was admitted to the hospital:

a. Each boy was to be placed in a room by himself.

b. Neither youngster was permitted visitors other than medical or nursing personnel who were to relate to them only in a "business-like" manner.

c. The patients were not allowed to leave their rooms to visit with other children.

d. The two boys were not allowed comic books and television sets in their rooms, although they might read books related to their school work.

e. The youngsters were to eat all meals in their rooms by themselves. They were permitted to leave their rooms only to go to the restroom.

These instructions were followed by the nursing staff for a period of six weeks. As depicted in Figure 5, the amount of time spent in the hospital during this condition decreased to 19% for Don and 7% for Jack.

3. *Reversal Condition (A).* In order to determine whether the time-out procedure (removing positive reinforcement) was the variable altering the malingering behavior, a reversal period was introduced. Essentially, this condition consisted of rescinding the instructions and permitting the nurses to manage Jack and Don as they managed them in condition A (baseline), and as they managed the other youngsters who were not malingering. The intent was to have the reversal period also last for 6 weeks; because of demands on the nursing staff, however, the time was cut in half.

Figure 5 shows that under the reversal condition hospital attendance and length of hospital stay increased for both boys: the percentage of time spent in the hospital rose to 38% for Don and to 24% for Jack, convincingly demonstrating that the reinstated reinforcers were maintaining the malingering behavior.

4. *Time-Out Condition (B).* The final condition involved the reinstatement of the time-out procedures for 8 weeks, a period of time that coincidentally terminated with the end of the academic year. Figure 5 shows that during this second time-out period Don spent 7% of his time in the hospital, while Jack's hospitalization time was reduced to 5%.

Use of time-out from positive reinforcement to manage malingering behavior was extended in another study reported by Creer *et al.*, (1974). Results from this single case study showed that the time-out procedure also was effective in reducing malingering when the technique was applied on a less regular, rate-dependent basis. These results are important because it is frequently difficult to sequester a patient for a prolonged length of time on pediatric wards. In such situations, the procedures would have to be applied on some sort of time-sampling or rate-dependent basis. Another important outcome of this study was that it demonstrated that a socially desirable class of behaviors could be increased coincident with the alteration (decrease) of the malingering responses. Academic performance was targeted, and the youngster acquired appropriate academic behaviors as the malingering behavior decreased. Finally, a program to insure generalization to the youngster's home situation upon discharge from the center was detailed for implementation by parents. A follow-up contact one year after discharge indicated that the malingering behavior had not returned and that high academic performance was still being maintained.

Recently, Shepard and Hockstadt (1978) reported on the application of time-out from reinforcement with asthmatic children admitted to a pediatric hospital. Seven youngsters took part in the investigation. These children were considered "overusers" of the hospital in that they (1) verbalized the intent to develop asthma symptoms in order to be admitted when a friend was hospitalized or a special event was planned; (2) stated a preference to remain in the hospital rather than return home; (3) predicted a quick return to the hospital following discharge; and (4) prolonged hospitalizations by developing symptoms of asthma when told of their impending discharge.

Intervention consisted of using a time-out procedure similar to that described by Creer (1979) each time a subject was admitted to the hospital. When a youngster's pulmonary functioning reached a prescribed level according to flow rates obtained with a peak flow meter, he was discharged from the hospital. The study, which was conducted over the course of one year, employed three dependent measures:

a. The average length of stay in the hospital
b. The average interval between hospitalizations
c. The latency required before a patient could blow a prescribed level on the peak flow meter

A statistically significant reduction from 18.3 days before intervention to 9.0 days after intervention was found in the mean length of stay in the hospital. The mean length of time between hospitalizations increased from 48.9 days to 85.3 days, and the mean time required by subjects to reach the prescribed level on the peak flow meter, that is, the time required for their respiratory obstruction to clear, decreased from 9.8 days to 4.8 days. These results not only demonstrate the effectiveness of the time-out procedure in both reducing time spent in the hospital and increasing the duration of time between hospitalizations but also provide an example of how multiple dependent measures can be used in assessing multiple effects of behavioral intervention (Creer, 1979).

PREVENTION AND HEALTH PROMOTION

The emphasis of the previous section has been on the behavioral treatment and management of children during asthma attacks. Such an emphasis reflects "emergency–fire" management procedures, which alone do little to prevent fires in the first place. Likewise, managing behavior during asthma attacks does little to prevent their occurrence. It would be ideal if patients could learn to prevent attacks and to live in a manner that would promote continued good health. It is in the area of health promotion and disease prevention where teaching self-management becomes a significant ingredient in any overall asthma treatment package. Space limitations restrict the discussion of self-management to a mere program outline.

The program to be described, funded by the Division of Lung Diseases of the National Institute of Heart, Lung, and Blood, has proved effective with children and their families (the program has been more fully discussed elsewhere [Creer, 1979; 1980a, 1980b]). The program focuses on four functions which constitute the essential ingredients of the self-care program: self-observation, self-instruction, decision-making, and self-induced stimulus or response change. Each of these functions will be discussed separately.

Self-Observation

There are two components to the self-observation routine (Creer, 1980a, 1980b): first, the patient or the physician (or both together) selects a specific behavior or behavioral pattern to be observed. In the case of asthma, for example, the patient may choose to track the number of asthma attacks he experiences, or the physician may ask the youngster to record medications taken because of asthma. The second component is the accurate monitoring and recording of the events of the behavior pattern targeted for observation. Hence, to continue our example, the patient must dutifully record the frequency of attacks or the exact amounts of asthma medications taken.

Two basic conclusions can be reached from the effort to teach self-observation procedures to asthmatic children (Creer, 1980a,b): first, most can learn to be reliable observers of their condition. Sometimes the acquisition of this skill involves behavior shaping. For instance, children may be initially asked to track (observe and record) the amount of water they drink daily. By recording each time they stop by a fountain for a drink or when they consume a glass of juice, these youngsters can learn to monitor their own behavior accurately. As a consequence, they can go a step further and be taught to keep records of behaviors more closely linked to asthma. Second, most children can supply knowledge regarding their asthma that, unless observed and recorded by them, would never become accessible. By using an asthma diary, for example, children can be taught to record information related to (a) their impressions of their own physical condition, for example, whether they experienced any symptoms of asthma; (b) their performance on the peak flow meter, that is, whether there was any objective evidence of pulmonary distress; (c) medications taken for asthma; (d) the times medications were taken, for instance, whether they took their medicine as prescribed; and (e) any information related to the socioeconomic aspects of asthma, for instance, how much school they missed because of the disease. Experiences in using an asthma diary on the project indicate that youngsters can become proficient at observing and recording their own actions and experiences.

Self-Instruction

Self-instructions are the verbal statements children make to themselves to prompt, direct, or maintain behavior (O'Leary and Dubey, 1979). It appears that self-instructions can assist children in controlling attacks provided that two conditions are met.

The first condition is sufficient knowledge about asthma and self-management procedures. This proviso was met in the self-management project through four weekly education sessions held with both children and their parents. The purpose of each session was as follows:

Session 1. During the initial session, the children and their families acquired a fundamental knowledge of asthma. A number of pamphlets and education materials, available from organizations such as the National Asthma Center and the American Lung Association, were used. In addition, a videotape depicting a physician describing the physiological aspects of asthma to a parent and child was shown. The content of the videotape, along with the written information, created the basis for group discussions. The materials were prepared and presented according to the ages and abilities of the participants. In the case of younger children, for example, this entailed the development of special reading and evaluative materials.

Session 2. In the second session, a videotape and assorted handouts described medications commonly used in the treatment of asthma. The asthma-controlling effects of the medications were outlined, followed by a discussion of the potential side-effects of each drug. This subject proved to be one of the more engaging topics in the entire program, judging from the increased participation in the discussion by both children and their parents.

Session 3. The third session was spent in presenting social learning or behavioral principles. The presentation was supported with handouts and written materials, including illustrations of how self-control techniques could be used by children to manage their illness and illness-related behaviors.

Session 4. Here, the focus shifted toward applying learning principles to asthma-related behaviors. Children were taught how to track their own behavior and how to apply behavioral principles to change specific problem behaviors related to their asthma.

The second prerequisite for effective self-instruction is to teach children the skills they need to use self-instruction procedures correctly. This was achieved by teaching children (1) to practice such skills as relaxing on self-cue, (b) to assume responsibility for taking asthma medications on schedule, and (c) to sharpen their observational competence so that they could track related responses and become aware of strategies for changing or managing them. Sessions 5 through 8 presented material covering antecedent and concurrent conditions, consequences of the disorder, and problem-solving strategies.

Session 5. If children can learn to recognize and react appropriately to certain stimuli or events, they may learn to keep their attacks from occurring or keep mild attacks from becoming more severe. Session 5 involved group discussion of antecedent conditions or events which occur prior to an asthma attack. Those children whose attacks were triggered by responses to anger (yelling or shouting) or anxiety (hyperventilating or holding breath) found this session to be invaluable. Their new-found skill in pinpointing and changing such antecedent events was a major step for them in establishing self-control over their asthma.

Session 6. The focus of session 6 was on those events or behaviors that occur concomitant with an asthma attack. Panic responses, medication-taking behavior, and breathing changes are examples of concomitant events. The children were taught skills which would help them to change such behavior and thus establish control over their asthma attack.

For example, they were taught relaxation skills which could be practiced during attacks, proper methods of breathing, and physiological monitoring procedures.

Session 7. The youngsters in the program and their families encountered many different consequences because of the disorder, some of which were problems. For example, the children often became the object of scorn or derision, were avoided, or were not permitted to engage in normal activities like other children. This session attempted to teach these children and their parents to respond to and resolve the various types of problems they encountered as a result of the illness, for example, how to be more assertive.

Session 8. The final session was spent discussing how to solve various types of problems that are often encountered by asthmatic patients. A representative situation used in teaching problem-solving strategy to them follows:

> You are out on a camping trip when you begin to suffer an attack. You seek out the medicine, but much to your dismay, you discover that you have left it at home. You are in a remote area and help is some distance away. What can you do in such a situation? (Creer, 1980a)

The children offered a number of solutions ranging from keeping calm and sending someone for help to having their parents brew some strong coffee for them to drink (remember that coffee is similar in chemical composition to theophylline). By this stage of the program, it was evident that the participants felt that they could solve many of the everyday problems they were likely to encounter because of their asthma.

Decision-Making

Creer (1980a) observed that

> The best medical attention and the best intentions of patients are wasted unless the patient makes the correct decisions with respect to his illness. Certainly, this is the case with asthma. A patient generates a large number of decisions each day that run the gamut from whether to avoid known precipitants to what medications should be taken to abort an attack. In short, the patient does have the ultimate control, in most instances, over the control of the illness.

Of particular significance are the decisions a patient must make during an attack. It is for this reason that an earlier section was devoted to detailing a chain of important behaviors that children might be taught to perform during an asthmatic episode (see Figure 2). Youngsters also have to make many other decisions that may directly affect their asthma. They must learn to make the correct decision, for example, if they are pressured by their peers to smoke. Not only may smoking precipitate an attack, but it may reduce the effectiveness of the medications they take to control asthma as well (Brandon, 1977). They must also learn what their limits are with regard to engaging in physical activity and then decide when they have reached those limits so as to halt activity and hopefully avoid an asthma attack. Finally, there are a number of decisions that asthmatic youngsters must make daily regarding such matters as when and how much medication is necessary for achieving control over the disorder. By closely monitoring their physical condition or status, they soon discover what decisions are appropriate given a wide choice of potential responses and varying consequences.

Self-Induced Stimulus or Response Change

Self-induced stimulus change means that asthmatic patients themselves are capable of changing conditions within their environment, in this case, those stimuli which may affect their asthma in some way. If they have knowledge of what precipitates their attacks, they attempt to alter the environment in such a way that they can avoid exposure to these precipitants. This may involve shunning pets or smoke-filled rooms (Creer, 1980a, 1980b). All persons have limited control over the environment within which they live, but it is imperative that asthmatic patients establish maximal control over those stimuli that trigger their attacks, even if it means initiating self-induced stimulus change procedures.

Self-induced change means that patients alter their own responses so as to avoid either an asthma attack or a worsening of the episode (Creer, 1980a, 1980b). Children with exercise-induced asthma face a particular dilema: on one hand, they recognize that exercise is often enjoyable as well as an ingredient of good health. On the other hand, they also realize that exercise may induce an asthma attack. To avoid the latter, they must engage in some self-induced response change procedure and not cross the boundary between healthy activity and harmful consequences to their physical well-being.

SUMMARY

The total management of a chronic respiratory disorder such as childhood asthma must include both medical and behavioral aspects, with patient self-reliance and responsibility a major goal. The application of the behavior analysis/diagnostic procedure and the powerful behavior change techniques described in this chapter by trained personnel can assist in this endeavor. It is only in this goal that asthmatic patients truly become partners with their physicians in the management of their disorder.

REFERENCES

Asa, K. Prognosis for asthmatic children. *Acta Paediatrica,* 1963, *140,* (Supplement) 87–88.

Alexander, A. B., Miklich, D. R., & Hershkoff, H. The immediate effects of systematic relaxation on peak expiratory flow rates in asthmatic children. *Psychosomatic Medicine,* 1972, *34,* 388–394.

American Lung Association. *Introduction to lung diseases* (6th ed.), 1975.

Baer, D. M., Wolf, M. M., & Risley, T. R. Some current dimensions of applied behavior analysis. *Journal of Applied Behavior Analysis,* 1968, *1,* 91–97.

Bharani, S. N., & Hyde, J. S. Chronic asthma and the school. *Journal of School Health,* 1976, *46,* 24–30.

Brandon, M. L. Newer medications to aid treatment of asthma. *Annals of Allergy,* 1977, *39,* 117–129.

Burns, K. L. An evaluation of two inexpensive instruments for assessing airway flow. *Annals of Allergy,* 1979, *43,* 246–249.

Chai, H. Intermediary mechanisms in asthma: Some informal comments on the pathogenesis of the breathing disturbance. *Clinical Pediatrics,* 1974, *13,* 409–412.

Chai, H. Management of severe chronic perennial asthma in children. *Advances in Asthma and Allergy,* 1975, *2,* 1–12.

Chai, H. Therapeutic modalities in the treatment of asthma. *Behavioral Medicine,* 1980, *7,* 45–46.

Chai, H., & Newcomb, R. W. Pharmacologic management of childhood asthma. *American Journal of Diseases of Children,* 1973, *125,* 757–765.

Chai, H., Purcell, K., Brady, K., & Falliers, C. J. Therapeutic and investigative evaluation of asthmatic children. *Journal of Allergy,* 1968, *41,* 23–36.

Chen, W. Y., & Horton, D. J. Heat and water loss from the airways and exercise-induced asthma. *Respiration,* 1977, *34,* 305–313.

Cooper, B. *The economic costs of selected respiratory diseases, 1972.* Unpublished report prepared for the Division of Lung Diseases Task Force on Prevention, Control and Education in Respiratory Diseases, 1976.

Creer, T. L. The use of a time-out from positive reinforcement procedure with asthmatic children. *Journal of Psychosomatic Research,* 1970, *14,* 117–120.

Creer, T. L. Behavior modification: New tools for rehabilitating the chronically ill. *Children's Asthma Research Institute and Hospital Research Bulletin,* 1971, *1* (1) 1–30.

Creer, T. L. Biofeedback and asthma. *Advances in Asthma and Allergy,* 1974, *1,* 6–11.

Creer, T. L. *Psychological impact of asthma.* Unpublished report prepared for the Asthma Committee, Task Force on Asthma and Other Allergic Diseases, National Institute of Allergy and Infectious Diseases, 1977.

Creer, T. L. Asthma: Psychologic aspects and management. In E. Middleton, Jr., C. E. Reed, & E. F. Ellis (Eds.), *Allergy: Principles and Practice.* St. Louis: Mosby, 1978, pp. 796–811.

Creer, T. L. *Asthma therapy: A behavioral health care system for respiratory disorders.* New York: Springer, 1979.

Creer, T. L. *Asthma: Self-management.* Paper presented at the National Jewish Hospital/

National Asthma Center Conference on Asthma, Keystone, Colorado, January 9, 1980. (a)

Creer, T. L. Self-management behavioral strategies for asthmatics. *Behavioral Medicine*, 1980, *7*, 14–24. (b)

Creer, T. L. Psychological aspects of asthma. *Current Views in Allergy and Immunology*, 1980, *4*, No. 2. (c) (Audiotape)

Creer, T. L. & Christian, W. P. *Chronically-ill and handicapped children: Their management and rehabilitation*. Champaign, Ill.: Research Press, 1976.

Creer, T. L., & Renne, C. M. *Panic in asthmatic children*. Unpublished manuscript, Children's Asthma Research Institute and Hospital, 1968.

Creer, T. L., & Yoches, C. The modification of an inappropriate behavioral pattern in asthmatic children. *Journal of Chronic Diseases*, 1971, *24*, 507–513.

Creer, T. L., Weinberg, E., & Molk, L. Managing a problem hospital behavior: Malingering. *Journal of Behavior Therapy and Experimental Psychiatry*, 1974, *5*, 259–262.

Davis, D. J. NIAID initiatives in allergy research. *The Journal of Allergy and Clinical Immunology*, 1972, *49*, 323–328.

Davis, M. H., Saunders, D. R., Creer, T. L., & Chai, H. Relaxation training facilitated by biofeedback apparatus as a supplemental treatment in bronchial asthma. *Journal of Psychosomatic Research*, 1973, *17*, 121–128.

Douglas, J. W. B., & Ross, J. M. The effects of absence on primary school performance. *British Journal of Educational Psychology*, 1965, *35*, 28–40.

Eney, R. D., & Goldstein, E. O. Compliance of chronic asthmatics with oral administration of theophylline as measured by serum and salivary levels. *Pediatrics*, 1976, *57*, 513–517.

Fry, J. *The catarrhal child*. London: Butterworth, 1961.

Ghory, J. E. Exercise and asthma: Overview and clinical impact. *Pediatrics*, 1975, *56*, 844–860.

Gordis, L. *Epidemiology of chronic lung diseases in children*. Baltimore: Johns Hopkins University Press, 1973.

Hahn, W. W. Autonomic responses of asthmatic children. *Psychosomatic Medicine*, 1966, *28*, 323–332.

Hahn, W. W., & Clark, J. A. Psychophysical reactivity of asthmatic children. *Psychosomatic Medicine*, 1967, *29*, 526–536.

Hartshorne, H., & May, M. A. *Studies in the nature of character. Vol. 1: Studies in deceit*. New York: Macmillan, 1928.

Jacobson, E. *Progressive relaxation*. Chicago: University of Chicago Press, 1938.

Johnstone, D. E. A study of the natural history of bronchial asthma in children. *American Journal of Diseases of Children*, 1968, *115*, 213–216.

Kotses, H., Glaus, K. D., Crawford, P. L., Edwards, J. E., & Scherr, M. S. Operant reduction of frontalis EMG activity in the treatment of asthma in children. *Journal of Psychosomatic Research*, 1976, *20*, 453–459.

McFadden, E. R., Jr., & Ingram, R. H., Jr. Exercise-induced asthma: Observations on the initiating stimulus. *New England Journal of Medicine*, 1979, *301*, 763–769.

Middleton, E., Jr., Reed, C. E., & Ellis, E. F. (Eds.). *Allergy: Principles and practice*. St. Louis: Mosby, 1978,

Miklich, D. R., Renne, C. M., Creer, T. L., Alexander, A. B., Chai, H., Davis, M. H., Hoffman, A., & Danker-Brown, P. The clinical utility of behavior therapy as an adjunctive treatment for asthma. *The Journal of Allergy & Clinical Immunology*, 1977, *60*, 285–294.

O'Leary, S. G., & Dubey, D. R. Applications of self-control procedures by children: A review. *Journal of Applied Behavior Analysis*, 1979, *12*, 449–465.

Porter, R., & Birch, J. (Eds.). Report of the working group on the definition of asthma. *Identification of asthma.* London: Churchill Livingston, 1971.

Purcell, K., & Weiss, J. H. Asthma. In C. C. Costello (Ed.), *Symptoms of psychopathology.* New York: Wiley, 1970, pp. 597–623.

Rackemann, F. M. Studies in asthma—Analysis of 213 cases in which patients were relieved for more than 2 years. *Archives of Internal Medicine,* 1928, 41, 346–355.

Rackemann, F. M., & Edwards, M. D. Medical progress: Asthma in children: Follow-up study of 688 patients after 20 years. *New England Journal of Medicine,* 1952, *246,* 815–858.

Reed, C. E., & Townley, R. G. Asthma: Classification and pathogenesis. In E. Middleton, Jr., C. E. Reed, & E. F. Ellis (Eds.), *Allergy: Principles and practice.* St. Louis: Mosby, 1978, pp. 659–677.

Renne, C. M., & Creer, T. L. The effects of training on the use of inhalation therapy equipment by children with asthma. *Journal of Applied Behavior Analysis,* 1976, *9,* 1–11.

Renne, C. M., Nau, E., Dietiker, K. E., & Lyon, R. *Latency in seeking asthma treatment as a function of achieving successively higher flow rate criteria.* Paper presented at the tenth annual convention of the Association for the Advancement of Behavior Therapy, New York, December 1976.

Rubinfeld, A. R., & Pain, M. C. F. Perception of asthma. *Lancet,* 1976, *1,* 882–884.

Scadding, J. G. Patterns of respiratory insufficiency, *Lancet,* 1966, *1,* 701–704.

Schiffer, C. G., & Hunt, E. P. *Illness among children* (Children's Bureau Publication No. 405). Washington, D.C.: U.S. Government Printing Office, 1963.

Segal, M. S. Death in bronchial asthma. In E. B. Weiss & M. S. Segal (Eds.), *Bronchial asthma: Mechanisms and therapeutics.* Boston: Little, Brown, 1976, pp. 1121–1142.

Shepard, J., & Hochstadt, N. J. *Reducing hospitalizations in children with asthma.* Paper presented to Western Psychological Association, San Francisco, April 19, 1978.

Slavin, R. *Prognosis in bronchial asthma.* Report prepared for the Asthma Committee, The Task Force on Asthma and Other Allergic Diseases, National Institute of Allergic and Infectious Diseases, 1977.

Snider, G. L. The interrelationships of asthma, chronic bronchitis, and emphysema. In E. B. Weiss & M. S. Segal (Eds.), *Bronchial asthma: Mechanisms and Therapeutics.* Boston: Little, Brown, 1976, pp. 31–41.

Speizer, F. E. Epidemiology, prevalence, and mortality in asthma. In E. B. Weiss & M. S. Segal (Eds.), *Bronchial asthma: Mechanisms and therapeutics.* Boston: Little, Brown, 1976, pp. 43–51.

Sublett, J. L., Pollard, S. J., Kadlec, G. J., & Karibo, J. M. Non-compliance in asthmatic children: A study of theophylline levels in a pediatric emergency room population. *Annals of Allergy,* 1979, *43,* 95–97.

Suess, W. *The effects of asthma medications upon the intellectual and motor performance of asthmatic children.* Paper presented at the National Jewish Hospital/National Asthma Center and Research Center, Denver, April 10, 1980.

Taplin, P. S., & Creer, T. L. A procedure for using peak expiratory flow rate data to increase the predictability of asthma episodes. *The Journal of Asthma Research,* 1978, *16,* 15–19.

Vance, V. J., & Taylor, W. F. Status and trends in residential asthma homes in the United States. *Annals of Allergy,* 1971, *29,* 428–437.

Weiss, E. B., & Segal, M. S. (Eds.). *Bronchial asthma: Mechanisms and therapeutics.* Boston: Little, Brown, 1976.

Wolpe, J. *Psychotherapy by reciprocal inhibition.* Stanford, Calif.: Stanford University Press, 1958.

Wright, B. M., & McKerrow, C. B. Maximum forced expiratory flow rates as a measure of ventilatory capacity. *British Medical Journal,* 1959, *2,* 1041–1046.

3

Pediatric Hematology/Oncology

Jonathan Kellerman and James W. Varni

Pediatric hematology/oncology encompasses a wide variety of diseases and disorders including cancer—both solid tumors and leukemias—bleeding disorders such as hemophilia, and the anemias. These present with a correspondingly varied range of symptoms, causes, treatments, and prognoses, as well as associated psychosocial issues. Although a comprehensive overview of these various diseases is beyond the scope of this chapter, disorders upon which the greatest amount of empirical research has been conducted will be discussed in detail.

CANCER: MEDICAL ASPECTS

Cancer in children is a rare disease, with an incidence of 11/10,000 (Sutow, Vietta, & Fernbach, 1973). Nevertheless, pediatric malignancy ranks second only to accidents as the most common cause of death in children following the neonatal period (Silverberg, 1979). The types of malignancies commonly encountered in children differ from those found in adults. Whereas most adult cancers are solid tumors of various organs—lung, breast, ovaries, gastrointestinal system, cervix, and uterus—the modal pediatric cancers are the acute leukemias, comprising

JONATHAN KELLERMAN • Childrens Hospital of Los Angeles; University of Southern California School of Medicine, Los Angeles, California 90027. JAMES W. VARNI • Orthopaedic Hospital; University of Southern California School of Medicine, Los Angeles, California 90007.

approximately 40% of all childhood malignancies. Tumors of the brain account for an additional 14%, with the remaining 46% made up of tumors of the kidney, lymphatic system, nervous system, bone, soft tissue, and eye. A summary of the distribution of specific cancer diagnoses has been presented by Siegel (1980), who traces the development of contemporary multimodal oncological therapy for children and emphasizes the dramatically improved prognosis for many types of pediatric cancer.

An example of such medical progress can be drawn from acute lymphoblastic leukemia (ALL), the single most common malignant disease of childhood. Progress in treating ALL over the last three decades has provided a model for the treatment of other malignancies of the hemopoietic system in both children and adults. Once a rapidly fatal disease whose victims could be expected to live only a few months, ALL can now be controlled to the point of remission in a majority of children with over 50% of patients alive five years after diagnosis. Concomitant with extended life expectancy, however, is an increase in problems of adjustment encountered by patients and families—problems of *living* with the disease and its treatment (Siegel, 1980; Kellerman, 1979a; Lansky, Lowman, Vats, & Gyulay, 1975; Koch, Hermann, & Donaldson, 1974; Katz, 1980; Kagen-Goodheart, 1977). This has created a significant challenge to multidisciplinary teams to develop psychosocial treatment approaches that address the various behavioral-medical aspects of long-term adjustment to a life-threatening disease.

CANCER: PSYCHOSOCIAL ASPECTS

Though the literature is replete with suggestions of higher rates of psychopathology in children with chronic or serious disease, empirical studies have generally failed to bear this out (Tavormina, Kastner, Slater, & Watt, 1976; Bedell, Giordani, Armour, Tavormina, & Boll, 1977; Kellerman, Rigler, & Siegel, 1977; Kellerman, Zeltzer, Ellenberg, Dash, & Rigler, 1980). This discrepancy may be explained when one examines the nature of references suggesting inevitable or highly probable psychopathology in physically ill children. These are usually anectodal case accounts by a mental health professional consulted *because* of suspected psychological or psychiatric disorder. Generalizing from such a clinical sample, which is likely to be skewed toward pathology, may lead to overestimation of problems. In addition, the consultant who is not familiar with the total context within which apparently problematic behavior arises may generate a high rate of false positives. For example, the mental health consultant who observes a specific patient demonstrating symptoms of withdrawal and who labels the child as hyperactive may not be taking into account that shortly prior to consultation the youngster underwent a bone marrow aspiration, a lumbar puncture, and a venipuncture. In addition, certain behavioral symptoms—listlessness, fatigue—occur as side effects of chemotherapy and radiotherapy and can be expected to cease once treatment has stopped. Kellerman (1980) has commented on the problems inherent in the mental health consultant model when applied to pediatric oncology and has suggested specific alternatives such as routine psychosocial care implemented soon after diagnosis.

This is not to suggest that children with cancer do not encounter adjustment problems, but rather that such problems are best viewed as reactions of *normal children under stress, rather than as psychiatric disorders.* This is supported by the findings of Kellerman, Zeltzer, Ellenberg, & Dash (1980), who found no differences in trait anxiety and self-esteem between healthy and seriously ill adolescents, including those with cancer. A concomitant study, however, did find various adaptive challenges, many of them specific to particular diseases, in the ill group (Zeltzer, Kellerman, Ellenberg, Dash, & Rigler, 1980). The major differences between teenagers with cancer and both healthy peers and adolescents with other diseases was a significant degree of problems related to medical treatment. Specifically, acute pain, nausea, emesis, and other forms of iatrogenic discomfort were viewed as major obstacles to adjustment. Supporting this, Katz, Kellerman, and Siegel (1980) found distress related to bone marrow aspirations to be virtually ubiquitous in a sample of 115 children with cancer. Similarly, Silberfarb, Philibert, and Levine (1980) found serious psychopathology absent in a sample of adult cancer patients.

Multidisciplinary team members working in pediatric oncology must acquaint themselves with the medical aspects of disease and treatment as well as the biopsychosocial context within which specific behaviors occur. As an example, while the alleviation of pain and discomfort provides a useful model for implementation of a behavioral pediatrics approach in oncology, understanding of several other topics should precede such work. Such areas include disease-related communication, school avoidance, and central nervous system effects of treatment. These will be discussed briefly, rather than in the form of comprehensive discourses, and the reader is urged to pursue the additional references cited.

Disease-Related Communication

The diagnosis of cancer in a child is a highly traumatic event for the entire family. Despite the improved prognoses of various childhood malignancies, the word *cancer* connotes inevitable death to many people. In fact, approximately 50% of children with cancer can be expected to die of their disease. It is, therefore, extremely important for practitioners to clarify what the child and family understand about the illness and its treatment. Spinetta and his colleagues have carried out a substantial body of research on illness and death-related communication (Spinetta, 1974, 1980; Spinetta & Maloney, 1975, 1978; Spinetta, Rigler, & Karon, 1973) and have shown that anxiety is not prevented by concealing relevant facts from children and in fact is likely to be exacerbated by incomplete or distorted communication. Despite the reticence of many adults to discuss specific details regarding disease and treatment, these findings, supported by others (Kellerman, Rigler, Siegel, & Katz, 1977; Vernick & Karon, 1965) point toward the value of an open approach. Pragmatically, it is unrealistic for the practitioner to attempt to treat specific behavioral problems in a patient who is unclear about the nature of his disease and why he is being treated. Furthermore, the practitioner can play an important role in facilitating communication about disease and treatment and in seeing that an open approach is used early in the diagnosis consistently by all who come into close contact with the child. It is essential for the clinician to be aware of how children perceive illness and death at differing developmental levels in order for such an intervention to be effective. The reader is referred to the work of Spinetta (especially Spinetta, 1980) for specific approaches to this area.

School Avoidance

In order to promote optimal psychosocial rehabilitation, many authors have emphasized the importance of having the child with cancer return to school and other premorbid activities as soon as is medically feasible (Katz, 1980; Karon, 1975; Kagen-Goodheart, 1977; Katz, Kellerman, & Siegel, 1980). The rationale for this approach is that optimal adjustment is likely to occur within the context of normal social and academic activities. Katz (1980) has noted that interference with age-appropriate, goal-directed behavior can lead to increased frustration, and he has extended this to include the child faced with imminent death.

Several authors (Katz, Kellerman, Rigler, Williams, & Siegel, 1977; Futterman & Hoffman, 1970; Lansky *et al.*, 1975; Katz, 1980) have discussed the impediment to school reintegration caused by inappropriate attitudes on the part of parents, health professionals, and school personnel. Whereas in some diseases—such as certain brain tumors and instances of central nervous system impairment, where illness impacts directly upon intellectual functioning—the child will require special or remedial education, the vast majority of pediatric cancer patients can be returned to a regular classroom within weeks after diagnosis. Katz (1980) has noted the value of early intervention with parents and school systems in easing the transition from hospital to school. This often takes the form of raising the issue of school return at a psychologically appropriate time, soon after diagnosis, determining which of the child's premorbid school experiences are likely to help or hinder quick return, and providing accurate illness-related information to child, family, and school.

Many patients exhibit anxiety about returning to school because of physical change brought about by disease and treatment (alopecia, weight gain or loss, nausea, emesis, surgical disfigurement—for a review see Siegel, 1980). Such obvious changes in appearance can elicit ridicule and teasing from peers. There is some evidence that girls are more adversely affected by this than are boys (Zeltzer, Kellerman, Ellenberg, Dash, & Rigler, 1980). Intervention can take place both with the child and with the offending peers. In the case of the latter, school visits that educate other children about childhood cancer and its treatment can often rapidly reduce the rate of teasing behavior. Patients, themselves, can be taught to emit appropriate responses to aversive comments. Should peer ridicule persist, children with cancer can be trained in the technique of extinction. By eliminating the social reinforcer (attention) that maintains teasing, the patient is able successfully to reduce its rate and eventually to eliminate it. When using extinction it is important to prepare patients for the likelihood of an initial rise of aversive comments following withdrawal of attention that precedes reduction.

Another cause of school-related anxiety and subsequent school avoidance is the high rate of absence brought about by rigorous regimens of treatment, some of which must be carried out on an inpatient basis. Rutter and Graham (1970) have described the disruption posed for chronically ill children by repeated short school absences that cause youngsters to fall behind their peers constantly. The use of in-house classroom programs that work directly in conjunction with the mainstream school is an important source of continuity for the child with cancer and is invaluable in mediating the psychological impact of repeated absence.

Combining a program of early school intervention, behavioral training, and parent education with hospital teaching programs reduces the risk of school avoidance described in the literature (Lansky *et al.*, 1975) to the point where virtually all children with cancer who are physically capable of participating in school can do so (Katz, 1980).

Central Nervous System Effects of Treatment

One of the major factors in increasing the survival rate of children with leukemia has been the use of prophylactic treatment aimed at preventing the spread of leukemic cells to the central nervous system. This has taken the form of either irradiation of the cranium, administration of the folic acid antagonist methotrexate (MTX) through the spine (intrathecally) or a combination of both of these modalities.

Recently, concern has been expressed regarding possible deleterious effects of such treatment upon the central nervous system. Moss and Nannis (1980) provide a comprehensive review of studies to date and organize this research into three groups: (1) studies evaluating structural changes within the brain through the use of histological or photographic techniques; (2) studies of personality change associated with CNS prophylaxis; and (3) studies assessing changes in intellectual, cognitive, and perceptual functioning. They note that although several authors suggest deleterious effects of CNS prophylaxis, controlled prospective studies are conspicuously lacking in the literature. Thus, further work is needed in this area, including the use of psychological and neuropsychological tests to examine longitudinal change associated with prophylactic chemotherapy and radiotherapy of the central nervous system and its potential impact on adjustment potential.

CANCER: MANAGEMENT OF PAIN AND DISCOMFORT

In contrast with the situation found in adults with cancer, chronic, debilitating pain is comparatively uncommon in pediatric oncology. An exception is the chronic pain experienced by some children with cancer during the terminal stage of illness, a period usually lasting no more than several weeks. The absence of long-term chronic pain problems is primarily due to the differing nature of cancers in adults and children mentioned before. For example, as opposed to carcinomas, leukemias and lymphomas are not extremely painful diseases *per se*, although they may be accompanied by joint aches, headaches, fatigue, and fever during periods of active disease.

A more common source of discomfort lies in *acute* distress, pain, nausea, emesis, anorexia, related to the *diagnosis and treatment* of cancer. For example, the bone marrow aspiration (BMA), the most frequently administered diagnostic procedure for leukemia, is associated with high levels of behavioral distress in a majority of pediatric patients (Katz *et al.*, 1980). Although such distress varies with age and sex, anxiety and pain are virtually ubiquitous in children undergoing this repeated procedure. Furthermore, a stable pattern of habituation to BMAs is not present. In addition to anxiety and pain encountered during the BMA itself, some patients experience anticipatory distress such as nausea, emesis, emotional withdrawal, and a general anxiety reaction to temporal and spatial stimuli associated with the procedure. The same is true of other procedures—lumbar punctures, venipunctures, chemotherapeutic injections.

Anticipatory symptoms are evidence of an apparently classically conditioned response pattern, in which the procedure is the unconditional stimulus and various other stimuli such as the waiting room, procedure room, sight of the nurse or doctor, are conditioned to elicit an anxious response. A similar pattern for adults was reported by Schulz (1980), who found approximately one third of a sample of adult cancer patients exhibiting classically conditioned nausea or emesis *prior* to the administration of chemotherapeutic injections.

Anticipatory distress adds to the discomfort of cancer patients by compounding nausea, emesis, and anorexia "normally" experienced as a physiological side effect of many of the chemotherapeutic agents used to treat cancer (Siegel, 1980). The impact of nausea and emesis is underscored by Schulz's (1980) finding that 69% of a sample of adult cancer patients receiving chemotherapy reported significant discomfort due to nausea and vomiting. Since many of the agents used to treat childhood cancer are similar to those employed with adults, it is not surprising to find similar patterns in pediatric cancer patients. It is common medical practice to utilize phenothiazines as antiemetic agents, but these vary widely in their effectiveness for specific patients and produce undesirable side effects such as drowsiness and lethargy. Recent suggestions have been made regarding the efficacy of tetrahydrocannabinol (THC) in reducing nausea and emesis (Sallan, Cronin, Zelen, & Zinberg, 1980); however, lack of baseline data and other methodological limitations lessen the impact of these findings (Zeltzer, Barbour, LeBaron, & Kellerman, 1980). Nevertheless, this is an area that bears further close scrutiny.

Anticipatory anxiety can be expected to respond to conditioning approaches that have been used successfully in the past to combat learned anxiety reactions, namely, systematic desensitization, relaxation training, hypnosis, biofeedback, meditation, guided imagery. It is our opinion that all of these modalities share more similarities than differences. Specifically, all of these autogenic techniques emphasize the following ingredients in varying proportions: classically conditioned counteranxious response(s), learning-enhanced self-control, integration of psychological-somatic responses, and operant reinforcement (therapist- and self-generated) of successful goal-attainment (either explicitly acknowledged as in biofeedback, or implicitly encouraged as in the case of suggestive methods). In addition to reducing anticipatory distress, there is evidence that these approaches can be effective in successfully mediating the perception of pain itself. For example, the clinical efficacy of hypnosis as a means for inducing analgesia has been documented over several centuries (Dash, 1980), despite the fact that little is known about the specific nature of mechanisms involved in this process (Hilgard, 1969). Theoretical models have included biochemical, sociological, and psychological variables, but none has served to explain hypnotic phenomena comprehensively. Despite clinical testimonies, acceptance of hypnosis in medical and psychological circles has historically waxed and waned. This may be due to the failure of clinicians using hypnosis to subject their methods to the rigors of experimental testing as well as to translate their findings into the lexicon of science.

The development of dependent measures poses methodological problems. Since there are no reliable physiological indexes with which to assess pediatric pain (Katz *et al.*, 1980), research in this area is dependent to a large extent upon self-report. There is evidence, however, that structured self-report techniques are more reliable than physiological measures and that they correlate highly with independent behavioral measures of distress (Katz *et al.*, 1980). Melzack (1975) has noted that pain is essentially a psychophysiological phenomenon strongly affected by environmental and other social variables and that research in pain reduction will—and should—rely heavily upon subjective ratings. However, when objective measures such as frequency of emesis and analgesic dosage do exist, all efforts should be made to utilize them. In view of the methodological limitations noted, many of the following studies should be viewed as exploratory in nature.

Burish and Lyle (1979) reported the case of a 30-year-old woman with thymic lymphoma who manifested extreme anxiety, nausea, and emesis prior to the intravenous administration of chemotherapy and who showed only a limited response to prochlorperazine given as an antiemetic. The level of this patient's anticipatory anxiety was such that she threatened to quit treatment. (This is not a rare occurrence. Zeltzer, 1980, notes the tendency for iatrogenic concerns to exacerbate an already heightened sense of powerlessness in adolescents with cancer who may respond to distress by dropping out of treatment or exhibiting other forms of noncompliance with medical regimens.) These authors used a 7-point self-rating of anxiety and nausea, pulse rate, blood pressure, and scores on the anxiety and depression components of the Multiple Affect Adjective Checklist as dependent measures and tested the efficacy of therapist- and patient-directed relaxation training in combating iatrogenic distress. Reduction in self-report indices was obtained during therapist-directed sessions. Though a positive response to patient-directed sessions did not occur initially, by the conclusion of 11 sessions the patient was able to achieve symptom reduction with minimal therapist intervention. However, little change was noted along the physiological measures. This disparity between subjective and "objective" measures is consistent with the literature on hypnotic analgesia (Hilgard, 1969; Dash, 1980) and with data on the measurement of anxiety in general (Katz *et al.*, 1980). Burish and Lyle did not report extensive results regarding postchemotherapy distress.

Kellerman (1979b) reported the successful use of a combined behavioral treatment plan utilizing parental relaxation training, anger as a counteranxious response, and operant reinforcement of appropriate sleep behavior eliminating night terrors in a 3-year-old girl with ALL. Behavioral analysis revealed sleep disturbance contiguous in time with bone marrow aspirations and exacerbated by separation anxiety. The hypothesis was made that the night terrors represented an anxiety reaction to medical procedures. (Similar results were reported by Roberts and Gordon, 1980, in treating nightmares secondary to burn trauma in a 5-year-old girl, using response prevention, extinction, and systematic desensitization.)

Ellenberg, Kellerman, Dash, Zeltzer, and Higgins (1980) reported the use of hypnotic training in treating multiple problems in an adolescent girl with chronic myelogenous leukemia who eventually died. Cue-controlled hypnosis was used to reduce frequency of chemotherapy-related emesis from baseline. The patient was trained in progressive relaxation and counteranxious imagery and was instructed to enter a relaxed state contingent upon initial sensations of nausea. Posthypnotic suggestions were given for improved tolerance of chemotherapeutic agents. The patient kept a prospective record of frequency and intensity (rated along a 1 to 10 scale) of emesis and nausea during a four-day course of vincristine, prednisone, 5-azacytidine, cystosine-arabinoside, and daunomycin prior to hypnotic treatment (baseline) and for a 4-day course of the same drugs following treatment. Frequency of emesis fell from 30 to 2 with no change in perceived intensity of symptoms. Frequency of nausea dropped from 30 to 4 and intensity fell from 10.0 to 7.5. These results were supported by independent ratings of nausea and emesis in the medical chart but were confounded by a reduction in the dosage of chemotherapeutic agents during the posthypnotic period. Despite the judgment of the attending oncologist that such a drastic reduction of symptoms was unlikely to be due solely to dosage reduction, these data must be considered cautiously.

Subsequently, Zeltzer, Kellerman, Ellenberg, and Dash (1981) have worked with a group of nine adolescents with cancer for whom baseline and posttreatment courses of chemotherapy and other medical factors have remained equivalent and have found a statistically significant reduction in both frequency and intensity of emesis (Table I). Longitudinal patterns of response have been obtained from three of these patients who went on to receive multiple equivalent treatments, and these data indicate maintenance of therapeutic gains over time. In grouping these data, rates of symptoms were weighed by length of chemotherapy course so as not to give undue bias to patients undergoing longer periods of treatment. Included in this group was one patient with a brain tumor (malignant astrocytoma) who was not receiving treatment of any kind and whose nausea and emesis were judged by her neurosurgeon to be related to tumor site. The ability of this patient to reduce successfully an apparently organically induced set of symptoms through the use of self-hypnosis raises important questions concerning the interaction be-

Table I. Changes in Emesis Due to Chemotherapy Following Hypnotic Training ($n = 9$)[a]

		Prehypnosis baseline	Posthypnosis	*p*
Frequency[b]				
	M	4.7	2.2	.01
	SD	2.3	1.4	
Intensity (1–10)[c]				
	M	7.3	6.4	.05
	SD	2.3	2.5	
Duration (hours)				
	M	7.0	3.1	NS
	SD	7.6	2.7	

[a]From Kellerman, Zeltzer, Ellenberg, & Dash (1980).
[b]Emetic episodes equivalent per chemotherapeutic course weighted by number of days of emesis-producing chemotherapeutic injections.
[c]1 = minimal discomfort; 10 = maximal discomfort.

tween physiological and psychological processes and supports the concept of biopsychosocial synthesis put forth by Schwartz (1980).

The patient reported by Ellenberg *et al.*, (1980) was also offered hypnotic treatment as a means of reducing distress associated with bone marrow aspirations and lumbar punctures. Reductions in pain and anxiety before, during, and after medical procedures from baseline were obtained. The hypnotic approach to acute procedural distress has been extended to a group of adolescents with leukemia (Kellerman, Zeltzer, Ellenberg, & Dash, 1980), with subsequent findings of significant reductions in distress (Table II). Multiple baselines across patients indicated no pattern of habituation to the procedure consistent with Katz *et al.* (1980). In addition, statistical tests applied to the multiple baseline revealed no drop in distress due to the passage of time, *per se*. Despite the successful use of hypnosis in reducing frequency of emesis and procedural distress in the patient reported by Ellenberg and associates (1980), this approach was not successful in modifying a period of disease-related anorexia occurring during the onset of the terminal phase of illness. Although relaxation training was followed by increased food

Table II. Changes in Acute Discomfort Due to Medical Procedures Following Hypnotic Training ($n = 16$)[a]

		Prehypnosis baseline	Posthypnosis	p
Anxiety (1–5)[b]				
Before	*M*	3.4	2.0	.001
	SD	1.0	.6	
During	*M*	3.8	1.9	.001
	SD	1.3	.7	
After	*M*	2.5	1.4	.01
	SD	1.3	.8	
Discomfort (1–5)[c]				
Before	*M*	2.8	1.7	.01
	SD	1.4	.6	
During	*M*	4.1	2.2	.001
	SD	.9	.8	
After	*M*	3.6	1.4	.001
	SD	1.3	.7	

[a]From Kellerman, Zeltzer, Ellenberg, & Dash (1980).
[b]1 = very calm; 5 = very anxious.
[c]1 = very comfortable; 5 = very uncomfortable.

intake, this was inevitably followed by emetic rejection of solid food. These authors note the presence of gastrointestinal ulcers at postmortem examination and suggest that these organic factors were sufficiently potent to override the effects of relaxation.

Cairns and Altman (1979) successfully treated posttreatment anorexia in a 6-year-old girl with malignant hemangiopericytoma using reinforcement of weight gain administered through a token economy. The treatment program was transferred from hospital to home and target weight was reached after approximately 3 months. These authors suggest that behavioral techniques offer a comparatively low-risk approach to cancer-related anorexia. In related work with adults, Redd (1980a) used systematic desensitization to eliminate gagging and simultaneously increase food intake in a 53-year-old woman who developed conditioned food aversion following successful surgery for gastrointestinal cancer. He suggested that anxiety related to recurrence of disease played a major role in the etiology of food aversion in this patient. Redd (1980b) reported the use of an operant approach combining social reinforcement and extinction in eliminating psychosomatic coughing and retching in two adult patients treated in protective isolation. The importance of carefully monitoring the psychological effects of treatment modalities such as isolation rooms, life islands, and laminar airflow units where sensory restriction is a by-product of the environment has been stressed by several authors (Holland, Plumb, Tates, Harris, Tuttolomondo, Homes, & Holland, 1977; Kellerman, Rigler, Siegel, McCue, Pospisil, & Uno, 1976a, 1976b). Further studies of behavioral modalities used in conjunction with more traditional (and hazardous) nutritional approaches, such as nasoesophageal tube-feeding and hyperalimentation, appear warranted.

In summary, behavioral techniques show promise as methods of alleviating distress due to cancer and its treatment. Findings are encouraging both with regard to anticipatory anxiety and in instances in which symptoms may be traced directly to physiological effects of disease and treatment. Further controlled studies are needed to elucidate specific components of intervention that bring about alleviation of symptoms. Although the researcher and practitioner must be aware of medical aspects of the disease as well as familiar with the biopsychosocial context of cancer as it affects the child and family, principles of learning that apply to healthy children are similarly applicable to the child with malignant disease. As medical advances continue to prolong life while concurrently subjecting the patient to treatment of growing intensity, the role of behavioral medicine in pediatric oncology can be expected to grow.

HEMOPHILIA: MEDICAL ASPECTS

Hemophilia represents a congenital hereditary disorder of blood coagulation, transmitted as an X-linked recessive trait by a female carrier to her male offspring. True hemophilia in a female is extremely rare, with carriers of the trait being protected by the production of sufficient clotting factor by cells under the control of the normal gene bearing X-chromosome. Because of its status as one of the most frequent spontaneous mutations known in medicine, about a third of all newly diagnosed patients evidence no known family history of hemophilia (Strauss, 1969). This life-long chronic disorder is characterized by recurrent, unpredictable internal hemorrhaging affecting any body part, especially the joints and the extremities. The severity of the disorder varies among individuals with the failure to produce one of the plasma proteins required to form a blood clot being the identifying characteristic. Severe hemophiliacs have 1% or less of normal activity of the clotting factor; moderate hemophiliacs evidence about 2–20% of normal clotting factor activity, and mild hemophiliacs show 20–60% of normal clotting factor. The incidence of hemophilia is estimated to be one in 10,000 of the male population, with the most common variety named hemophilia A (or classic hemophilia), represented by a plasma protein Factor VIII deficiency; hemophilia B (or Christmas disease) is the next most common, with a Factor IX deficiency in the coagulation chain (Hilgartner, 1976). There are other blood clotting disorders (e.g., Von Willebrand's disease), but they are considerably rarer.

Since in hemophiliacs one of the clotting proteins is at a diminished circulatory level, a blood clot takes a substantially longer period of time to form, and it may not be an effective clot for the ruptured vessel (Kasper, 1976). Small external cuts on the skin can usually be treated by pressure bandages; however, untreated bleeding (typically indicated by pain, loss of motion, and/or swelling) may be quite prolonged. Although internal bleeding episodes can occur as a result of obvious physical trauma, many bleeding episodes occur spontaneously, that is, without clear precipitating cause but usually as a result of unrecognized mild trauma (Arnold & Hilgartner, 1977). Spontaneous bleeding episodes are more common in severe hemophiliacs, whereas moderate to mild hemophiliacs may experience excessive internal bleeding only as a result of obvious physical injury. Whereas severe hemophiliacs may hemorrhage internally as often as once or twice a week as children, moderate hemophiliacs may have as few as one or two bleeding episodes a year, while mild hemophiliacs may bleed excessively only during a surgical intervention. As the hemophilic child grows older, there appears to be in general a substantial decrease in bleeding frequency, even in severe hemophiliacs (Aronstam, Rainsford, & Painter, 1979).

A major breakthrough in the medical care and management of hemophilia came less than 15 years ago with the discovery and separation of the essential clotting factors from the plasma of normal blood. Freeze-dried concentrates of the clotting factor proteins made from pooled plasma are now available, and are reconstituted with sterile water. Dosage of factor deficiency replacement is calculated on the basis of the patient's weight, the assumed plasma volume, and the severity of the bleeding episode. The intravenous infusion of the factor concentrate temporarily replaces the missing clotting factor and converts the clotting status to normal, allowing a functional blood clot to form. As a result of home-based administration and self-infusion of factor replacement, early treatment is now possible, consequently reducing the severity of bleeding episodes. Recently, Sergis-Deavenport and Varni (1981a, 1981b) developed a treatment package consisting of a behavioral assessment checklist of associated infusion behaviors, modeling, behavioral rehearsal, and feedback techniques successful in teaching the parents of hemophilic children the necessary skills to be placed on the home care program for factor replacement therapy. The advantages of early treatment of each hemorrhage in the home environment as well as the independence from hospital-based treatment are readily apparent and warrant further research.

Even though muscle and soft tissue hemorrhages (hematomas) may be quite serious, the most frequent problem and major cause of disability in severe hemophilia is repeated bleeding into the joint areas (hemarthroses) (Arnold & Hilgartner, 1977). The large weight-bearing joints are affected most frequently, with hemarthroses of the knees accounting for about half of all joint hemorrhages and ankle and elbow hemarthroses each accounting for about 20% (Stuart, Davies, & Cummings, 1960). With repeated hemarthroses, a pathophysiological condition similar to osteoarthritis develops, marked by articular cartilage destruction, pathological bone formation, and impaired function (Sokoloff, 1975). Although prophylactic administration of factor replacement to prevent hemarthrosis has been investigated, the expense of this form of treatment, the amount of factor replacement required, and the difficulties of multiple transfusions on an ongoing basis have made it unrealistic for most patients (Kisker, Perlman, & Benton, 1971). It is even questionable whether in fact prophylaxis will prevent hemarthroses totally and the resultant degenerative arthropathy (Kasper, Dietrich, & Rapaport, 1970). An estimated 75% of hemophilic adolescents and adults are affected by degenerative hemophilic arthropathy, with chronic arthritic pain representing the most frequent problem in the care of adolescent and adult hemophiliacs (Dietrich, 1976). Since pain and pain management represent such an outstanding characteristic of the hemophilia disorder, a section devoted wholly to this problem follows.

HEMOPHILIA: PAIN

Pain has been conceptualized as containing both physical and psychological components, with their relative contribution to perceived pain varying across individuals (Melzack, 1973). Thus, pain perception involves not only sensing painful stimuli but also behavioral and cognitive components which may intensify or lessen the pain experience (Varni, Bessman, Russo, & Cataldo, 1980). Whereas acute pain in the hemophiliac is associated with a specific bleeding episode, chronic arthritic pain represents a sustained condition over an extended period of time (Dietrich, 1976). Thus, pain perception represents a complex psychophysiological event, further complicated in the hemophiliac by the existence of both acute bleeding and chronic arthritic pain, requiring differential treatment strategies (Varni, 1981a). More specifically, acute pain of hemorrhage provides a functional signal, indicating the necessity of intravenous infusion of factor replacement. Arthritic pain, on the other hand, represents a potentially debilitating chronic condition which may result in impaired life functioning and analgesic dependence (Varni & Gilbert, in press). A substantial number of analgesics and anti-inflammatory medications are of limited usefulness since they inhibit platelet aggregation and prolong the bleeding time in hemophiliacs (Arnold & Hilgartner, 1977). Consequently, the challenge has been to develop an effective intervention in the reduction of perceived arthritic pain while not interfering with the essential functional signal of acute bleeding pain.

Chronic Arthritic Pain

Medical observations have suggested the possible therapeutic value of warming and heat application in the management of arthritic joints (Lehman, Warren, & Scham, 1974). White (1973) reported data from a questionnaire administered to 30 osteoarthritic and rheumatoid arthritic patients, with 27 associating pain relief with past experiences of warmth and massage, with the application of a counterirritant (10% menthol and 15% methyl salicylate) producing active tissue hyperemia and a sensation of heat accompanied by decreased pain perception. Citing earlier findings by Wasserman (1968) which demonstrated abnormal electromyographic readings in muscles adjacent to arthritic joints, White proposed that the counterirritant-produced hyperemia reduced the ischemia in the contracted muscles surrounding the arthritic joint. Swezey (1978) reviewed the evidence on the relative efficacy of superficial heating in articular disorders, suggesting that the threshold for pain can be raised by superficial heating, but cautioned that the increases in superficial circulation in relationship to pain relief and muscle relaxation had to be experimentally determined. Hilgard's (1975) laboratory research on hypnosis-controlled experimental pain has indicated the role of distraction and refocusing of attention, anxiety reduction, suggestions of pain relief, and the imagination of past experiences that were incompatible with pain as potential cognitive variables in the reduction of pain perception.

Varni (1981a, 1981b) reported on the successful self-regulation of arthritic pain perception by hemophiliacs involving a treatment strategy consisting of progressive muscle relaxation exercises, meditative breathing, and guided imagery techniques. In these studies, the patient was instructed to imagine himself actually in a scene previously identified or associated with past experiences of warmth and arthritic pain relief. The scene was initially evoked by a detailed multisensory description by the therapist, and then further details of the scene were subsequently described out loud by the patient. Once the scene was clearly described and visualized by the patient, the therapist's suggestions included imagining the gentle flow of blood from the forehead down all the body parts to the targeted arthritic joint, images of warm colors such as red and orange, and the sensations of warm sand and sun on the joint in the context of a beach scene. Further suggestions consisted of statements indicating reduction of pain as the joint progressively felt warmer and more comfortable. Each patient had previously identified one joint as the site of greatest arthritic pain, and this joint was targeted for active warming during clinic and home practice sessions.

In the initial investigation, preliminary findings with two hemophiliacs suggested the potential utility of these self-regulation techniques in the reduction of arthritic pain (Varni, 1981a). The next investigation was designed to extend these earlier findings systematically to three additional hemophiliacs under improved methodological conditions, including longer baseline and follow-up assessments and measurement of multiple subjective and medical parameters in relationship to arthritic pain (Varni, 1981b). Each patient was instructed in the use of a daily self-monitoring multidimensional pain assessment instrument. In addition, each patient's subjective evaluation of the overall impact of the intervention was determined by a comparative assessment inventory administered at the final follow-up session. An analysis of each patient's medical chart and pharmacy records provided data on the number of analgesics ordered, factor replacement units for bleeding episodes, and the number of hemorrhages. These records were analyzed retrospectively at the completion of the investigation to represent the patient's status six months prior to the self-regulation intervention and during the follow-up period (averaged monthly for both pre- and post-assessments). Finally, a thermal biofeedback unit served as a physiological assessment device rather than as a training technique, with the patient instructed to attempt actively to increase the temperature at the joint site during the self-regulation conditions without the benefit of incremental feedback.

An analysis of Figure 1 demonstrates that the self-regulation techniques were effective in significantly reducing the number of days of perceived arthritic pain per week for all three patients, maintained over an extended follow-up assessment. Further analysis of the daily pain assessment instrument showed that on the 10-point scale (1 = mild pain, 10 = most severe pain), arthritic pain perception decreased from an average of 5.1 during baseline to 2.2 on those follow-up days in which arthritic pain occurred. Average bleeding pain was essentially unchanged from baseline to follow-up conditions (6.9 to 6.8). Each patient's evaluation at the final follow-up session on the comparative assessment inventory indicated substantial positive changes in arthritic pain, mobility, sleep, and general overall functioning, with no reported changes in bleeding pain perception. Assessment of the surface skin temperature over the targeted arthritic joint showed increased thermal readings from baseline to self-regulation conditions, averaging an increase of 4.1°F. These findings were consistent across patients and were maintained over the follow-up period (Table III). The analysis of the medical charts and pharmacy records demonstrated that the units of factor replacement prescribed covaried with the number of bleeding episodes for all three patients, with the average number of bleeding episodes and units of factor replacement remaining essentially constant across pre- and post-assessment periods. The number of analgesic tablets showed a consistent decrease across patients (Table III).

These data on both the medical parameters of bleeding frequency and factor replacement units, as well as the subjective pain ratings, support the proposal that these self-regulation techniques did not affect bleeding pain perception. Substantial decreases in reported bleeding frequency and factor replacement units would have been expected had bleeding pain perception been affected. It is essential to emphasize that these techniques are not meant to supercede proper and correct medical care. Nevertheless, in the case of chronic degenerative arthropathy, the

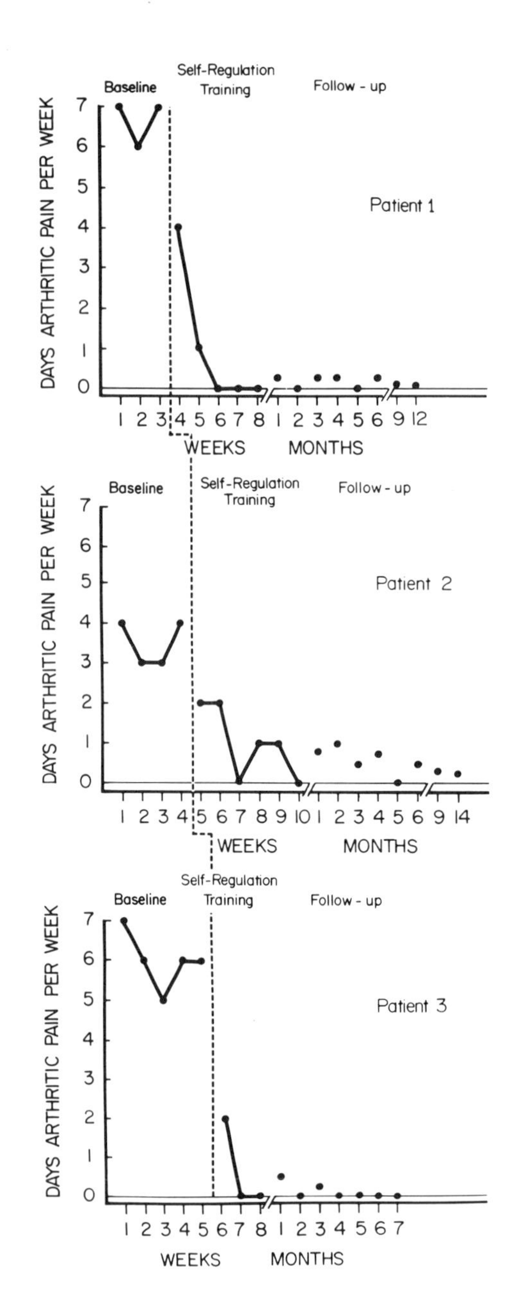

Figure 1. Days of total arthritic pain perception per week during baseline, self-regulation training, and follow-up for three patients. From "Self-Regulation Techniques in the Management of Chronic Arthritic Pain in Hemophilia" by J. W. Varni, *Behavior Therapy*, 1981, *12*, 185–194. Copyright 1981 by the Association for Advancement of Behavior Therapy. Reprinted by permission.

Table III. Physiological and Medical Parameters[a]

Patient	Arthritic joint skin temperature (°F)[b]		Analgesic tablets (#)[b]	Factor replacement (units)[b]	Bleeding frequency[b]
1	Pre	84.4	4.3	10667	5.1
	Post	87.9	0	9893	4.8
2	Pre	83.1	15.8	6264	3.7
	Post	88.3	7.9	8919	4.6
3	Pre	85.8	2.7	1166	.33
	Post	89.3	0	858	.29
$\bar{x}$	Pre	84.4	7.6	6032	3.0
	Post	88.5	2.6	6557	3.2

[a]From "Self-Regulation Techniques in the Management of Chronic Arthritic Pain in Hemophilia" by J. W. Varni, *Behavior Therapy*, 1981, *12*, 185–194. Copyright 1981 by the Association for Advancement of Behavior Therapy. Reprinted by permission.
[b]For skin temperature, premeasures were recorded during the baseline condition, with postmeasures recorded during the follow-up condition. For the other parameters, the medical charts and pharmacy records were analyzed retrospectively to yield a monthly average 6 months prior to the baseline condition and during the follow-up condition.

arthritic pain, in contrast to the acute bleeding pain, exceeds its functional intent, potentially resulting in limitations in normal activities, with the possibility of analgesic dependence always of concern. Thus, within the context of a detailed consideration of each patient's physical and psychological status, these techniques represent a noninvasive and supportive therapeutic intervention which, when applied within an interdisciplinary hemophilia center, provides an additional treatment modality in the comprehensive management of hemophilia.

Pain and Analgesia Management Complicated by Factor VIII Inhibitor

The treatment of children with hemophilia has evidenced significant advances in recent years as a result of the development of antihemophilic concentrate, home infusion programs, and the opportunity for comprehensive care through regional hemophilia centers. Unfortunately, approximately 10% of hemophilic children develop an inhibitor to Factor VIII, presenting a serious problem in the management of bleeding episodes (Buchanan & Kevy, 1978). Although the frequency of hemorrhage is no different, the neutralization of Factor VIII replacement by an inhibitor (antibody) makes the control of bleeding ineffective. A recent advance in the treatment of hemophilic patients with inhibitors has been the use of prothrombin-complex concentrates (PCC), effecting some level of hemostasis through apparent activated by-products present in the reconstituted preparation. However, recent clinical reports have indicated decreased effectiveness of these products (Parry & Bloom, 1978; Scranton, Hasiba, & Gorenc, 1979). The pain associated with uncontrolled hemorrhage can be extremely severe, with narcotic analgesics traditionally prescribed (Dietrich, 1976). Thus, although the acute pain of hemorrhage provides a functional signal indicating the necessity of factor replacement therapy, in the hemophilic child with Factor VIII inhibitor the intensity of the pain supercedes its functional intent, with analgesic dependence of constant concern. Consequently, an effective alternative to analgesic dependence in the reduction of perceived pain in the patient with an inhibitor has been greatly needed.

Varni, Gilbert, and Dietrich (1981) recently reported on a study involving a 9-year-old hemophilic child with Factor VIII inhibitor. At 4 years of age, when the inhibitor developed and subsequent Factor VIII replacement therapy became impossible, the patient began to require narcotics in order to tolerate the pain of each hemorrhage. Progressively, the need for pain medication increased for both bleeding pain and for arthritic pain in his left knee secondary to degenerative arthropathy. Since the arthritic pain eventually occurred almost daily, the requests for analgesics further increased so that the acute pain of hemorrhage required ever larger doses for pain relief, even though PCC therapy and joint immobilization continued for the management of bleeding episodes. As a consequence of bleeding and arthritic pain in the lower extremities, the patient was wheelchair-bound nearly 50% of the time, had been hospitalized 16 times in the 4½-year period prior to the study for a total of 80 days after the development of the inhibitor, with analgesic medication also kept at his school for pain control. The final precipitating event in this steadily worsening cycle occurred during an evening visit to the emergency room because of a very painful and severe left knee hemorrhage which had not responded to home PCC therapy, with the administration of an adult dose of meperidine hydrochloride and intravenous diazepam providing no pain relief.

Training in the self-regulation of pain perception consisted of techniques developed earlier by Varni (1981a, 1981b), with modifications in the guided imagery techniques required for the bleeding pain intensity. The patient recorded the severity of his pain on the 10-point scale for a 2½-week baseline prior to self-regulation training. The average score for both arthritic and bleeding pain during this period was 7, indicating rather intense pain. At 1-year follow-up after the self-regulation training, the patient reported that both arthritic and bleeding pain were reduced to 2 on the scale when he engaged in the self-regulation techniques. In addition to this measure of pain perception, the patient's evaluation at the 1-year follow-up session on the comparative assessment inventory indicated substantial positive changes in arthritic and bleeding pain, mobility, sleep, and general overall functioning. As seen in Table IV, once the patient began using the self-regulation techniques for pain management, there were no further requests for meperidine hydrochloride during the 1-year posttreatment assessment, with substantially decreased amounts of acetaminophen with codeine required. The table also shows that significant improvements in other areas of functioning, including improved mobility as evidenced by the physical therapy measures on his arthritic left knee in comparison with his normal right knee on the dimensions of range of motion (0–150° = normal) and quadricep strength (1 = no joint motion; 5 = complete range of motion against gravity with full resistance). Normalization of psychosocial activities is suggested by increased school attendance, decreased hospitalizations, and parental report, noting a distinct elevation of the child's overall

Table IV. Parameters Associated with Pain Intensity[a]

Parameters	1 year before initiation of self-regulation training	1 year after initiation of self-regulation training
Pain intensity (1 = mild; 10 = severe)	7[b]	2[c]
Meperidine	74 tablets (50mg/ea.)	0 tablets
Acetaminophen/codeine elixir	438 doses (24mg codeine/dose)	78 doses (24mg codeine/dose)
Physical therapy measures		
Range of motion	Normal right knee 0°–150° Arthritic left knee 15°–105°	Right knee 0°–150° Left knee 0°–140°
Quadricep strength (0–5 scale)	Normal right knee 4– Arthritic left knee 3+	Right knee 4+ Left knee 4
Girth (knee joint circumference)	Not available	Right knee 26 cm Left knee 25.8 cm
Ambulation on stairs	2–3 maximum	No limitation
School days missed	33	6
Hospitalizations		
Total days	11	0
Number of admissions	3	0

[a]From "Behavioral Medicine in Pain and Analgesia Management for the Hemophilic Child with Factor VIII Inhibitor" by J. W. Varni, A. Gilbert, and S. L. Dietrich, *Pain*, 1981, *11*, 121–126. Copyright 1981 by Elsevier/North-Holland Biomedical Press B.V. Reprinted by permission.
[b]2½-week preassessment during pain episodes just prior to self-regulation training.
[c]1-year average rating during pain episodes when using self-regulation techniques.

mood and the fact that he was considerably less depressed during pain episodes now that he had the skills to reduce his pain perception actively without having to depend on pain medication.

The analysis of the various parameters assessed in this study suggests a significant improvement across a number of areas. As envisioned by the authors, a deteriorating cycle was evident prior to the intervention, schematically represented as: hemorrhage → pain → analgesics/joint immobilization → atrophy of muscles adjacent to the joints/joint deterioration → hemorrhage. Thus, as has been previously suggested (Dietrich, 1976), pain-induced immobilization results in muscle weakness surrounding the joints, setting the occasion for future hemorrhaging. By breaking this deteriorating cycle at the point of pain severity, the patient was offered the opportunity to decrease immobilization and increase therapeutic activities such as swimming, subsequently improving the strength and range of motion in the left knee, and with this improved ambulatory status, school attendance and his general activity level were consequently increased. The possibility that this early intervention may have prevented or reduced the likelihood of later drug abuse must also be considered (Varni & Gilbert, in press). Finally, it is important to reiterate that these procedures were used for a child with an inhibitor. For the hemophiliac without an inhibitor, bleeding pain serves a functional signal and is best managed with factor replacement therapy. However, in the present case, no effective medical procedure was available to control bleeding pain other than powerful narcotic analgesics, clearly an undesirable therapy modality.

HEMOPHILIA: PSYCHOSOCIAL ASPECTS

Prior to advances in factor replacement therapy, hemophilia medical care was most concerned with keeping the hemophilic child alive through the control of severe hemorrhages. Frequent and extensive hospitalizations were often required during childhood to manage internal bleeding, with whole plasma the only source of factor replacement. Although psychosocial dysfunction had been well recognized in a number of clinical observations (Agle, 1964; Browne, Mally, & Kane, 1960; Mattsson & Gross, 1966; Salk, Hilgartner, & Granich, 1972; Spencer, 1971), few well-controlled treatment studies exist (Mattsson & Agle, 1972). Problems in self-concept, familial relationships, perceived relationship to others outside the family, academics, and maladaptive response to stress have been suggested (cf. Agle, 1977, for review), but only in recent years as a result of the development of antihemophilic concentrate and home infusion programs has the opportunity for truly comprehensive multidisciplinary care through regional hemophilia centers been available for the hemophiliac (Boone, 1976). These advances in hemophilia treatment have significantly reduced the frequency of hospitalizations and the accompanying severe disruptive effects on normal psychosocial development. Nevertheless, typical of other chronic medical disorders with an inherently unstable clinical course (cf. Cataldo, Russo, Bird, & Varni, 1980), hemophilia may be accompanied by secondary emotional dysfunction; that is, the hemophiliac may be at risk for demonstrating psychological adjustment difficulties (Agle & Mattsson, 1976). Epidemiological research has further suggested that children with chronic illness and disabilities are at risk for developing behavioral disorders secondary to their chronic condition (Rutter, Tizard, & Whitmore, 1970), with Yule (1977) pointing out the potential of behavioral techniques in the secondary prevention of further and more severe behavioral/emotional dysfunction. From this perspective, a secondary prevention strategy would be directed toward the early identification of familial and individual risk factors in the hemophilic child such as unassertiveness and social skills deficits, anxiety reactions, and behavior problems (Varni & Russo, 1980). This early identification in the hemophilic child of a behavioral disorder and subsequent behavioral treatment may quite reasonably contribute to the child's later adjustment potential.

Behavioral Disorders

The behavioral approach in the modification of behavioral disorders consists of training the child's parents and teachers, when indicated, in behavioral management techniques (cf. Patterson, 1975), and teaching the older child self-regulation skills (Varni & Henker, 1979). Recently, Varni (1980a) reported on a case investigation designed to assess the efficacy of behavioral techniques in reducing persistent behavior problems and increasing age-appropriate behaviors in both the home and school settings with a 4½-year-old child with severe classical hemophilia. At the time of referral, the child was residing in a foster home consisting of both foster parents and four children. Foster home placement was necessitated by chronic parental neglect in his hemophilia care. His hemophilia status was further complicated by a high titer Factor VIII antibody, superceding traditional factor replacement therapy, resulting in numerous hospitalizations and intensive ambulatory medical care. Inappropriate behaviors in the home consisted of noncompliance, temper tantrums, threatening language, answering back, and aggression toward his peers. Deficits in age-appropriate behaviors included household chores such as making his bed and putting away his clothes. Classroom behavioral goals were identified as following directions, not physically disturbing others, staying in seat, not talking inappropriately, paying attention, and similar behaviors necessary for successful classroom participation (cf. Schumaker, Hovell, & Sherman, 1977). The home intervention consisted of a point system for the daily contingency manage-

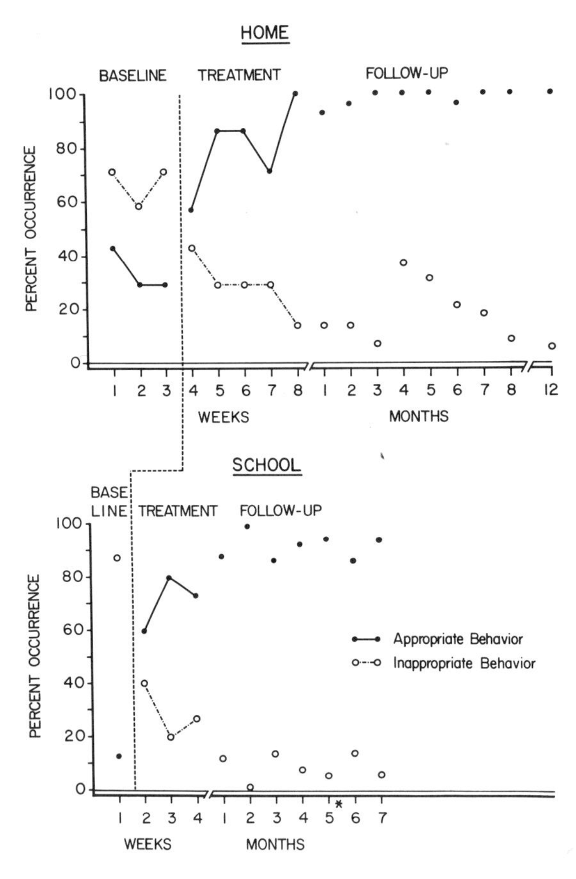

Figure 2. Percent occurrence of appropriate and inappropriate behaviors across baseline, treatment, and follow-up in the home and school. The asterisk on the school graph represents a $2\frac{1}{2}$-month summer vacation and beginning of first grade (6- and 7-month follow-up). From "Behavior Therapy in the Management of Home and School Behavior Problems with a $4\frac{1}{2}$-year-old Hemophilic Child" by J. W. Varni, *Journal of Pediatric Psychology*, 1980, *5*, 17–23. Copyright 1980 by the *Journal of Pediatric Psychology*. Reprinted by permission.

ment of behavior, with the occurrence of appropriate behaviors resulting in plus points which could be exchanged for previously agreed-upon preferred activities from a reinforcement list. Inappropriate behaviors resulted in a loss of points and privileges. A home-based contingency system was developed for the school intervention whereby the teacher completed a daily checklist of the classroom behaviors which were then consequated by the foster parents in the same manner as the home behaviors. As shown in Figure 2, the behavioral program resulted in clinically significant positive changes in both the home and school, with a 12-month follow-up in the home and a 7-month follow-up in the school demonstrating maintenance of the therapeutic gains. Although the potential of these techniques as a component in the secondary prevention of later adjustment problems in the hemophiliac awaits their application to a larger sample of hemophilic children with subsequent long-term follow-up, the findings from this case study are most encouraging.

Disease-Related Chronic Insomnia

Whereas in hemophilic children the stress and anxiety which may be associated with their illness can often be observed as a behavioral disorder, adolescent and young adult hemophiliacs exhibit topographically different responses to their potentially life-threatening condition. Chronic insomnia is often indicative of such a stress reaction. A recent report from the Institute of Medicine has delineated the many contraindications of sleep-inducing medications for chronic insomnia, while emphasizing the development and evaluation of psychosocial and behavioral interventions (Solomon, White, Parron, & Mendelson, 1979). Within the behavioral approach to insomnia, various techniques have been investigated, including progressive muscle relaxation (Borkovec, Kaloupek, & Slama, 1975), stimulus control procedures (Bootzin, 1972; Zwart & Lisman, 1979), and cognitive refocusing strategies (Mitchell, 1979). These studies have focused on patients demonstrating insomnia as a primary symptom, rather than as a secondary dysfunction accompanying a medical illness or disability. Varni (1980b) reported on a case investigation in which chronic insomnia was a secondary emotional reaction to hemophilia-related anxiety. Because of the development of an anticoagulant inhibitor (high titer Factor VIII antibody), treatment of this patient's bleeding episodes with Factor VIII replacement was not possible, further complicating his medical care with increased severity and duration of each internal hemorrhage. Additional medical complications included severe hemophilic arthropathy, hypertension, obesity, and borderline adult-onset diabetes. The patient evidenced a history of chronic insomnia which was further aggravated 1 year prior to the investigation when his older hemophilic brother died as a result of a severe internal hemorrhage. Also at that time, the patient's antibody titer was found to be higher than previously noted, resulting in even greater feelings of his own vulnerability. At the time of referral, the patient reported that these factors resulted in his constant worrying at bedtime about the potential consequences of a severe hemorrhage.

During the initial session, four measures were identified for assessment: (1) total number of hours of uninterrupted sleep per night averaged to the half hour, (2) daily tension rating on a scale from 1 (relaxed) to 10 (most stressed), (3) number of days per week in which bleeding was evident, and (4) bleeding pain intensity on a scale from 1 (mild) to 10 (most severe). A daily questionnaire checklist was developed for the self-recording of these parameters. After an analysis of the information provided from the initial evaluation and a 3-week self-recording baseline, three areas were targeted for intervention: (1) presleep and daily tension, (2) presleep intrusive cognitions, and (3) sleep-incompatible activities while in bed. Subsequently, training in the self-management of insomnia and chronic tension consisted of a 4-component treatment package: (1) a 25-step progressive muscle relaxation sequence; (2) meditative breathing; (3) cognitive refocusing techniques involving a detailed multisensory image of a previously experienced pleasant scene; (4) stimulus control techniques consisting of delaying bedtime until very sleepy, using the bed only for sleep-compatible behaviors (no television viewing or reading), getting out of bed within 15 minutes if unable to sleep and engaging in an activity such as reading with the return to bed only when very sleepy, and no daytime naps. As shown in Figure 3, implementation of the treatment package significantly increased the daily number of hours spent in uninterrupted sleep, as well as decreasing the daily tension ratings. Analysis of the data shows that the number of hours of sleep was maintained over a 27-week follow-up even during weeks when the bleeding frequency and pain intensity were quite high. On the other hand, daily tension ratings covaried with the frequency of bleeding episodes, but at substantially lower levels than during baseline. The asterisk represents a 4-week hospitalization during weeks 16 to 20 as a result of a severe hemorrhage, with an increase of daily tension specific to the prehospitalization and posthospitalization periods.

The clinical significance of these findings is even more evident in consideration of the patient's high titer Factor VIII antibody, precluding traditional factor replacement therapy. Chronic insomnia and tension alone can result in the debilitation and limitation of healthy physical and

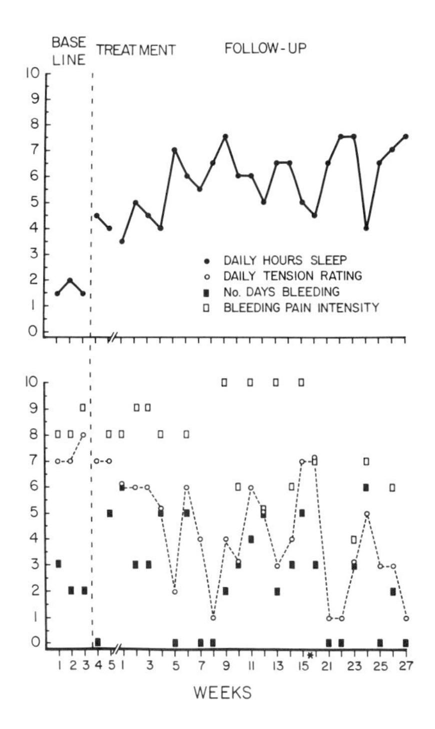

Figure 3. Daily number of hours spent in uninterrupted sleep, daily tension ratings, bleeding frequency per week, and bleeding pain intensity across baseline, treatment, and follow-up. The asterisk represents a 4-week hospitalization. From "Behavioral Treatment of Disease-Related Chronic Insomnia in a Hemophiliac" by J. W. Varni, *Journal of Behavior Therapy and Experimental Psychiatry*, 1980, *11*, 143–145. Copyright 1980 by Pergamon Press, Ltd. Reprinted by permission.

psychological functioning. Given the patient's numerous physical problems in addition to hemophilia, including hypertension, obesity, and borderline adult-onset diabetes, one might assume that a continuation of his chronic tension and severe insomnia in synergistic action with the patient's other pathophysiological conditions, could quite reasonably represent in itself a life-threatening potential for coronary heart disease (Margolis, 1977). Although the patient's status remains precarious, the behavioral control of these two potent health care variables (chronic tension and severe insomnia) may represent a case example in preventive medicine as related to coronary heart disease (Margolis, 1977), with the findings further suggesting the effectiveness of behavioral techniques in the reduction of stress and anxiety reactions secondary to hemophilia.

HEMOPHILIA: ADHERENCE TO FACTOR REPLACEMENT TECHNIQUES

The recognition of the advantages of prompt factor replacement therapy for each bleeding episode has served as a stimulus for the creation of the home care program in an attempt to minimize or prevent the crippling effects of internal hemorrhaging (Dietrich, 1976). The home care program consists of training the hemophiliac and his parents in the techniques necessary for the administration of factor replacement products. Although the home care program has been adopted nationally through hemophilia comprehensive care centers, adherence to proper factor replacement procedures has been a previously uninvestigated component in the home care program.

Adherence to long-term treatment regimens in pediatric chronic disorders represents a serious and continuous problem (see Chapter 11). A review of the literature suggests that the instructional strategies employed may influence adherence by improving the patient's comprehension and memory of the regimen. These instructional strategies include providing the information in incremental quantities over time, organizing the information into specific categories, and combining verbal and written instructions. Although these strategies have shown some initial encouraging results with patients on short-term regimens for acute illness, their effectiveness with chronic disorders has not previously been systematically investigated.

Sergis-Deavenport and Varni (1981a, 1981b) investigated the effectiveness of behavioral techniques and instructional strategies in the teaching of and the therapeutic adherence to proper factor replacement procedures for the hemophilia home care program. The instructional strategies included the techniques described above, with the behavioral techniques consisting of the modeling of correct factor replacement procedures by the pediatric nurse practitioner (PNP), with the parent of a hemophilic child observing these modeled behaviors (observational learning). Three classifications of behaviors were delineated: (1) reconstitution of factor replacement products, (2) syringe preparation, and (3) infusion. The parent's behavioral rehearsal of the observed techniques was recorded on a behavioral checklist, with proper sequencing required for the correct responding of certain techniques. A minimum score of 80% correct performance was required before learning the next classification.

In addition to a treatment group of 5 parents who received factor replacement training for the first time, a comparison group of 7 parents who had been on the home care program for an average of 5 years was also tested on the behavioral checklist. The performance of the treatment group increased from 15% correct during baseline to 92% by the end of the treatment condition and maintained at 97% correct adherence over long-term follow-up assessment. A representative multiple-baseline analysis of the training process for a parent in the treatment group is shown in Figure 4. In contrast, the comparison group showed only 65% correct adherence when tested. These findings suggest the necessity and

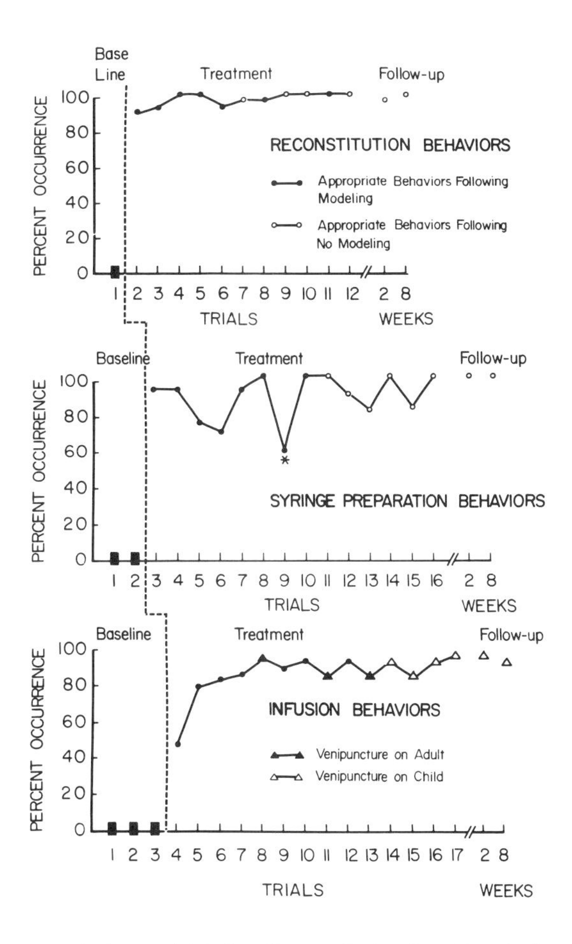

Figure 4. A representative multiple-baseline analysis of training in factor replacement therapy. The asterisk represents a 22-day delay in practicing syringe preparation steps. From Sergis-Deavenport & Varni (1981a).

the potential of these techniques in improving long-term adherence to a complex treatment regimen for a chronic disorder like hemophilia.

CONCLUSION

Results to date on the use of biobehavioral approaches in pediatric hematology/oncology are encouraging, though numerous research and clinical questions exist. As medical treatment for many hematological and oncological disorders grows increasingly intensive, sophisticated, and effective, biopsychosocial issues related to iatrogenicity and chronicity can be expected to increase concomitantly. The emphasis placed by a behavioral pediatrics approach upon prospective collection of data and rigorous evaluation of results should prove valuable in further defining problem areas as well as in providing avenues for effective clinical care. The potential for optimizing health care in children with these diseases, through the use of an interdisciplinary approach, is great and warrants further investigation.

REFERENCES

Agle, D. P. Psychiatric studies of patients with hemophilia and related states. *Archives of Internal Medicine*, 1964, *114*, 76–82.

Agle, D. P. (Ed.). *Mental health services in the comprehensive care of the hemophiliac*. New York: National Hemophilia Foundation, 1977.

Agle, D. P., & Mattsson, A. Psychological complications of hemophilia. In M. W. Hilgartner (Ed.), *Hemophilia in children*. Littleton, Mass.: Publishing Sciences Group, 1976.

Arnold, W. D., & Hilgartner, M. W. Hemophilic arthropathy: Current concepts of pathogenesis and management. *Journal of Bone and Joint Surgery*, 1977, *59*, 287–305.

Aronstam, A., Rainsford, S. G., & Painter, M. J. Patterns of bleeding in adolescents with severe haemophilia A. *British Medical Journal*, 1979, *1*, 469–470.

Bedell, J. R., Giordani, B., Armour, J. L., Tavormina, J., & Boll, T. Life stress and the psychological and medical adjustment of chronically ill children. *Journal of Psychosomatic Research*, 1977, *21*, 237.

Boone, D. C. (Ed.). *Comprehensive management of hemophilia*. Philadelphia: Davis, 1976.

Bootzin, R. A stimulus control treatment for insomnia. *Proceedings of the American Psychological Association*, 1972, 395–396.

Borkovec, T. D., Kaloupek, D. G., & Slama, K. M. The facilitative effect of muscle tension-release in the relaxation treatment of sleep disturbance. *Behavior Therapy*, 1975, *6*, 301–309.

Browne, W. J., Mally, M. A., & Kane, R. P. Psychosocial aspects of hemophilia: Study of 28 hemophilic children and their families. *American Journal of Orthopsychiatry*, 1960, *30*, 730–740.

Buchanan, G. R., & Kevy, S. V. Use of prothrombin complex concentrates in hemophiliacs with inhibitors: Clinical and laboratory studies. *Pediatrics*, 1978, *62*, 767–774.

Burish, T. G., & Lyle, J. N. Effectiveness of relaxation training in reducing the aversiveness of chemotherapy in the treatment of cancer. *Journal of Behavior Therapy and Experimental Psychiatry*, 1979, *10*, 357–361.

Cairns, G. F., & Altman, K. Behavioral treatment of cancer-related anorexia. *Journal of Behavior Therapy and Experimental Psychiatry*, 1979, *10*, 353–356.

Cataldo, M. F., Russo, D. C., Bird, B. L., & Varni, J. W. Assessment and management of chronic disorders. In J. M. Ferguson & C. B. Taylor (Eds.), *Comprehensive handbook of behavioral medicine. Vol 3: Extended applications and issues*. New York: Spectrum, 1980, pp. 67–95.

Dash, J. Hypnosis for symptom amelioration. In J. Kellerman (Ed.), *Psychological aspects of childhood cancer*. Springfield, Ill.: Charles C Thomas, 1980, 215–230.

Dietrich, S. L. Musculoskeletal problems. In M. W. Hilgartner (Ed.), *Hemophilia in children*. Littleton, Mass.: Publishing Sciences Group, 1976.

Ellenberg, L., Kellerman, J., Dash, J., Zeltzer, L., & Higgins, G. Multiple use of hypnosis for an adolescent girl with chronic myelogenous leukemia. *Journal of Adolescent Health Care*, 1980, *1*, 132–136.

Hilgard, E. R. Pain as a puzzle for psychology and physiology. *American Psychologist*, 1969, *24*, 103–113.

Hilgard, E. R. The alleviation of pain by hypnosis. *Pain*, 1975, *1*, 213–231.

Hilgartner, M. W. (Ed.). *Hemophilia in children*. Littleton, Mass.: Publishing Sciences Group, 1976.

Holland, J., Plumb, M., Yates, J., Harris, S., Tuttolomondo, A., Homes, J., & Holland, J. F. Psychological response of patients with acute leukemia to germ-free environments. *Cancer*, 1977, *36*, 871–879.

Kagen-Goodheart, L. Reentry: Living with childhood cancer. *American Journal of Orthopsychiatry*, 1977, *47*, 651.

Karon, M. Acute leukemia in childhood. In H. F. Conn (Ed.), *Current therapy*. Philadelphia: Saunders, 1975.

Kasper, C. K. Hematologic care. In D. C. Boone (Ed.), *Comprehensive management of hemophilia*. Philadelphia: Davis, 1976.

Kasper, C. K., Dietrich, S. L., & Rapaport, S. I. Hemophilia prophylaxis with Factor VIII concentrate. *Archives of Internal Medicine*, 1970, *125*, 1004.

Katz, E. R. Illness impact and social reintegration. In J. Kellerman (Ed.), *Psychological aspects of childhood cancer*. Springfield, Ill.: Charles C Thomas, 1980.

Katz, E. R., Kellerman, J., Rigler, D., Williams, K., & Siegel, S. E. School intervention with pediatric cancer patients. *Journal of Pediatric Psychology*, 1977, *2*, 72.

Katz, E. R., Kellerman, J., & Siegel, S. E. Behavioral distress in children with leukemia undergoing bone marrow aspirations. *Journal of Consulting and Clinical Psychology*, 1980, *48*, 356–365.

Kellerman, J. Psychological intervention in pediatric cancer: A look toward the future. *International Conference of Psychology and Medicine*, University of Swansea, Wales, 1979(a).

Kellerman, J. Behavioral treatment of pavor nocturnus in a child with acute leukemia. *Journal of Nervous and Mental Disease*, 1979, *167*, 182–185. (b)

Kellerman, J. Comprehensive psychosocial care of the child with cancer. In J. Kellerman (Ed.), *Psychological aspects of childhood cancer*. Springfield, Ill.: Charles C Thomas, 1980.

Kellerman, J., Rigler, D., Siegel, S. E., McCue, K., Pospisil, J., & Uno, R. Psychological evaluation and management of pediatric oncology patients in protected environments. *Medical and Pediatric Oncology*, 1976, *2*, 353–360. (a)

Kellerman, J., Rigler, D., Siegel, S. E., McCue, K., Pospisil, J., & Uno, R. Pediatric cancer patients in reverse isolation utilizing protected environments. *Journal of Pediatric Psychology*, 1976, *1*, 21–25. (b)

Kellerman, J., Rigler, D., & Siegel, S. E. The psychological effects of isolating patients in protected environments. *American Journal of Psychiatry*, 1977, *134*, 563–565.

Kellerman, J., Rigler, D., Siegel, S. E., & Katz, E. Disease-related communication and depression in pediatric cancer patients. *Journal of Pediatric Psychology*, 1977, *2*, 52–53.

Kellerman, J., Rigler, D., & Siegel, S. E. Psychological response of children to reverse isolation in a protected environment. *Journal of Behavioral Medicine*, 1979, *2*, 263–274.

Kellerman, J., Zeltzer, L., Ellenberg, L., & Dash, J. Hypnotic reduction of distress associated with adolescent cancer and treatment (1): Acute pain and anxiety due to medical procedures. Unpublished manuscript, 1980.

Kellerman, J., Zeltzer, L., Ellenberg, L., Dash, J., & Rigler, D. Psychological effects of illness in adolescence: Anxiety, self-esteem and the perception of control (Part 1). *Journal of Pediatrics*, 1980, *97*, 126–131.

Kisker, C. T., Perlman, A. W., & Benton, C. Arthritis in hemophilia. *Seminars in Arthritis and Rheumatism*, 1971, *1*, 220–235.

Koch, C. R., Hermann, J., & Donaldson, M. H. Supportive care of the child with cancer and his family. *Seminars in Oncology*, 1974, *1*, 81.

Lansky, S. B., Lowman, J. T., Vats, T., & Gyulay, J. School phobias in children with malignant neoplasms. *American Journal of Diseases of Children*, 1975, *129*, 42.

Lehmann, J. F., Warren, C. G., & Scham, S. M. Therapeutic heat and cold. *Clinical Orthopedics and Related Research*, 1974, *99*, 207–245.

Margolis, S. Physician strategies for the prevention of coronary heart disease. *Johns Hopkins Medical Journal*, 1977, *141*, 170–176.

Mattsson, A., & Agle, D. P. Group therapy with parents of hemophiliacs: Therapeutic process and observations on parental adaptation to chronic illness in children. *Journal of American Academy of Child Psychiatry*, 1972, *11*, 558–571.

Mattsson, A., & Gross, S. Adaptional and defensive behavior in young hemophiliacs and their parents. *American Journal of Psychiatry*, 1966, *122*, 1349–1356.

Melzack, R. *The puzzle of pain*. Harmondsworth, England: Penguin, 1973.

Melzack, R. The McGill pain questionnaire: Major properties and scoring methods. *Pain*, 1975, *1*, 277–299.

Mitchell, K. R. Behavioral treatment of presleep tension and intrusive cognitions in patients with severe predormital insomnia. *Journal of Behavioral Medicine*, 1979, *2*, 57–69.

Moss, H. A., & Nannis, E. D. Psychological effects of central nervous system treatment on children with acute lymphocytic leukemia. In J. Kellerman (Ed.), *Psychological aspects of childhood cancer*. Springfield, Ill.: Charles C Thomas, 1980, pp. 171–183.

Parry, D. H., & Bloom, A. L. Failure of Factor VIII inhibitor bypassing activity (Feiba) to secure hemostasis in hemophilic patients with antibodies. *Journal of Clinical Pathology*, 1978, *31*, 1102–1105.

Patterson, G. R. *Families: Applications of social learning to family life*. Champaign, Ill.: Research Press, 1975.

Redd, W. H. In vivo desensitization in the treatment of chronic emesis following gastrointestinal surgery. *Behavior Therapy*, 1980, *11*, 421–427. (a)

Redd, W. H. Stimulus control and extinction of psychosomatic symptoms in cancer patients in protective isolation. *Journal of Consulting and Clinical Psychology*, 1980, *48*, 448–455. (b)

Roberts, R. N., & Gordon, S. B. Reducing childhood nightmares subsequent to a burn trauma. *Child Behavior Therapy*, 1980, *1*, 373–381.

Rutter, M., & Graham, P. Psychiatric aspects of intellectual and educational retardation. In M. Rutter, J. Tizard, & K. Whitmore (Eds.), *Education, health and behaviour*. London: Longmans, 1970.

Rutter, M., Tizard, J., & Whitmore, K. (Eds.). *Education, health, and behaviour*. London: Longmans, 1970.

Salk, L., Hilgartner, M., & Granich, B. The psycho-social impact of hemophilia on the patient and his family. *Social Science and Medicine*, 1972, *6*, 491–505.

Sallan, S. E., Cronin, C., Zelen, M., & Zinberg, N. E. Antiemetics in patients receiving chemotherapy for cancer: A randomized comparison of delta-9-tetra- hydrocannabinol and prochlorderazine. *New England Journal of Medicine*, 1980, *302*, 135–138.

Schulz, L. S. Classical conditioning of nausea and vomiting in cancer chemotherapy. *Abstracts of American Society of Clinical Oncology*, 1980.

Schumaker, J. B., Hovell, M. F., & Sherman, J. A. An analysis of daily report cards and parent-managed privileges in the improvement of adolescents' classroom performance. *Journal of Applied Behavioral Analysis*, 1977, *10*, 449–464.

Schwartz, G. E. Behavioral medicine and systems theory: A new synthesis. *National Forum*, 1980, *60*, 25–30.

Scranton, P. E., Hasiba, U., & Gorenc, T. J. Intramuscular hemorrhage in hemophiliacs with inhibitors. *Journal of the American Medical Association*, 1979, *241*, 2028–2030.

Sergis-Deavenport, E., & Varni, J. W. *Emphasizing behavioral techniques in teaching factor replacement procedures for the hemophilia home care program*. Unpublished manuscript, 1981. (a)

Sergis-Deavenport, E., & Varni, J. W. *Behavioral techniques and therapeutic adherence to proper factor replacement procedures in hemophilia*. Unpublished manuscript, 1981. (b)

Siegel, S. E. The current outlook of childhood cancer: The medical background. In J. Kellerman (Ed.), *Psychological aspects of childhood cancer*. Springfield, Ill.: Charles C Thomas, 1980, pp. 5–13.

Silberfarb, P. M., Philibert, D., & Levine, P. M. Psychosocial aspects of neoplastic disease: Affective and cognitive effects of chemotherapy in cancer patients II. *American Journal of Psychiatry*, 1980, *137*, 597–601.

Silverberg, E. Cancer statistics. *Cancer*, 1979, *29*, 6.

Sokoloff, L. Biochemical and physiological aspects of degenerative joint disease with special reference to hemophilic arthropathy. *Annals of New York Academy of Science*, 1975, *240*, 285–290.

Solomon, F., White, C. C., Parron, D. L., & Mendelson, W. B. Sleeping pills, insomnia, and medical practice. *New England Journal of Medicine*, 1979, *300*, 803–808.

Spencer, R. F. Psychiatric impairment versus adjustment in hemophilia: Review and five case studies. *Psychiatry and Medicine*, 1971, *2*, 1–12.

Spinetta, J. J. The dying child's awareness of death: A review. *Psychological Bulletin*, 1974, *81*, 256.

Spinetta, J. J. Disease related communication: How to tell. In J. Kellerman (Ed.), *Psychological aspects of childhood cancer*. Springfield, Ill.: Charles C Thomas, 1980, 257–269.

Spinetta, J. J., & Maloney, J. L. Death anxiety in the outpatient leukemic child. *Pediatrics*, 1975, *56*, 1034.

Spinetta, J. J., & Maloney, L. J. The child with cancer: Patterns of communication and denial. *Journal of Consulting and Clinical Psychology*, 1978, *46*, 1540.

Spinetta, J. J., Rigler, D., & Karon, M. Anxiety in the dying child, *Pediatrics*, 1973, *52*, 841–845.

Strauss, H. S. The perpetuation of hemophilia by mutation. *Pediatrics*, 1969, *39*, 186.

Stuart, J., Davies, S. H., & Cummings, E. A. Haemorrhagia episodes in haemophilia: A 5-year prospective study. *British Medical Journal*, 1960, *3*, 1624.

Sutow, W. W., Vietta, T. J., & Fernbach, D. J. *Clinical Pediatric Oncology*, St. Louis: Mosby, 1973.

Swezey, R. L. *Arthritis: Rational therapy and rehabilitation*. Philadelphia: Saunders, 1978.
Tavormina, J., Kastner, L. S., Slater, P. M., & Watt, S. L. Chronically ill children: A psychologically and emotionally deviant population? *Journal of Abnormal Child Psychology*, 1976, *4*, 99.
Varni, J. W. Behavior therapy in the management of home and school behavior problems with a 4½-year-old hemophilic child. *Journal of Pediatric Psychology*, 1980, *5*, 17–23. (a)
Varni, J. W. Behavioral treatment of disease-related chronic insomnia in a hemophiliac. *Journal of Behavior Therapy and Experimental Psychiatry*, 1980, *11*, 143–145. (b)
Varni, J. W. Behavioral medicine in hemophilia arthritic pain management: Two case studies. *Archives of Physical Medicine and Rehabilitation*, 1981, *62*, 183–187. (a)
Varni, J. W. Self-regulation techniques in the management of chronic arthritic pain in hemophilia. *Behavior Therapy*, 1981, *12*, 185–194. (b)
Varni, J. W., & Gilbert, A. Self-regulation of chronic arthritic pain and long-term analgesic dependence in a hemophiliac. *Rheumatology and Rehabilitation*, in press.
Varni, J. W., & Henker, B. A self-regulation approach to the treatment of three hyperactive boys. *Child Behavior Therapy*, 1979, *1*, 171–192.
Varni, J. W., & Russo, D. C. Behavioral medicine approach to health care: Hemophilia as an exemplary model. In M. Jospe, J. E. Nieberding, & B. D. Cohen (Eds.), *Psychological factors in health care*. Lexington, Mass.: Lexington Books, 1980.
Varni, J. W., Bessman, C., Russo, D. C., & Cataldo, M. F. Behavioral management of chronic pain in children: Case study. *Archives of Physical Medicine and Rehabilitation*, 1980, *61*, 375–379.
Varni, J. W., Gilbert, A., & Dietrich, S. L. Behavioral medicine in pain and analgesia management for the hemophilic child with Factor VIII inhibitor. *Pain*, 1981, *11*, 121–126.
Vernick, J., & Karon, M. Who's afraid of death on a leukemia ward? *American Journal of Diseases of Children*, 1965, *109*, 393.
Wasserman, R. R. Electromyographic, electrodiagnostic, and motor nerve conduction observations in patients with rheumatoid arthritis. *Archives of Physical Medicine and Rehabilitation*, 1968, *49*, 90–95.
White, J. R. Effects of a counterirritant on perceived pain and hand movement in patients with arthritis. *Physical Therapy*, 1973, *53*, 956–960.
Yule, W. The potential of behavioral treatment in preventing later childhood difficulties. *Behavioural Analysis and Modification*, 1977, *2*, 19–31.
Zeltzer, L. The adolescent with cancer. In J. Kellerman (Ed.), *Psychological Aspects of Childhood Cancer*. Springfield, Ill.: Charles C Thomas, 1980.
Zeltzer, L., Barbour, J., LeBaron, S., & Kellerman, J. Delta-9-tetrahydro- cannabinol as an anti-emetic. *New England Journal of Medicine*, 1980, *302*, 1363.
Zeltzer, L., Kellerman, J., Ellenberg, L., Dash, J., & Rigler, D. Psychological effects of illness in adolescence: Crucial issues and coping styles. *Journal of Pediatrics*, 1980, *97*(1), 132–138.
Zeltzer, L., Kellerman, J., Ellenberg, L., & Dash, J. *Hypnotic reduction of distress associated with adolescent cancer and treatment (II): Nausea and emesis*. Unpublished manuscript, 1981.
Zwart, C. A., & Lisman, S. A. Analysis of stimulus control treatment of sleep-onset insomnia. *Journal of Consulting and Clinical Psychology*, 1979, *47*, 113–118.

4

Behavioral Interventions in Pediatric Neurology

Bruce L. Bird

INTRODUCTION

Pediatric neurology encompasses a broad continuum of disorders with great diversity of prevalences, etiologies, symptom categories, and severities. For example, the prevalence of neurological learning disabilities is estimated to be as high as 5% of the population (Tarnopol, 1971). Rare disorders such as dystonia musculorum deformans, a genetic degenerative disease, afflict a very small subset of the population but produce extreme physical incapacity (Eldridge & Fahn, 1976). Etiologies of neurological disorders include trauma, genetic, metabolic, or developmental disorders, diseases specific to the nervous system (NS) or central nervous system (CNS), or diseases which produce byproducts which in turn damage the NS or CNS. Symptom categories include all brain and NS functions, selectively or in combination. Severities range from mild aberrations in selective functions (i.e., written letter perception) to complete physical incapacity, coma, and death (Ford, 1973).

In the past decade, progress has occurred in understanding the pathophysiological bases and improving treatments of many neurological disorders, with exemplary results in neurovirology (Johnson & Hern-

BRUCE L. BIRD • University of New Orleans; Associated Catholic Charities of New Orleans, New Orleans, Louisiana 70092.

don, 1974) and selected biochemical disorders (Moser, 1975). However, pathophysiologies of many neurological disorders remain, at best, poorly understood. The relationships among etiology, pathophysiology, symptoms, and treatment are therefore often poorly formulated, resulting in problems of therapeutic efficacy. Surgical and pharmacological treatments have not shown in neurology the increases in benefits shown in other areas of medicine. Because current medical treatments are deficient both in curative and palliative effects, a large number and variety of neurological disorders are considered chronic (Cataldo, Russo, Bird, & Varni, 1980). In neurology, traditional therapies are heavily supplemented with other modalities such as physical and occupational therapy and with a variety of special educational and behavioral techniques.

A related problem for both the researcher and the practicing neurologist concerns diagnosis. Despite impressive improvements in diagnostic technology, such as computerized axial tomography, neurological diagnoses in many disorders are determined by the patient's history and presenting behavior. Differential diagnosis of neurological as opposed to behavioral/psychiatric etiology is difficult for a number of disorders, such as hyperactivity, urinary incontinence, and headache (cf. Diamond & Dalessio, 1978).

The practicing pediatric neurologist, therefore, encounters a number of disorders for which medical treatments are prescribed for palliative effects with uncertain promise of efficacy. Other disorders present with symptoms which may have behavioral/psychiatric contributions to etiology. Increasingly throughout the last decade, researchers and practitioners have attempted learning-based therapeutic interventions for disorders which resist traditional therapies or which may have behavioral components in their etiologies. These attempts have been prompted by evidence of successful behavioral treatments for selected medical disorders and by increasing evidence from basic research indicating significant relationships among physical and emotional/behavioral variables (Basmajian, 1978).

Subsequent sections of this chapter summarize literature and provide case examples of behavioral evaluations and treatments for selected pediatric neurological disorders. Emphasis is given to neuromuscular and paroxysmal disorders, which have received substantial attention from researchers but less attention from reviewers than other disorders. The recurring theme of poorly understood neuropathophysiologies and potential contributions of behavioral pediatrics to this problem are discussed in the final section of the chapter.

OVERVIEW OF BEHAVIORAL TREATMENTS

As reviewed in Chapter 1 of this text, a variety of learning-based behavioral interventions may be applied to medical disorders. Before reviewing literature and providing clinical examples for selected disorders, we will review briefly major behavioral treatments for pediatric neurological disorders. Behavioral treatments share a common goal—normalizing a measurable behavioral or physiological symptom through learning. As discussed throughout this chapter, understanding mechanisms of these effects should provide important insights into the nature of neurological disorders related with behavioral methods.

Behavior Modification

Behavior modification, contingency management, or operant conditioning, involves: (1) identifying and defining a target behavior to be changed (increased/decreased in rate or strength, changed in form or topography); (2) obtaining reliable observations of behavior before, during, and after treatment; (3) dispensing rewarding or punishing events or activities contingent upon some number or rate of responding (Bandura, 1969). For example, Sachs and Mayhall (1972) have demonstrated that rewards may be used to improve muscle control in a child with cerebral palsy. Reinforcement systems using token economies in which points, chips, or other tangible items are dispensed for appropriate responses and are later cashed in for privileges or treats have been widely used with great success to build skills in severely retarded populations (Wetherby & Baumeister, 1981).

The social environment has long been recognized by medicine as a potent influence on symptoms (Peabody, 1927). The principles of social learning theory (Bandura, 1969), sufficiently studied by experimental psychology, have been applied with increasing success to problems in clinical neurology. For example, Fordyce (1976) has demonstrated that a patient's natural social environment can maintain undesired complaints and poor functional activity in chronic pain patients. The social environment can be rearranged so that social reinforcers shape adaptive functions and minimize maintenance of unnecessary complaints. Social reinforcement used in behavior modification programs has also been reported to be influential in seizures (Gardner, 1967) and in a variety of developmental disabilities (Creer & Christian, 1976).

Behavior modification as a body of scientific knowledge has only begun to be applied to medical disorders. For example, principles of situational or stimulus control of behavior and generalization, or transfer of learned behavior across situations may offer important tools for analyzing and improving symptoms in neurological disorders (Bird, Cataldo, & Parker, 1981). Moreover, all behavioral interventions utilize some components of behavior modification, including direct treatments such as biofeedback (Bird & Cataldo, 1978) and adjunctive treatments such as programs designed to increase compliance with medication (Epstein & Masek, 1978). The explanatory principles for behavior modification are parsimonious and may be proposed at the behavioral level of analysis (infrequent, low-strength behaviors are increased and maladaptive behaviors decreased) without proposing any changes in neuropathophysiology, as the organism learns to use remaining intact neurophysiology more effectively.

Biofeedback

Biofeedback principles and procedures have been thoroughly reviewed elsewhere (Miller, 1969, 1978). Briefly, biofeedback uses electronic apparatus to monitor physiological responses and display them to subjects using nonnormal sensory modalities, as subjects are motivated to practice a certain type of control of responses. For example, in electromyographic (EMG) biofeedback, a patient with cerebral palsy may watch his or her trapezius EMG change on an oscilloscope while attempting to reduce undesired spasms and relax the muscles.

Several types of EMG biofeedback have been applied to muscular disorders, including relaxation of forehead muscles to achieve whole-body relaxation (Cataldo, Bird, & Cunningham, 1978); increases in targeted muscle strength, such as anterior tibial contractions in post-CVA foot drop (Basmajian, Takebe, Kukulka, & Narayan, 1976); and matching abnormal with contralateral normal muscle groups (Jankel, 1978). Principles and procedures of EMG biofeedback (Basmajian, 1978) and reviews of EMG biofeedback applications to muscle disorders (Bird, Cataldo, & Parker, 1981) have been presented in detail elsewhere. The literature is generally very positive, indicating effective improvements in a variety of muscular disorders. However, additional evidence is needed to separate the contributions of EMG biofeedback from other potentially active procedures such as behavior modification. Generalization and maintenance of therapeutic effects have also been reported to be generally good, but additional documentation is needed for strong claims of efficacy.

Two other biofeedback techniques, providing force or position feedback to subjects, have been used successfully with neuromuscular patients. For example, Woolridge and Russell (1976) have used head position feedback to improve control of head posture in cerebral palsy. Engel, Nikoomanesh, and Schuster (1974) trained patients with diverse disorders to improve anal sphincter control, using a balloon-pressure feedback system for sphincter contractions. Wannstedt and Herman (1978) have taught post-CVA patients to improve weight-bearing of afflicted limbs using feedback from pressure-sensitive footpads. As with EMG biofeedback, force and position feedback techniques currently appear promising but merit additional research attention on transfer of effects and long-term durability.

Electroencephalographic (EEG) biofeedback provides nonnormal (usually visual or auditory) sensory feedback on selected EEG responses to subjects practicing to control or change the EEG in some way. Popular reports of EEG biofeedback training producing desirable altered states of consciousness in the early 1970s were followed by scientific studies of EEG feedback conditioning in humans and animals (Chase, 1974).

Two neurological disorders to which EEG biofeedback has been applied are paroxysmal disorders (Sterman, 1977) and hyperkinesis (Lubar & Shouse, 1976). In both cases, therapeutic efficacy has been proposed to result from the patient's learning neural responses which are both desirable and antagonistic to undesired EEG and behavioral symptoms of the disorder. The work in EEG biofeedback for hyperkinesis is not yet sufficiently extensive to be either well received or controversial. The literature on EEG biofeedback for paroxysmal disorders, as discussed below, is quite controversial, but also promising.

As with EMG and EEG responses, a growing literature indicates that humans can learn to modify skin temperature through biofeedback (Taub & Stroebel, 1978). The proposed mechanism for learned increases and decreases in skin temperature is modulation of regional blood flow. Photoplethysmographic responses which indicate regional blood flow have also been trained in humans (Feuerstein & Adams, 1977).

As discussed in detailed reviews, training patients to increase finger skin temperature and to decrease plethysmographic measures of blood flow in extracranial arteries have both been shown to be effective in treating migraine headaches (Diamond & Dalessio, 1978).

Behavior Therapy, Relaxation Training, and Self-Control

Behavior therapy includes a variety of techniques based on operant conditioning, classical conditioning, and related principles and methods for changing behaviors in the clinic. It may be considered a set of principles and procedures which overlaps with behavior modification (Calhoun & Turner, 1981). Three techniques related to behavior therapy which have been applied to neurological disorders include relaxation training, cognitive behavior therapy, and self-control.

A variety of procedures for training patients in muscular relaxation have been applied to a diverse set of medical disorders and have been popularized as preventitive exercises for stress-related medical problems (Benson, 1975). Many of the current "packaged" protocols are based on the clinical work of Jacobsen (1938, 1967), who developed a technique called *progressive relaxation training*. In this protocol, a patient is taught to sense tension and produce relaxation in muscle groups throughout the body. This is accomplished by regimented practice sessions in which the patient proceeds through major muscle groups, alternately tensing and relaxing them, until all major muscle groups are relaxed. Jacobsen's (1967) technique suggested that long practice sessions (30 minutes) over long periods (months) were necessary to acquire skill in relaxation.

More recent relaxation protocols involve shortening the individual practice sessions (cf. Haynes, Griffin, Mooney, & Parise, 1975), emphasizing breathing techniques (Benson, 1975), and/or using variants of meditation or cognitive strategies with physical exercises (Schwartz, 1977). Stroebel (1980) has developed a packaged relaxation program which appears to be particularly well suited to children. Another popular technique is Luthe's (1969) autogenic training, which attempts to train patients in a hierarchy of responses which produce relaxation and parasympathetic autonomic nervous system states. In general, these techniques propose clinical benefits from practicing skeletomuscular and proposed autonomic relaxation exercises, although mechanisms for such effects are speculative (Stoyva, 1977).

Relaxation training has been applied to such diverse neurological disorders as cerebral palsy (Ortega, 1978), paroxysmal disorders (Johnson & Myer, 1974), and both tension and migraine headaches (Diamond & Dalessio, 1978). In addition to potential utility for disorders of muscle tension, proposed skeletomuscular-autonomic antistress effects may have desireable impact on biochemical factors influencing many neurological disorders (Stoyva, 1977).

Cognitive behavior therapy proposes that behavioral procedures useful for changing overt behavior may also be applied to covert behavior, or thoughts (Meichenbaum, 1976). A related literature has developed on self-control programs totally implemented by patients, who seek to modify their own behavior using a variety of behavioral techniques (Thorsen & Mahoney, 1974). The rationales for applying cognitive behavior therapy and self-control techniques to psychophysiological problems have been cogently presented (Epstein & Blanchard, 1977;

Meichenbaum, 1976). Although there are virtually no data documenting the exclusive use of cognitive behavior therapy or self-control for neurological disorders, many reports of behavioral interventions either indicate that these techniques were components of the intervention or at least allow the possibility that they were used. For example, patients in biofeedback studies are often urged to practice learned control in their natural environment and/or think the same way they have learned to think in the clinic to achieve learned benefits (cf. Brudny, Grynbaum, & Korein, 1974).

In the following sections, the success of behavioral interventions will be reviewed for selected disorders of pediatric neurology. Case examples will then be presented to illustrate each of the three major types of intervention in pediatric neurology.

NEUROMUSCULAR DISORDERS

Traditional Classifications and Treatments

Pediatric neuromuscular disorders as a class are prevalent, seriously debilitating, and difficult to treat (Keats, 1965; Menkes, 1974). Etiologies include trauma, hereditary disorders of metabolism, degenerative disorders, tumors, cerebrovascular disorders, and infections, among others (Ford, 1973). Cerebral palsy, which encompasses a variety of nonprogressive motor disorders, has been purported to be caused by prenatal or perinatal anoxia or trauma (Keats, 1965). Neuropathophysiologies also vary with proposed etiology and clinical presentation but are poorly understood, especially for most chronic CNS disorders (cf. Eldridge & Fahn, 1976). Unfortunately, both clinical neurologists (Myers, 1952) and neuroscience researchers (Kornhuber, 1974) have indicated serious problems in the simplicity of traditional models of the motor system. Clinical presentation is usually categorized by type of abnormal movement or motor activity and body parts affected. Table I defines common clinical presentation of CNS motor problems and purported sites of problems (Ford, 1973; Keats, 1965; Menkes, 1974).

Traditional medical treatments for pediatric neuromuscular disorders include psychotherapy, pharmacology, neurosurgery, orthopedic surgery, and treatments by related disciplines such as occupational therapy, special education, and physical therapy. The goals, methods, and risk/benefit ratios of these modalities differ greatly, as do reports of their efficacy in the various disorders (Landau, 1974; Wolf, 1969).

Although psychotherapy has received little research attention for

Table I. Definitions of CNS Motor Disorders and Purported Sites of Problems

Term	Clinical presentation	Purported site of problem
Quadriplegia	All 4 limbs affected	Cortical/pyramidal (C/P), spinal
Hemiplegia	Ipsilateral arm and leg affected	C/P
Paraplegia	Legs affected	Spinal
Diplegia	Legs affected most, arms slightly	C/P
Athetosis	Slow, writhing, twisting of limbs or body parts	Basal ganglia/extrapyramidal (BG/E)
Ataxia	Disrupted coordination	Cerebellum
Chorea	Rapid, jerky, purposeless movements of body parts	BG/E
Choreoathetosis	Combined athetosis and chorea	BG/E
Dystonia	Slow twisting and sustained abnormal postures	BG/E
Dyskinesia	Abnormal tone/movement	Any site
Dysmetria	Inability to perform rapid alternating movements	Cerebellum
Myoclonus	Rapid jerking of body parts, flexors dominating	Any site
Spasm	Rapid involuntary tensing of muscles	Any site
Spasticity (cerebral)	Hypertonia, hyperreflexivity, of arm flexors and leg extensors and adductors, poor voluntary control	C/P
Spasticity (spinal)	Hypotonia, hyporeflexivity, weakness/paralysis of body below damage, loss of voluntary control	Spinal
Tics	Rapid, sudden, purposeless, brief verbal or motor responses	BG/E
Torticollis	Sustained or spasmodic lateral deviation of neck/head	BG/E
Tremor	Rapid alternating activation of antagonist muscles	Cerebellum and/or BG/E

efficacy of either psychological or related motor symptoms, children with neurological disorders are often referred for psychotherapy if they present with secondary psychiatric problems or if their disorder is considered functional or hysterical (Arieti, 1975). As has been the case with a number of disorders, pediatric neuromuscular disorders such as dystonia musculorum deformans (DMD) have received increasing support as neurological rather than psychiatric entities, and cautions have been issued against diagnosing symptoms as hysterical on the basis of the absence of neuropathophysiology (cf. Eldridge, Riklan, & Cooper, 1969).

Cooper (1969) and his colleagues have reported promising neurological treatments for movement disorders. Dentatotomy and ventrolateral thalamotomy (lesions of selected CNS nuclei) have been reported to be successful for DMD and athetoid cerebral palsy (Cooper, Amin, Riklan, Walz, & Poon, 1976). Unfortunately, their clinical reports offer little quantitative evidence. Electronic stimulators with implants in anterior and posterior cerebellar cortical lobes have been reported to reduce spasticity and athetosis and increase adaptive functioning in cerebral palsy (Cooper, Riklan, & Snider, 1974). However, a recent double-blind study failed to find significant benefits (Gahn, Russman, Cerciello, Fiorentino, & McGrath, 1981).

Over the past 20 years, prominent researchers have repeatedly noted the lack of scientifically acceptable data on therapeutic efficacy and have called for better assessments of physical therapies (Wolf, 1969; Wright & Nicholson, 1973). Recent developments in motor assessment, including using single-subject research designs and reliably quantified motor behavior, should produce much needed data on these therapies (Martin & Epstein, 1976; Molnar & Alexander, 1974). The goals of physical therapies are similar to the goals of behavioral therapies. That is, physical therapies attempt to teach patients to make maximum use of residual CNS motor functions. The differences in behavioral and traditional physical therapies are the hallmarks of any behavior therapy: reliable data, principles of learning developed by scientific inquiry, and replicable procedures.

Pharmacological approaches to pediatric neuromuscular disorders have been a widely accepted part of medical management for over 20 years, despite serious questions about the rationales and the amount and quality of empirical support for such agents (Landau, 1974). An extended discussion of this area elucidates problems in clinical diagnosis and traditional treatment of these disorders.

Central relaxants such as diazepam (Valium), which affect spinal mechanisms (reduce neural impulse transmission) to reduce muscle tone, have long been used for spasticity in cerebral palsy (Engle, 1966).

Diazepam has been reported by Phelps (1963) to reduce spasticity and involuntary movement in cerebral palsy, by Keats (1965) to improve general function and reduce spasticity in a variety of cerebral palsied children, and by Engle (1966) to have mildly beneficial effects in a variety of cerebral palsied children. Engle's (1966) study, which included a double-blind condition and a placebo, found clinically disappointing but statistically significant results.

Diazepam has been generally accepted as the agent of choice in treating spasticity with diverse etiologies (Nathan, 1970). Similarly, clinical researchers have reported some benefits in motor control in dystonics receiving diazepam (Barrett, Yahr, & Duvoisin, 1970). However, because of numerous inconsistencies in past reports, the continued use of diazepam in managing movement disorders merits additional research attention (Engle, 1966).

Other agents have been found to have similarly conflicting success in treating neuromuscular disorders. For example, although sodium dantrolene has received significant attention in recent years, the results of trials in cerebral palsy have not been conclusive. Denhoff, Feldman, Smith, Litchman, and Holden (1975) concluded that benefits were not significantly greater than risks to continue therapy; Haslam, Walcher, and Lietman (1974) reported generally worthwhile benefits; and Joynt's research (1976) was inconclusive. Levodopa has been proposed as potentially beneficial for choreoathetoid forms of cerebral palsy, also thought to be caused by dysfunctions of dopamine systems in the basal ganglia (Cooper, 1969).

The literature on the pharmacological management of DMD also demonstrates complexity and confusion. Preliminary studies of DMD have suggested beneficial effects of levodopa (Barrett, Yahr, & Duvoisin, 1970) and haloperidol (Marsden & Harrison, 1974) in some patients, but worsening of dystonia symptoms in others. The reported positive benefits of both haloperidol, a dopamine antagonist, and levodopa, a dopamine agonist, in these studies are problematic. Other pharmacological agents used in management of dystonia included diazepam (Batshaw & Haslam, 1976) and carbamazepine (Eldridge & Fahn, 1976). However, the neuropharmacology of dystonia appears to be quite complex and the general utility of any particular agent questionable (Marsden & Harrison, 1974).

A discussion of drug therapies for DMD during an international meeting in June, 1975, was most instructive in the matter of problems of the drug management of neuromuscular disorders (Eldridge & Fahn, 1976). Several participants reported that although no particular drug had been found generally useful for DMD, remarkable improvement in motor control occurred with one or two selected drugs in one or a few individual patients. The discussion provided important insights into the current problems in making sense of individual differences in biochemistry and neuropathophysiology in DMD, and possibly in other pediatric neuromuscular disorders. In addition to providing a rationale for procedures which objectively assess treatment effects for each patient, the proposal of important individual differences in biochemical and neuropathophysiological mechanisms suggests a need for improved quantitative analyses of muscular and behavioral aspects of DMD, and for alternative treatments.

Landau (1974) has critically discussed problems in the rationales for pharmacological treatments for spasticity. Two of his points are worth noting here: (1) that, following Jackson's (1931) model, researchers should focus on negative effects (functional losses produced by the disorder) instead of the frequently studied positive effects (undesired involuntary movements, or tone, produced by the disorder) in assessing therapeutic benefits; and (2) that the neurons in the final pathway of voluntary muscle control are the same neurons participating in undesired spastic reflexes and involuntary movements. This suggests that pharmacological treatments which reduce transmission in this path in order to reduce spasticity must also reduce voluntary movement and/or tone, therefore reducing the benefits of therapy.

Pediatric neuromuscular disorders are therefore particularly difficult to treat, and traditional medical management is usually supported by a variety of behaviorally based therapies (Vining, Accardo, Rubenstein, Farrel, & Roizen, 1976).

Behavioral Treatments

A variety of neuromuscular disorders have been approached using behavioral treatments. Table II lists selected studies in the literature which exemplify reported successful behavioral or biofeedback treatment of neuromuscular disorders. The number of disorders treated and reported successes are impressive in this relatively new field.

The majority of studies to date have utilized EMG biofeedback. It is important to note that there are a variety of permutations of EMG biofeedback training procedures, considering such variables as muscles selected for training, type of training (e.g., increases or decreases in EMG signals), modality of feedback signal (e.g., auditory, visual), and additional treatment procedures, such as home practices, rewards for

Table II. Selected Studies of Behavioral Treatments of Neuromuscular Disorders

Author	Date	Disorder	Treatment
Basmajian *et al.*	1976	Cerebral spasticity	EMG biofeedback and physical therapy
Bird & Cataldo	1978	Dystonia	EMG Biofeedback
Brierly	1967	Torticollis	Head position biofeedback and aversive contingency
Brudney *et al.*	1976	Cerebral spasticity Peripheral nerve injuries Spinal spasticity Torticollis	EMG biofeedback
Cataldo *et al.*	1978	Cerebral palsy	EMG biofeedback
Cleeland	1973	Torticollis	EMG biofeedback and aversive contingency
Engel *et al.*	1974	Fecal incontinence	Pressure biofeedback
Finley *et al.*	1977	Cerebral palsy	EMG biofeedback
Ince *et al.*	1976	Neurogenic bladder Spinal injuries	EMG biofeedback
Jankel	1978	Bell's palsy	EMG biofeedback
Johnson & Garton	1973	Cerebral spasticity	EMG biofeedback
Kohlenberg	1973	Fecal incontinence	Pressure biofeedback
Marinacci	1968	Peripheral nerve injuries	EMG biofeedback
Netsell & Cleeland	1973	Parkinsonism	EMG biofeedback
Nusselt & Legewie	1975	Parkinsonism	Position feedback
Ortega	1978	Cerebral palsy	Relaxation
Peck	1977	Blepharospasms	EMG biofeedback
Rosen & Wesner	1973	Tics	Contingency management
Sachs & Mayhall	1971	Cerebral palsy	Contingency management
Swaan *et al.*	1974	Poliomyelitis	EMG biofeedback
Thomas, Abrams, & Johnson	1971	Tics	Relaxation and self-monitoring
Varni *et al.*	1978	Tics	Behavior therapy/self-control
Wannstedt & Herman	1978	Cerebral palsy	Force biofeedback

performance, and/or relaxation training procedures. A second point worth noting is that the literature has generally been characterized by clinically promising but poorly documented results.

For example, Basmajian *et al.* (1976) trained patients with foot-drop due to stroke to increase anterior tibial contractions, raising the foot. A group trained with EMG biofeedback and physical therapy performed generally better than a group trained with physical therapy alone, but results were not statistically reported. Finley, Niman, Standley, and Wansley (1977) have used EMG biofeedback to reduce forehead muscle tension and produce general relaxation and have shown improved behavioral performance in children with cerebral palsy. This group's procedures included using behavior modification, dispensing small objects (toys, etc.) contingent upon appropriate biofeedback responses. The relative contributions of the biofeedback and behavior modification components of treatment were not clearly identified. Using yet another EMG biofeedback technique, Jankel (1978) taught a patient with Bell's palsy to match EMG levels on the normal side of the face with muscles on the damaged side, with considerable success.

The simplest rationale for improved behavioral/muscular performance in central neuromuscular disorders proposes that residual brain and spinal systems allow patients to learn to increase the rate, duration, or intensity of desired responses. Although proposals of neurochemical or neurophysiological changes are interesting, there are currently no data to support such proposals. The rationale for improved performance after spinal injury is problematic and difficult to sustain. Little evidence exists on behavioral interventions for muscular problems in injuries. For peripheral injuries, the rationale is plausible, stating that recovered neuroanatomical paths are often not correlated with recovered function and that retraining the brain/spinal system to use the reconnected paths may be important (and difficult) in these disorders (Marinacci, 1968).

Whatever the neurological mechanisms involved, they remain poorly understood. The literature has indicated successful treatments of behavioral/muscular symptoms of these disorders with increasing improvements in rigor of evidence (Bird, Cataldo, & Parker, 1981). Issues of larger-scale studies documenting effects, contributions of behavioral treatment components, and interactions with other therapies are currently critical.

The issues of transfer and maintenance of effects deserve some special attention. In our research on neuromuscular disorders (Bird, Cataldo, & Parker, 1981; Cataldo, Bird, & Cunningham, 1978), we have attempted to study generalization of effects of biofeedback training. A number of variables may adversely affect transfer of learned effects, including transfer across changes in: head position, body posture, muscles trained, particular responses, cognitive and motor activities, and social situations. That is, learned beneficial motor responses in one set of muscles may or may not affect other muscles and/or complex adaptive responses, and may or may not be maintained when the neuromuscular patient changes head position, stands up, tries to solve mathematics problems, walks, or sees a person enter the training room. We have found surprisingly good generalization across muscles, responses, and situations in patients who have successfully learned motor control (Bird, Parker, & Cataldo, 1979).

In patients who do not display sufficient transfer of learned effects, the therapist may use a variety of techniques to attempt facilitating transfer, including training the patient while performing the targeted response or in the desired situation, prescribing no-feedback practice for the patient in the targeted situation, or "fading" the patient into the targeted situation, in which the environmental situation is changed in small steps over sessions as successful performance continues.

Although the literature on EMG biofeedback for neuromuscular disorders is not definitive, the evidence is promising and suggests a success rate at least equivalent to other traditional therapies (Basmajian & Fernando, 1978).

PAROXYSMAL DISORDERS

Traditional Classifications and Treatments

Seizures are abnormal neural responses which disrupt normal brain functions, causing aberrant homeostatic, sensory, cognitive, and/or motor behaviors (Gastaut & Broughton, 1972). Epilepsy refers to chronically recurring seizures and may occur in a variety of conditions termed paroxysmal disorders. Estimates of prevalence of paroxysmal disorders vary but are as high as 10 in 1,000 persons (So & Penry, 1981). Paroxysmal disorders accompany and/or are produced by a variety of neurological disorders, are associated with anoxia, trauma, or cerebral diseases, but may be idiopathic (Schain, 1977).

Etiologies include genetic disposition, inborn errors of metabolism, brain damage, infectious diseases, tumors, abcesses, cerebrovascular disorders, systemic metabolic disorders (e.g., renal insufficiency), toxic poisons, anoxia, or febrile episodes (Niedermeyer, 1974). In the past two decades, a number of neuropathophysiological mechanisms of seizures have been identified and have received considerable attention (Brazier, 1974). In addition to purported etiology, paroxysmal disorders are classified according to the proposed neuronal systems and behavioral symptoms. Many authors consider certain movement disorders as *subcortical* paroxysmal disorders (Mostofsky, 1981). The International Classification of Epileptic Seizures proposes three major classes of partial seizures, which begin locally and have simple sensory and/or motor aberrations, with minimal impairment of consciousness; partial complex seizures, which begin locally and usually impair consciousness with cognitive, affective, sensory, motor, or combined symptoms; and partial seizures which generalize. Other major classes of seizures are generalized (including petit mal, atonic, akinetic, and grand mal or tonic-clonic, among others), unilateral, and unclassified seizures (Niedermeyer, 1974).

Complicating the understanding of etiological-pathophysiological-symptom relationships is the fact that one subclass of seizures may have several etiologies. For example, myoclonic seizures, which include usually bilateral synchronous movements of extremities (flexors dominating), may be due to such varying conditions as: CNS lipidosis, acute encephalitis, subacute sclerosing panencephalitis, acute cerebral anoxia, hyperuremia due to renal insufficiency, Tay–Sachs disease, and encephalopathies, among others (Niedermeyer, Fineyre, Riley, & Bird, 1979). There is also considerable overlap among subclasses. For example, the Lennox–Gastaut syndrome presents with almost all types of seizures (Niedermeyer, 1974).

Traditional medical treatments rely on anticonvulsant medication, which has a successful record for many types of seizures (Pippenger, Penry, & Kutt, 1978). With the development of sodium valproate, (Rodenbaugh, Sato, Penry, Dreifuss, & Kupferberg, 1980), an anticonvulsant for the historically resistant petit mal subclasses, it has been estimated that seizures could significantly be controlled in 80% of all patients with epilepsy if monitoring of therapeutic levels and patient compliance were improved (Pippenger *et al.*, 1978). However, there are significant identified risks for both anticonvulsants and less well documented neurosurgical techniques (Glaser, 1980; Niedermeyer, 1974). Prognosis depends upon specific type and etiology but continues to be uncertain for many patients (So & Penry, 1981).

Behavioral Treatments

The rationale for behavioral treatments of seizures includes: (1) learning procedures present lower risks than do surgical and pharmacological treatments; (2) observations and reports in clinical neurology indicate that emotions and learning may be important factors in seizures; and (3) evidence from basic and clinical research indicates that CNS electrophysiology and biochemistry may be manipulated by learning.

In addition to potentially valuable direct treatments, indirect or adjunctive behavioral programs for problems such as medication compliance (Epstein & Masek, 1978) and psychosocial consequences (Nordan, 1976) may offer significant benefits to this population.

Efron (1957a, 1957b) has reported using counterconditioning to reduce seizures reliably preceded by an olfactory aura or cue, by sequentially replacing a competing olfactory stimulus with a visual cue (a bracelet) and then a cognitive cue. Forster (1972) has reduced a variety of reflex seizures with procedures which may be characterized as habituation or classical conditioning. This does not suggest a primary learning etiology, as his basic research has shown that seizures could not be acquired with classical conditioning unless animals were pharmacologically predisposed (Forster, 1966).

Gardner (1967) has reported successful reduction of seizures in a child by manipulating parental attention. Aversive contingencies have been reported successful in reducing responses which induced seizures in a retarded child (Wright, 1973). However, Stevens, Milstein, and Dodds (1967) reported failure to reduce seizures with aversive contingencies, although seizure awareness increased. Cataldo, Russo, and Freeman (1980) have recently reduced rates of myoclonic and grand mal seizures in a 4-year-old girl by requiring a brief rest contingent on high myoclonic rates. The methodology in this case study was notable for controlled data and calibration of behavioral and EEG measures of seizures.

A combined EEG and EMG biofeedback relaxation training procedure has reportedly reduced seizures in an adolescent female (Johnson & Meyer, 1974). Kaplan (1975) suggested that successful reductions in seizure rates with slow frequency EEG biofeedback training may have been due to reduced arousal or relaxation. Similarly, Kuhlman's (1976) therapeutic success with 12–15 Hz EEG biofeedback but lack of EEG and seizure correlations suggested a reduced arousal effect. A recently reported study with crossover design indicated that EEG biofeedback enhanced a relaxation-training effect (Cabral & Scott, 1976). Poor controls in this study make interpretations of results difficult. The relationship of relaxation and arousal to seizures is currently complicated. For example, experimental studies have shown that increased attention and problem-solving may abort seizures (Davidoff & Johnson, 1964). Wyler's basic (Wyler, Fetz, & Ward, 1974) and clinical (Wyler, Lockard, Ward, & Finch, 1976) studies on effects of increased EEG desynchronization suggest a beneficial arousal effect. Specification of seizure type and related behavioral symptoms may improve the understanding of arousal/relaxation relationships with seizure frequencies and intensities.

In some forms of seizures, such as Jacksonian psychomotor epilepsy, a reliable chain of neural-behavioral responses precedes a generalized seizure. Zlutnick, Mayville, and Moffat (1975) have reported four cases in which interrupting seizure response chains not only stopped the sequence but also reduced the future occurrence of the entire chain. Arousing and aversive properties of interruptive stimuli were not clearly separable.

The value of electroencephalographic (EEG) analysis in diagnosis of epilepsy has been well-documented for many years (Neidermeyer,

1974). In programmatic research, Sterman and his colleagues have been studying biofeedback training of a particular EEG frequency which may inhibit seizure rates in patients for whom traditional drug treatments have been unsuccessful. Early studies indicated a prominent 12–16 Hz rhythmic EEG activity over sensorimotor cortex in cats during absence or inhibition of movement (Roth, Sterman, & Clemente, 1967; Sterman & Wyrwicka, 1967). Sterman (1977) subsequently demonstrated that operant conditioning of this sensorimotor rhythm (SMR) in cats increased their resistance to monomethylhydrazine-induced seizures. Sterman and Friar (1972) then reported that increasing SMR by biofeedback conditioning over a three-month period reduced seizures in a patient not well-controlled by medication. Subsequent to their initial case report, Sterman, MacDonald, and Stone (1974) reported clinically successful seizure reductions in four epileptic patients trained in SMR biofeedback over periods from 6 weeks to 18 months. Finley (1976) has reported a case in which SMR training was successful in reducing seizures in a 13-year-old male by a factor of 10. Finley's study also included a reversal phase, wherein SMR training was briefly discontinued, with a reported concurrent increase in seizures, subsequently followed by a period of SMR increases and seizure decreases. Lubar and Bahler (1976) have also reported successful results of SMR training with eight patients with seizures of varying type and severity. All authors have been cautiously optimistic about the potential of SMR biofeedback for seizure reduction.

The neurophysiological mechanisms mediating the effects of SMR increases on seizure activity are currently not clear. Sterman (1977) has also reported on possible neurophysiological mechanisms of SMR, which involve EEG–single neuron correlates in the nucleus ventralis posterolateralis of the thalamus. Moreover, during SMR production, firing rates of large neuronal units in the red nucleus were reportedly suppressed while somatosensory-evoked responses in the dorsal column nuclei have been reportedly facilitated. Sterman's model, therefore, suggests some alterations of the sensorimotor cortex-ventral posterolateral thalamic circuits which increase the refractoriness of those circuits to recruitment in seizure activity. On the basis of power spectral analysis of EEG during sleep, Sterman has also more recently speculated that SMR biofeedback, which increased sleep spindles, may reduce seizures by improving sleep. Poor sleep has been found to exacerbate seizures in clinical populations (Pratt, Mattson, Weikers, & Williams, 1968). In a recent double blind crossover study, Sterman and MacDonald (1978) have reported that biofeedback for increasing SMR was more effective in reducing seizures than less effective 18–23 Hz activity and a nontherapeutic 6–9 Hz activity.

The basic research of Wyler *et al.* (1974) has suggested an interesting alternative proposal for the effects of EEG biofeedback on epilepsy. A sophisticated system monitored rhesus monkeys for seizures for 24 hours a day. Monkeys were also operantly conditioned for alternating increases and decreases in firing rates of cortical neurons in eleptogenic foci induced by alumina cream. Results showed that the number of findable epileptic units (defined by firing rate and burst characteristics) and the number of behavioral seizures decreased over training days. Wyler *et al.* (1976) have also reported that reductions in single-unit seizure activity were accompanied not necessarily by SMR activity, but by desynchrony in the EEG. However, Wyler has also pointed out that his work involves attempting to change electrical activity in neurons within the immediate area of seizure activity, whereas Sterman has attempted to change the EEG in the sensorimotor cortex. In a subsequent clinical report, Wyler *et al.* (1976) have reported reductions in seizures with EEG biofeedback for desynchronization. Therefore, although based on different neurophysiological rationales, both treatment models may be clinically effective in reducing seizures.

Kaplan (1975) has reported that training two patients to increase slower (about 10 Hz) EEG frequencies reduced seizures. Gastaut (1975) reinforced her discussion of problems in relating EEG frequency to seizure control. Kuhlman (1976) has also noted poor correlation of learned EEG activity and seizures.

Much more research is obviously needed in this area, to answer several critical questions, such as: (1) what types of seizures might benefit most or least from particular types of behavioral treatment; (2) whether the location of EEG changes might prove to be an effective variable for particular types of seizures; (3) whether through EEG training patients simply learn to discriminate more readily seizure-related changes in their EEG and to make any avoidance responses in their repertoire which prove to be successful; (4) what the mechanisms of these reported benefits are and how EEG biofeedback benefits compare with alternative medical or behavioral treatments.

The literature on behavioral and biofeedback treatments for seizures has been criticized on several points. In addition to general problems of experimental control related to collaboration with neurology, a few deficiencies have been particularly evident, including: (1) lack of neurological definition, verification, or measurement of seizures in the subjects studied; (2) presentation of rationales for behavioral or biofeedback treatments which have not fit easily with traditional neurology; (3) lack of large controlled studies on patients with well-defined seizure subtypes; and (4) lack of consideration or manipulation of ongoing pharmacological regimens.

From the viewpoint of behavioral medicine, the most disconcerting problem has been the development of research on behavioral treatments for seizures in settings independent from clinical neurology. This trend stands in contrast with behavioral medicine for a variety of medical disorders, such as chronic pain, headache, and cardiovascular disorders, which have developed in the setting of clinical services. The evidence on behavioral and biofeedback treatments for seizures is most promising. In a recent conference on behavioral interventions in neurology, Freeman (1980) reviewed the potential of individualized behavioral approaches to particular symptoms displayed by seizure patients. Increasing collaboration among behavioral and neurological researchers and clinicians is clearly predicted for this area in this decade.

DEVELOPMENTAL DISABILITIES

The incidence of mental retardation and associated developmental disabilities has been estimated to be greater than 200,000 in the United States (President's Committee on Mental Retardation, 1975). As recent reviewers have stressed, retardation is in part a cultural definition of deviation from expected norms, with major criteria including significant deficits in intellectual functioning and adaptive behavior displayed before the 18th birthday (Wetherby & Baumeister, 1981). Only within the past 25 years has retardation acquired status as a medical entity (Salam & Adams, 1975). In the past two decades, great progress has been made in identifying etiologies and pathophysiologies of selected disorders which produce retardation (Moser, 1975). The pledging of enormous national resources toward prevention of and rehabilitation for retardation and developmental disabilities, culminating in Public Law 94–142 (the Right to Education of the Handicapped), indicates a strong commitment on the part of society and government to the developmentally disabled population (Cataldo & Russo, 1979).

During the past 25 years, the number and sophistication of behavioral interventions for behavior problems and skill deficits of retardation have grown tremendously. Techniques for problems ranging from decreasing self-injurious behaviors (Romanczyk & Goren, 1975) to step-by-step building of job-related skills (Neef, Iwata, & Page, 1978) have been reported. Selected behavioral techniques have been "packaged" for dispensation and are now extensively used throughout the health care system (Azrin & Foxx, 1971). Recent reviews have indicated that behavioral interventions have been successfully applied to almost every problem imaginable in a variety of settings (Gardner, 1971; Wetherby & Baumeister, 1981). The current emphasis in application of behavioral techniques is toward facilitating adaptive home- or community-based living instead of institutionalization (Cataldo & Russo, 1979).

Throughout the past 25 years, an exploding literature on hyperactivity, learning disabilities (Tarnopol, 1971), and minimal brain dysfunction (Wender, 1971), has produced complicated and confusing pictures of these disorders and their interrelationships with other developmental disabilities.

Most recently, three areas of treatment have received most recognition as potentially effective for children with some forms of hyperactivity or attentional disorder: pharmacotherapy, using amphetamine, methylphenidate, or pemoline (Liberman & Davis, 1975); alterations in diet, which have received more popular than scientific support (Feingold, 1975; Smith, 1976); and behavioral interventions (Lahey, Delamater, & Kupfer, 1981).

Behavioral interventions have progressed from simple behavior modification programs aimed at reducing activity and/or increasing attention span in single environments to more sophisticated programs which are aimed at selected adaptive functions (Ayllon, Layman, & Kandel, 1975) and include components of self-control (O'Leary & Dubey, 1979) and/or combination with drug therapy to facilitate effects (Whalen, Henker, Collins, Finck, & Dotemoto, 1979).

Recent reviews have confirmed the efficacy of behavioral interventions applied in a variety of settings (Lahey *et al.*, 1981). The efficacy of biofeedback, training for increases in EEG indices of motor quiescence (Lubar & Shouse, 1976) or for decreases in EMG activity (Hughes, Henry, & Hughes, 1980) has been also promising.

A related issue in the literature on hyperactivity concerns the increasing emphasis on initially training the desired end-product skill, for example, by rewarding performance on mathematics tasks, rather than training for quiescence or increased attention span prior to training for selected skills or performance (cf. O'Leary & Dubey, 1979). This approach contrasts with that of biofeedback practitioners, who view motor quiescence (Hughes *et al.*, 1980) or improvement of CNS attention processes (Lubar & Shouse, 1976) as important first steps toward improving skills. Although both approaches have been noted to be successful, comparative and interactive studies would be most informative on this issue.

The case study which follows the section on mental retardation involves a child with hyperactivity and exemplifies assessment and treatment techniques useful for these disorders.

CASE EXAMPLES

Behavior Modification in a Retarded Hyperactive Child

Although recent books in the popular press have advocated proposals that diet may adversely affect hyperactivity, the literature remains controversial (Feingold, 1975; Smith, 1976). In this case study, behavioral techniques were employed to assess the effects of a dietary component (wheat gluten) on the behavior of a child with a suspected food allergy (Crook, 1975). This study has been reported in detail elsewhere (Bird, Russo, & Cataldo, 1977).

The patient was a retarded 9-year-old male (IQ = 40) referred for diagnosis and treatment of several behavior problems which were reportedly aggravated by certain food substances. The child, the product of a complicated pregnancy and labor, had experienced repeated episodes of infant diarrhea attributed to an allergy to wheat gluten and milk products (celiac profile). The child's developmental problems were first noted between 2 and 3 years of age when his word usage declined and he began to withdraw from the social environment. Over the next few years, withdrawal and word usage worsened, and tantrums, perseverative and autistic behaviors, and other inappropriate behaviors developed. Overall reports of the child's behavior suggested that he was unmanageable for long periods each day, disrupting the home and school environments and frequently endangering himself through recklessness or pica. An initial evaluation by an interdisciplinary team diagnosed the child as retarded with autistic behaviors. The parents reported that behavior problems worsened considerably within a few hours of the child's ingestion of gluten-containing foods. The parents felt strongly that the child's behavior was a result of an allergic reaction to wheat gluten, that his behavior had improved slightly as a result of a dietary restriction program suggested by a pediatrician, and that such a program should be the treatment of choice for their child.

Two somewhat different inpatient settings were selected for observing and treating the child. In one setting, called an *individual environment*, the child was observed while with one therapist in a rather bare room. In a second setting, called a *group environment*, the child was observed while in a much larger room containing many chairs, tables, toys, and games, with a group of children who were supervised by two or three adults. The group environment simulated the classroom and supervised recreation programs which the child regularly attended.

Four behaviors were selected for measurement and modification in the study: (1) *pica*—touching nonfood objects or body parts to the mouth, teeth, lips, or tongue; (2) *inappropriate vocalizations*—any vocal noise which was unrecognizable as a word; (3) *cooperation*—a correct nonverbal response to any of three simple requests ("Come here;" "Sit down;" "Hand me that———," with the name of a familiar object inserted) within 10 seconds; and (4) *locomotor activity*—any movement about the room, such as walking or noticeably propelling the body across horizontal space (crawling, slithering, lying on back and pushing across floor with feet, etc.)

The frequencies of pica and inappropriate vocalizations and cooperation with requests are recorded. Cooperation was measured only as a response to one of the three requests defined above. Activity was measured by a time-sampling, which involved noting the absence or occurrence of locomotion in a brief 2-second period sampled each minute (Hersen & Barlow, 1976).

Throughout the study the child was observed in each environment for a 15-minute period each day. In the individual environment, the therapist handed objects to the child at an approximate rate of one per minute, thus allowing the child frequent opportunities to exhibit pica. Also about once each minute, the therapist made one of the three requests of the child, giving him opportunities to exhibit cooperation. No specific regimen of requests or pica opportunities occurred in the group environment. Reliability or percentage of agreement of independent observers ranged from 67% to 100% throughout the study, averaging 92% for all behaviors. All gluten-containing foods were introduced or excluded from the child's three daily meals, depending on the condition. Adherence to dietary conditions was monitored by selected staff. Changes in diet were not made known to observers. The child was fed a gluten-containing diet during 7 daily observation sessions, and then was switched to a gluten-free diet for 8 daily sessions, and back to a gluten-containing diet for 11 daily sessions. The different durations for the three conditions were determined by stability of the data rather than by an absolute, preset number of days. The effects of gluten on the child's behavior had been estimated by the parents to occur within hours of ingestion.

After the dietary assessment had been completed, behavioral treatment was begun for cooperation in the individual setting. During this phase, the therapist played with the child only if the child cooperated with requests. Noncooperation resulted in the child's being required to sit away from the activity area for 30 seconds, after which the request was repeated. Occurrences of the other two problem behaviors were recorded by observers but ignored (not consequated) by the therapist. After 8 sessions of cooperation training, behavioral treatment was expanded to include inappropriate vocalizations. During this second phase, activities were made contingent upon both cooperation with the therapist's request and the absence of inappropriate vocalizations. After 19 more sessions, behavioral treatment was extended to include pica, for 13 more sessions. During the last phase of the behavioral treatment, the entire package for increasing cooperation, decreasing inappropriate vocalizing, and decreasing pica was taught to a hospital employee and used by her in the group setting for 5 sessions. During the entire study, locomotor activity was monitored but not treated.

In addition, the therapist prepared the child's mother to use the treatment program at home. The mother was trained with explanation, modeling, and feedback, during three $\frac{1}{2}$-hour sessions in the individual environment. When the child, who lived a considerable distance away, was discharged, he was followed by occasional telephone contacts. After 6 months of using the treatment program at home, the mother and child visited for a follow-up session.

Figure 1 presents the results of dietary manipulations and behavioral treatment across individual and group environments for cooperation, vocalizations, and pica. The data are variable across dietary conditions, and differences emerged as a result of gluten introduction. However, with the introduction of behavioral treatment in session 28, cooperation immediately increased and remained high. Behavioral treatment of inappropriate vocalization was instituted during session 36, and data indicate that variability was reduced. The data for pica behavior show a decrease in variability and rate over time, with no apparent dietary effect and no conclusive behavioral effect.

Data on the effects of dietary manipulations in the group environment are

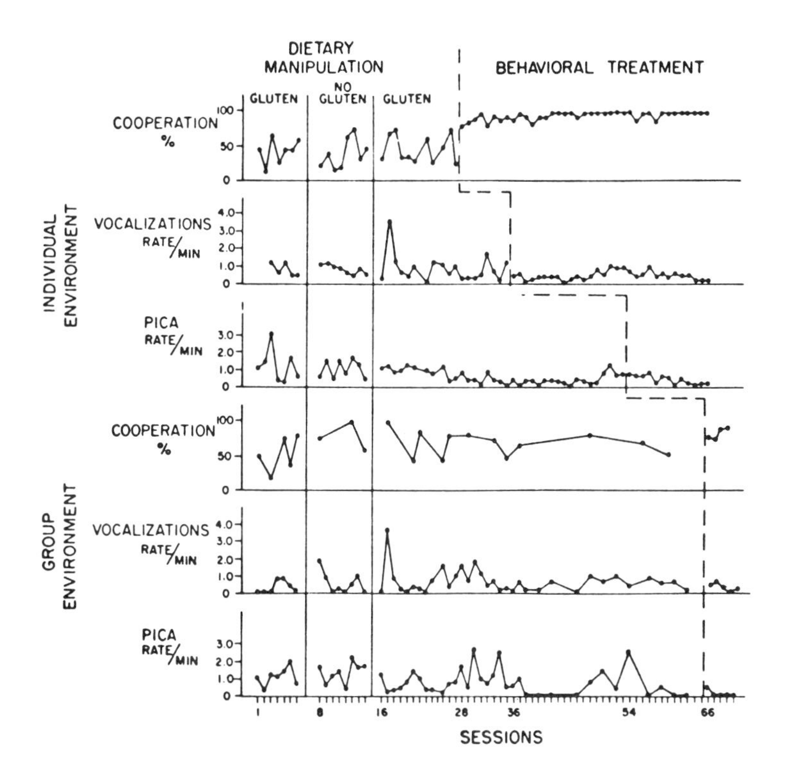

Figure 1. Effects of dietary manipulation and behavioral treatment program on the child's cooperation, vocalization, and pica in the individual and group environments. From "Considerations in the Analysis and Treatment of Dietary Effects on Behavior Disorders" by B. L. Bird, D. C. Russo, and M. F. Cataldo, *Journal of Autism and Childhood Schizophrenia*, 1977, *7*, 373–382. Copyright 1977 by Pergamon Press. Reprinted by permission.

presented in the lower section of Figure 1. There are no indications that diet affected behavior, and there is a general indication that cooperation increased when the treatment was added in session 66.

Figure 2 presents data on locomotion, across both individual and group environments for the course of the investigation. No dramatic differences were found between dietary conditions. The initiation of behavioral treatment for the three other target behaviors is labeled on the graph with arrows. As treatment for each was introduced, the percentage of locomotion declined in the individual environment. This drop in locomotion during behavioral treatment was not the result of the timeout treatment. As the child became more cooperative and exhibited fewer problem behaviors, his time spent in timeout decreased, and he spent more time in constructive play with the therapist. Data on locomotion in the group environment showed similar changes.

Six months after the child's discharge, follow-up observations were made to assess maintenance on the program. Measures in the individual environment, with the mother acting as therapist, indicated a 100% compliance rate, a low rate of 0.1 vocalizations per minute, and 0.0 pica responses per minute.

Four points are worth noting from this study. First, the behavioral technology was useful in demonstrating that the suspected dietary agent was not affecting behavior. Second, the behavioral intervention dramatically improved behavior. Third, the procedures used were reason-

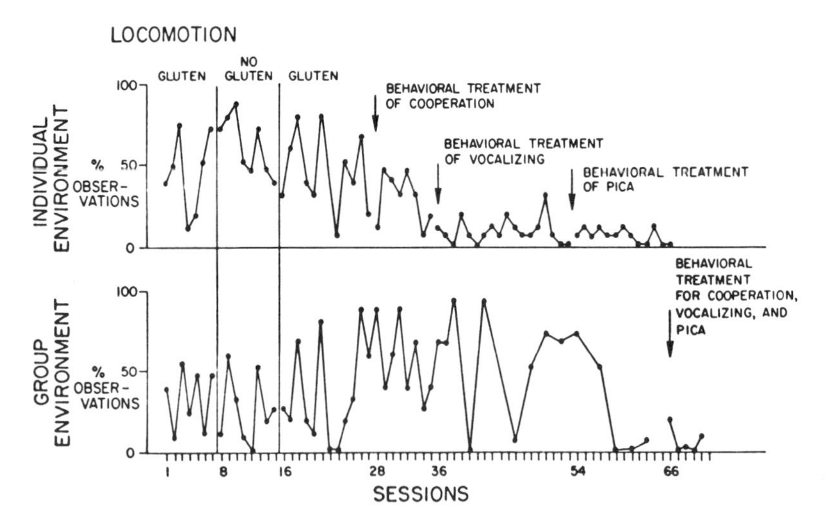

Figure 2. Changes in the child's locomotion during dietary manipulation and behavioral treatment in the two environments. From "Considerations in the Analysis and Treatment of Dietary Effects on Behavior Disorders" by B. L. Bird, D. C. Russo, and M. F. Cataldo, *Journal of Autism and Childhood Schizophrenia*, 1977, *7*, 373–382. Copyright 1977 by Pergamon Press. Reprinted by permission.

ably simple and were taught to the mother, who used them successfully in the home. The empirical demonstrations of no dietary and potent behavioral effects were critical in convincing the family to implement the successful behavioral program at home.

The fourth point is worth noting since it concerns a principle which has relevance to all areas of behavioral pediatrics. It involves the observation that increasing cooperation also produced beneficial reductions of other undesirable behavior. The phenomenon has been discussed by Wahler (1975), who has proposed that learned increases in the prevalence or strength of a subset of responses (such as cooperation with a few requests) will generalize or transfer to related responses. The results of training a few simple cooperation responses may include generally increased cooperation and a correlated decrease in undesirable responses.

Biofeedback Treatment for Dystonia

The patient in this case was a 20-year-old male who presented with normal intelligence and a 3-year progression of symptoms of dystonia. The patient had experienced numerous neurological evaluations and was diagnosed as having idiopathic DMD. After the study was completed and published, the patient's brother developed similar symptoms, and the diagnosis was changed to autosomal recessive DMD. At the time of this study, the patient presented with moderate to severe involvement of neck, facial, and esophageal muscles. Eating and speaking were severely impaired, and the patient's face was "uncontrollably" contorted throughout the waking hours. Distal portions of the limbs (forearm flexors/extensors, gastrocnemius/anterior tibialis, etc.) were moderately symptomatic. During sleep and infrequent, brief periods of the day, the patient appeared to be generally relaxed. Fine motor skills were significantly impaired due to increased tone in hand and forearm muscles. The patient displayed an awkward gait and occasionally fell backward due to excessive plantar flexion of both feet, caused by excessive gastrocnemius contractions.

The patient had been previously treated with pharmacological agents including diazepam, carbamazepine, and haloperidol, without significant improvements in the symptoms. A notable loss of function during the 6 months prior to this study had precipitated his admission to the hospital.

Based on a thorough assessment of clinical problems by a physical therapist and the behavior therapists, three sites were chosen for electrode placement: right forearm (flexors), left jaw (recording activity from several facial muscles), and frontalis muscles. For forearm and jaw muscles, silver-chloride disc electrodes filled with conductive paste were attached with adhesive collars to alcohol-cleaned skin. For frontalis muscles, a rubber head band held in place by an adhesive strip at the rear of the head contained small metal discs inside small cups filled with conductive paste which were placed over alcohol-cleaned skin.

For all placements, two active electrodes were separated by a reference electrode led to ground.

The biofeedback apparatus included three commercially available EMG amplifiers with low and high pass filters set at 1,000 and 10 Hz, respectively, and a gain of 1,000, feeding into three integrators, which in turn fed into a PDP-8 computer. The computer-controlled visual and auditory feedback signals printed out digitized EMG pulse rates for 20-second periods for each muscle group. Visual feedback consisted of three-digit electronic numbers which updated continuously and was inversely proportionate to EMG pulses from a selected muscle. A second three-digit number, set by the therapist, displayed a criterion which the patient was asked to exceed, by reducing tension in a selected muscle, producing higher numbers in the feedback display. If the numbers in the feedback display fell below criterion, indicating increased tension in the muscle, the patient heard a tone varying in pitch proportionately to EMG level, which remained on until the patient reduced tension and met the criterion.

A within-subject experiment assessed the effects of feedback training by comparing within-session EMG data across conditions of no-feedback and no-instructions baselines, no-feedback relaxation periods, and feedback relaxation-training periods. Prior to feedback training, 6 sessions were conducted in which the baseline condition was compared with relaxation without feedback, in order to assess the patient's ability to relax without feedback.

During daily sessions lasting about one hour, experimental conditions were alternated. On day 7, EMG feedback signals for frontalis muscles were activated during feedback relaxation periods. The frontalis muscle was chosen for training because this mildly symptomatic group was proposed as potentially easier to train and potentially likely to provide transfer of relaxation to jaw muscles.

On day 16 of training, a systematic attempt to facilitate transfer of learned relaxation began. The therapist entered the training room from time to time and conversed with the patient during relaxation without feedback. The conversation was found to be necessary because in selected instances the therapist's entry into the room was followed by loss of relaxation.

Figure 3 displays the results of training over days. Prior to the start of feedback training on day 7, in selected sessions the patient was tenser during attempted relaxation than during baseline periods. After feedback was begun, frontalis and jaw EMG data significantly decreased during feedback conditions. By day 11, EMG reductions were occurring during relaxation without feedback. As generalization training began, the patient initially demonstrated higher EMG with the therapist in the room, but regained relaxation control as training continued. Data from forearm flexors showed more variability than other muscles in all conditions, although there was a tendency for EMG levels to be low during all relaxation periods.

Figure 4 illustrates data from within-session reversals of conditions from the last two no-feedback daily sessions, and the first, third, and fifth sessions in which feedback was instituted. Careful analysis of the data in Figure 4 suggested three important points: (1) prior to feedback training, high variability and levels of muscle tension occurred in both baseline and relaxation periods, and

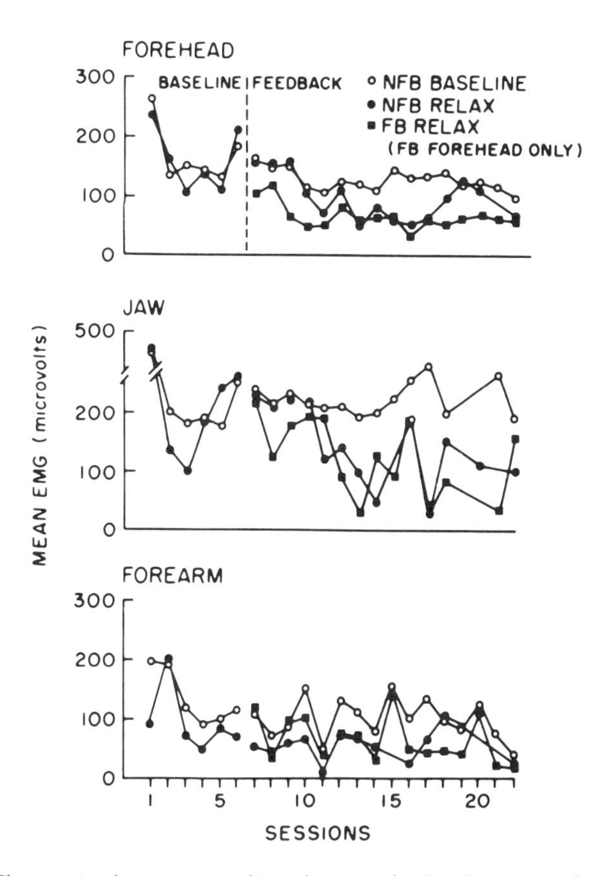

Figure 3. Changes in electromyographic voltages in forehead, jaw, and forearm muscles in experimental conditions over 22 sessions. Conditions were no-feedback baseline (NFB Baseline), no-feedback relaxation (NFB Relax), and relaxation with feedback (FB Relax). Feedback was provided for forehead relaxation only. From "Experimental Analysis of EMG Feedback in Treating Dystonia" by B. L. Bird and M. F. Cataldo, *Annals of Neurology*, 1978, *3*, 310–315. Copyright 1978 by Little, Brown. Reprinted by permission.

no difference occurred in frontalis and jaw average EMG in baseline and relaxation periods; (2) as training progressed, greater reductions occurred in both average levels and variability of EMG for the frontalis muscle during feedback periods, until by day 12 significant reductions were evident during feedback periods; (3) the relaxation evident in the frontalis muscles apparently generalized to jaw and forearm muscles, and from feedback periods to no-feedback periods.

Observations from the pretraining and posttraining videotapes which were scored with a reliable coding method indicated that the patient improved lip closure from 2% of the intervals scored before training to 100% of the intervals after training. Interobserver agreement was 97%, indicating that the coding

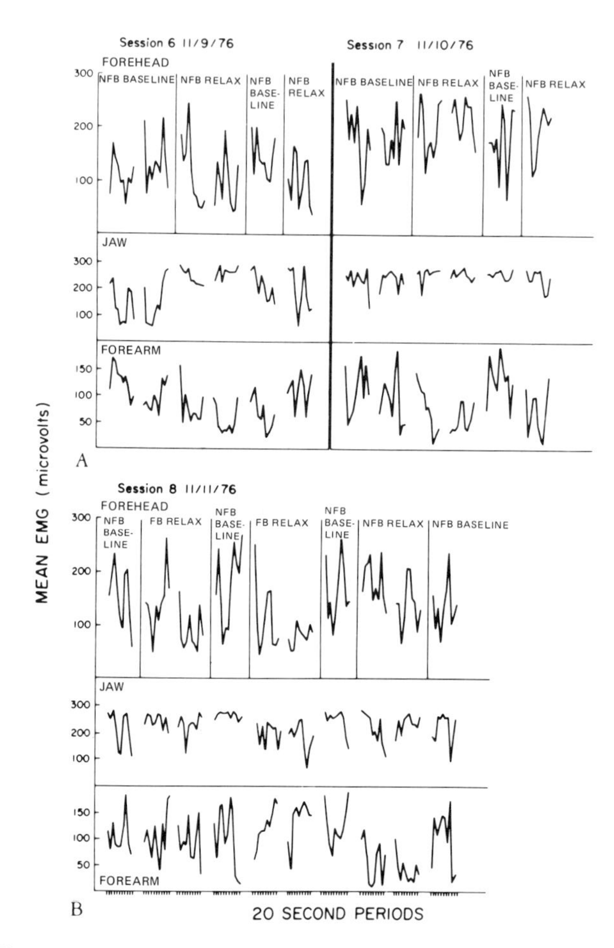

Figure 4. Changes in EMG of forehead, jaw, and forearms over experimental conditions during: (A) the last 2 no-feedback sessions; (B) the 1st feedback session; (C) the 3rd feedback session; (D) the 5th feedback session. From "Experimental Analysis of EMG Feedback in Treating Dystonia" by B. L. Bird and M. F. Cataldo, *Annals of Neurology*, 1978, 3, 310–315. Copyright 1978 by Little, Brown. Reprinted by permission.

system was reliable. During the posttraining videotaping, the patient maintained normal facial relaxation while sitting, moving arms and head, standing, walking, and listening to several people speak to him. At the end of the taping session, when told he could stop exerting control, he immediately relapsed to abnormal dystonic facial posture.

After the period of intensive inpatient training reported in this study, the patient was given instructions to practice relaxation at home for about 30 minutes each day and was seen biweekly in the clinic for 3 months. During that time,

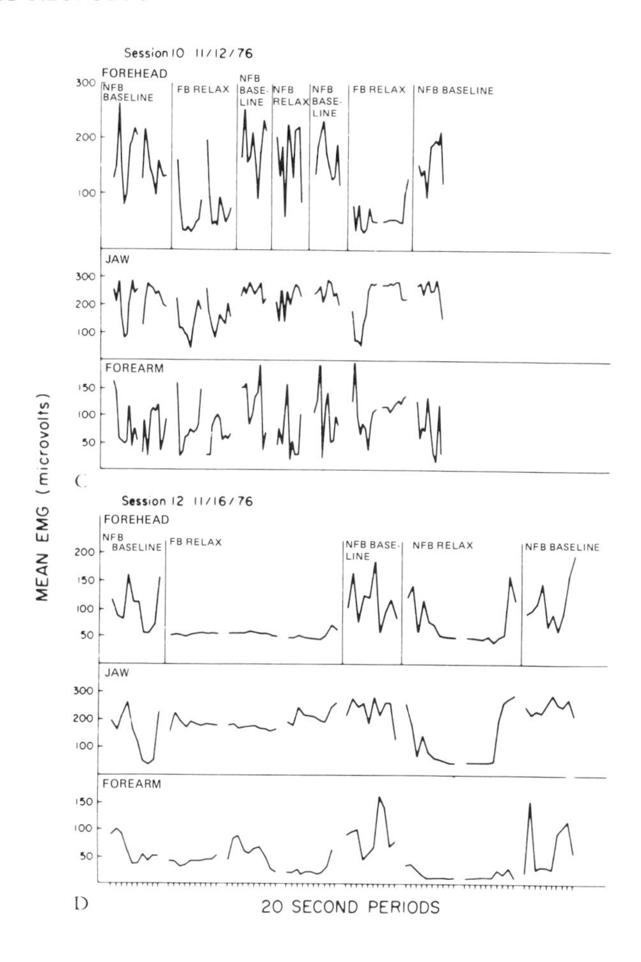

the patient continued to relax when prompted and reportedly was able to relax in a variety of situations, including previously stressful social situations. However, between 3 and 6 months after the study ceased, the patient's overall neurological condition worsened, and these gains were considerably reversed.

In discussing this case, several points are worth noting. First, substantial symptomatic relief was provided to a patient with a severe neurological disorder which was refractive to traditional therapies. Second, although the procedures and equipment used were somewhat more sophisticated in data collection than that routinely used in biofeedback clinics, a simpler, commercially available EMG biofeedback training unit would have provided equally effective training capabilities with much simpler procedures. Thirdly, the neurophysiological mechanisms by which these beneficial effects occurred were, and have remained, unclear. The issue of mechanisms of behavioral interventions in neurology will be discussed in detail in a subsequent section. Briefly, the most parsimonious explanation for the benefits in this study is that an infrequent or weak response in this patient's repertoire was increased. There was no reason to propose any change in neuropathophysiology. Fourth, although the treatment transferred to the natural environment and persisted for a period of months, when the disease again became progressive, the symptomatic benefits were considerably lost.

Behavior Therapy and Self-Control for Tics

Tics are sudden, unexpected, purposeless, stereotypic, usually brief, rapid, and frequent motor or verbal responses, occurring singly, in "bursts," or in groups of different responses, with no readily determined etiology or pathophysiology (Shapiro, Shapiro, Bruun, Sweet, Wayne, & Soloman, 1976). In the past decade, organic etiologies have been proposed increasingly, as the disorder has been related to the dystonias (Eldridge & Fahn, 1976) and shown to be often responsive to pharmacotherapy (Barrett *et al.*, 1970). However, both problems in efficacy and risks in drug treatments such as haloperidol (Weiss & Santelli, 1978) have encouraged clinicians to seek alternative treatments.

A few studies have reported successful behavioral treatment of tics, using both aversive (Rafi, 1962) and positive consequences contingent upon tics (Doleys & Kurtz, 1974). The following case study was reported in detail by Varni, Boyd, and Cataldo (1978).

A 7-year-old male presented with a history of normal intelligence, social immaturity, noncompliance, distractibility, hyperactivity, and multiple tics since age 2. The child's tics included facial grimaces (eyes squinting and mouth distorted), shrugging of both shoulders, protruding the rump (with knees bent) and vocalizing (a grunting "huh"). The child had not previously experienced medical treatment for his disorder. For purposes of the study, all tics were defined operationally to allow reliable behavioral recording once every 10 seconds for three 5-minute sessions once each week in the clinic. The child's mother noted whether or not there was any occurrence of a tic every half-hour, averaging 18 observations each day, during which the mother followed her normal routine.

During a baseline period in the clinic, the child simply played with toys. During baseline at home, the entire family was instructed to follow normal routine, but the mother was asked to collect observations of tics.

The behavioral treatment included the following procedures, in sequence: (1) the child was initially asked to watch himself in a mirror and was rewarded with praise by the therapist when he accurately reported and with feedback if he did not report a facial tic; (2) the mirror was faded out, and the child was rewarded with praise for accurate reporting of facial tics; (3) the child was rewarded by access to toys in a corridor if he evinced fewer than 10 facial tics for 5 minutes; (4) if he evinced more than 10 facial tics in a 5-minute period, he was left alone in the room with no toys; (5) the child was given a stopwatch and

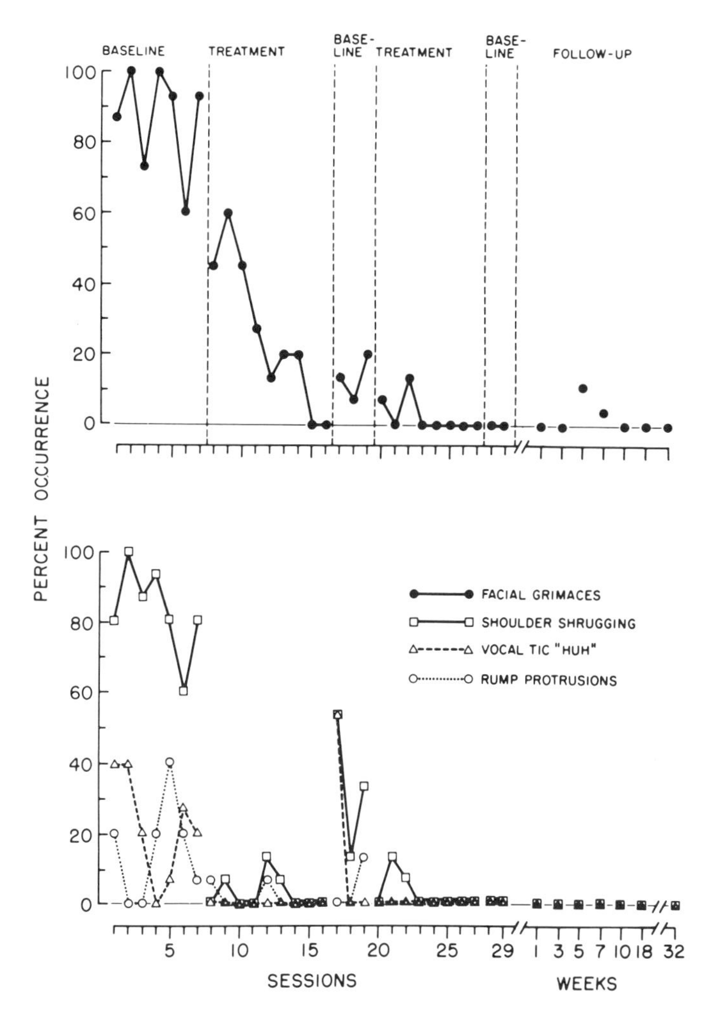

Figure 5. Effects of behavioral treatment program on four types of tics across varying conditions in the clinic. Only facial grimace tics were treated. From "Self-Monitoring, External Reinforcement, and Timeout Procedures in the Control of High Rate Tic Behaviors in a Hyperactive Child" by J. W. Varni, E. F. Boyd, and M. F. Cataldo, *Journal of Behavior Therapy and Experimental Psychiatry*, 1978, *9*, 353–358. Copyright 1978 by Pergamon Press. Reprinted by permission.

asked to monitor and record his own facial tics; (6) as tic frequency decreased, the definitions of facial tics were also changed (for example, from intense bilateral squints to slight unilateral squints or mouth twitches); (7) the child's verbal labeling of facial tics was rehearsed before each session; (8) after session 11, the criteria were changed so that the child's access to the play area required his

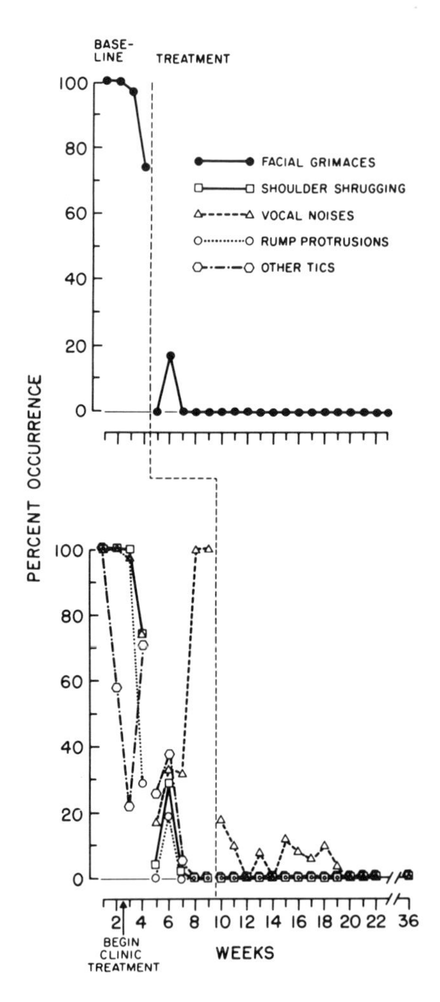

Figure 6. Changes in tics during baseline and treatment conditions in the home. Facial grimaces were treated first, then vocal noises. Other types of tics were only measured, not treated. From "Self-Monitoring, External Reinforcement, and Timeout Procedures in the Control of High Rate Tic Behaviors in a Hyperactive Child" by J. W. Varni, E. F. Boyd, and M. F. Cataldo, *Journal of Behavior Therapy and Experimental Psychiatry*, 1978, *9*, 353–358. Copyright 1978 by Pergamon Press. Reprinted by permission.

emitting less than 5 facial tics; (9) after session 23, play area required 5 minutes of no facial tics. Twice during the study, the child was informed and the contingency was removed, returning to the no-treatment baseline condition.

Home treatment started 2 weeks after treatment in the clinic began. Each time the mother observed a facial tic, she placed the child in timeout for 5 minutes. The child was given a gold star, later exchanged for desirable activities, each time he did not have a facial tic for one half-hour period. After week 9, the timeout/gold-star contingency was also placed on vocal noises.

Figure 5 displays data on tics observed in the clinic and Figure 6 data on tics observed at home. The results of the study indicate an effective suppression of tics first in the clinic, then in the home environment. These suppressions lasted through a 32-week follow-up.

Several features of this case are worth noting. First, the treatment procedures were relatively cost-effective and also potentially therapeutic. Second, although only facial tics were treated in the clinic and facial and vocal tics at home, learned suppression of all tics occurred. The theoretical notion that learned control of key responses in a class or category of functional responses may produce reduction of the entire class has been evoked by other authors to account for such phenomena (Wahler, 1975). Finally, the procedures produced durable effects.

CONCLUSIONS

The literature summarized in this chapter indicates great potential for behavioral treatments in pediatric neurology. The areas of application and issues reviewed here are by no means comprehensive. Other areas of productivity in behavioral neurology include chronic pain (Fordyce, 1976), migraine (Diamond & Dalessio, 1978) and tension headache (Budzynski, 1978), rehabilitation (Ince, 1976), and pediatric intensive care (Cataldo, Bessman, Parker, Pearson, Reid, & Rogers, 1979).

It is interesting to note two divergent but ultimately compatible trends in neurological research and treatment. One trend involves the explosive development of basic knowledge of genetic (Brady & Rosenberg, 1978) and cellular (McKhann, 1978) mechanisms of neurological diseases, and prevention and therapies based on this knowledge. Related improvements in medical diagnostics such as computerized axial tomography (Bachman, Hodges, & Freeman, 1977) and success in medically treating previously untreatable disorders (cf. Rodenbaugh *et al.*, 1980) indicate increasing potency in medical treatments.

The second trend involves increasing interest in basic neurosciences and clinical neurology in the role of emotions, stress, and behavior in neurological disorders and their treatments. This trend is evident in renewed interest in CNS regeneration (Research News, 1980), functional recovery from brain damage (Yu, 1976), and neuroendocrine-systems–behavior–disease relationships (Mason, 1980). It is also evident in the extent and quality of clinical research in behavioral neurology, as indicated by the studies reviewed in this chapter.

Two major issues appear critical to progress in behavioral neurology in this decade. The first concerns the obvious need for more evidence on durable and generalized treatment effects, applied to larger groups of subjects in tightly controlled studies. A corollary of this issue is that studies of behavioral treatment have infrequently been collaborative endeavors with colleagues in neurology. Collaborative, interdisciplinary work will be vital to this area.

A second issue concerns the poor understanding of the relationships among etiology, pathophysiology, and symptoms of many neurological disorders. It has been proposed elsewhere that features of the technology of behavioral assessment may be helpful in improving diagnoses of disorders in which behavioral symptoms are complex and poorly correlated with physical variables (Russo, Bird, & Masek, 1980). These features of reliable and valid quantification of difficult to measure responses and single-subject designs allowing establishment of functional relationships among variables should contribute to behavioral neurology. Currently, as has been obvious throughout this chapter, diagnostic specificity of neurological as opposed to behavioral etiology and specificity of neurological pathophysiology have not been shown to contribute significantly to the outcome of behavioral interventions. This situation does not make rational sense to practitioners of scientific medicine.

The "borderland" of neurology, in which disorders of questionable psychiatric/behavioral versus organic/neurological etiology and pathophysiology reside, has been with us for some time (Gowers, 1907). During the past 25 years, this terrain has expanded, as researchers and practitioners have applied behavioral interventions to a variety of neurological disorders. During the coming decade, it is hoped that this area might achieve some definition of topography, with the establishment of specific treatment–symptom–pathophysiology–etiology relationships. It is clearly predicted that the practitioner in pediatric neurology will travel this area with a growing array of valuable behavioral assessment and treatment techniques.

Acknowledgment

The author gratefully acknowledges the editorial guidance of Dennis C. Russo and the help of Sylvia Giaise in preparing the manuscript.

REFERENCES

Arieti, S. (Ed.). *American handbook of psychiatry, Vol. 4: Organic disorders and psychosomantic medicine*. New York: Basic Books, 1975.

Ayllon, T., Layman, D., & Kandel, H. J. A behavioral-educational alternative to drug control of hyperactive children. *Journal of Applied Behavior Analysis*, 1975, *8*, 137–146.

Azrin, N., & Foxx, R. A rapid method of toilet training the institutional retarded. *Journal of Applied Behavior Analysis*, 1971, *4*, 89–99.

Bachman, D. S., Hodges, F. J., & Freeman, J. M. Computerized axial tomography in neurologic disorders of children. *Pediatrics*, 1977, *59*, 352–363.

Bandura, A. *Principles of behavior modification*. New York: Holt, Rinehart, & Winston, 1969.

Barrett, R. E., Yahr, M. D., & Duvoisin, R. C. Torsion dystonia and spasmodic torticollis: Results of treatments with L-dopa. *Neurology*, 1970, *20*, 107–113.

Basmajian, J. V. Research or retrench: The rehabilitation professions challenged. *Physical Therapy*, 1975, *55*, 607–610.

Basmajian, J. V. *Biofeedback: Principles and practice for clinicians*. Baltimore: Williams & Wilkins, 1978.

Basmajian, J. V., & Fernando, C. K. Biofeedback in physical medicine and rehabilitation. *Biofeedback and Self-Regulation*, 1978, *3*, 435–455.

Basmajian, J. V., Takebe, K., Kukulka, C. G., & Narayan, M. G. Biofeedback treatment of foot drop after stroke compared with standard rehabilitation technique. *Archives of Physical Medicine and Rehabilitation*, 1976, *57*, 9–11.

Batshaw, M. L., & Haslam, R. H. Multidisciplinary management of dystonia misdiagnosed as hysteria. In R. Eldridge & S. Fahn (Eds.), *Advances in neurology, Vol. 14: Dystonia*. New York: Raven, 1976.

Benson, H. *The relaxation response*. New York: Avon, 1975.

Bird, B. L., & Cataldo, M. F. Experimental analysis of EMG feedback in treating dystonia. *Annals of Neurology*, 1978, *3*, 310–315.

Bird, B. L., Russo, D. C., & Cataldo, M. F. Considerations in the analysis and treatment of dietary effects on behavior disorders. *Journal of Autism and Childhood Schizophrenia*, 1977, *7*, 373–382.

Bird, B. L., Parker, L. H., & Cataldo, M. F. *Experimental analysis of EMG biofeedback in cerebral spasticity*. Paper presented at the tenth meeting of the Biofeedback Society of America, San Diego, Calif., February 1979.

Bird, B. L., Cataldo, M. F., & Parker, L. H. Muscular disorders. In S. M. Turner, K. S. Calhoun, & H. E. Adams (Eds.), *Handbook of clinical behavior therapy*. New York: Wiley, 1981.

Brady, R. O., & Rosenberg, R. N. Autosomal dominant neurological disorders. *Annals of Neurology*, 1978, *4*, 548–552.

Brazier, M. A. B. The search for the neuronal mechanisms in epilepsy: An overview. *Neurology*, 1974, *24*, 903–911.

Brierly, H. The treatment of hysterical spasmodic torticollis by behavior therapy. *Behavior Research and Therapy*, 1967, *5*, 139–142.

Brudny, J., Grynbaum, B., & Korein, J. Spasmodic torticollis: Treatment by feedback display of the EMG. *Archives of Physical Medicine and Rehabilitation*, 1974, *55*, 403–408.

Budzynski, T. Biofeedback in the treatment of muscle contraction (tension) headache. *Biofeedback and Self-Regulation*, 1978, *3*, 409–434.

Cabral, R. J., & Scott, D. F. Effects of two desensitization techniques, biofeedback and relaxation, on intractable epilepsy: Follow up study. *Journal of Neurology, Neurosurgery, and Psychiatry*, 1976, *39*, 504–507.

Calhoun, K. S., & Turner, S. Historical perspectives and current issues in behavior therapy.

In S. Turner, K. Calhoun, & H. Adams (Eds.), *Handbook of clinical behavior therapy*. New York: Wiley, 1981.

Cataldo, M. F., & Russo, D. C. Developmentally disabled in the community: Behavioral/medical considerations. In L. A. Hamerlynck (Ed.), *Behavioral systems for the developmentally disabled. II: Institutional, clinic and community environments*. New York: Bruner/Mazel, 1979, pp. 105–143.

Cataldo, M. F., Bird, B. L., & Cunningham, C. Experimental analysis of EMG feedback in treating cerebral palsy. *Journal of Behavioral Medicine*, 1978, *1*, 311–322.

Cataldo, M. F., Bessman, C. A., Parker, L. H., Pearson, J. E., Reid, J. E., & Rogers, M. C. Behavioral assessment for pediatric intensive care units. *Journal of Applied Behavior Analysis*, 1979, *12*, 83–97.

Cataldo, M. F., Russo, D. C., Bird, B. L., & Varni, J. W. Assessment and management of chronic disorders. In J. Ferguson & C. B. Taylor (Eds.), *Comprehensive handbook of behavioral medicine* (Vol. 3), New York: Spectrum, 1980, 76–95.

Cataldo, M. F., Russo, D. C., & Freeman, J. M. Behavioral treatment in myoclonic and grand mal seizures. *Journal of Autism and Developmental Disorders*, 1980, *9*, 413–427.

Chase, M. H. (Ed.). *Operant control of brain activity*. Los Angeles: Brain Information Service/Brain Research Institute, 1974.

Cleeland, C. S. Behavior techniques in modification of spasmodic torticollis. *Neurology* (Minneapolis), 1973, *23*, 1241–1247.

Cooper, I. S. *Involuntary movement disorders*. New York: Harper & Row, 1969.

Cooper, I. S., Riklan, M., & Snider, R. S. (Eds.). *The cerebellum, epilepsy, and behavior*, New York: Plenum Press, 1974.

Cooper, I. S., Amin, I., Riklan, M., Walz, J. M., & Poon, T. P. Chronic cerebellar stimulation in epilepsy: Clinical and anatomical studies. *Archives of Neurology*, 1976, *33*, 559–570.

Creer, T. L., & Christian, W. F. *Chronically ill and handicapped children*. Champaign, Ill.: Research Press, 1976.

Crook, W. G. Food allergy: The great masquerader. *Pediatric Clinics of North America*, 1975, *22*, 227–238.

Davidoff, R. A., & Johnson, L. C. Paroxysmal EEG activity and cognitive-motor performance. *Electroencephalography and Clinical Neurophysiology*, 1964, *16*, 343–354.

Denhoff, E., Feldman, S., Smith, M. G., Litchman, H., & Holden, W. Treatment of spastic cerebral palsied children with Sodium Dantrolene. *Developmental Medicine and Child Neurology*, 1975, *17*, 736–742.

Diamond, S., & Dalessio, D. J. *The practicing physician's approach to headache*. Baltimore: Williams & Wilkins, 1978.

Doleys, D. M., & Kurtz, P. S. A behavioral treatment program for Gilles de la Tourette syndrome. *Psychological Reports*, 1974, *35*, 43–48.

Efron, R. The effect of olfactory stimuli in arresting uncinate fits. *Brain*, 1957, *79*, 267–281. (a)

Efron, R. The conditioned inhibition of uncinate fits. *Brain*, 1957, *80*, 251–262. (b)

Eldridge, R., & Fahn, S. (Eds.). *Advances in neurology, Vol. 14: Dystonia*. New York: Raven Press, 1976.

Eldridge, R., Riklan, M., & Cooper, I. S. The limited role of psychotherapy in torsion dystonia: Experience with 44 cases. *Journal of American Medical Association*, 1969, *210*, 705–712.

Engel, B. T., Nikoomanesh, P., & Schuster, M. M. Operant conditioning of rectosphincteric responses in the treatment of fecal incontinence. *New England Journal of Medicine*, 1974, *290*, 646–649.

Engle, H. A. The effect of diazepam (Valium) in children with cerebral palsy: A double-blind study. *Developmental Medicine and Child Neurology*, 1966, *8*, 661–669.

Epstein, L. H., & Blanchard, E. B. Biofeedback, self-control, and self-management. *Biofeedback and Self-Regulation*, 1977, *2*, 201–211.
Epstein, L. H., & Masek, B. T. Behavioral control of medicine compliance. *Journal of Applied Behavior Analysis*, 1978, *11*, 1–9.
Feingold, B. F. *Why your child is hyperactive*. New York: Random House, 1975.
Feuerstein, M., & Adams, H. E. Cephalic vasomotor feedback in the modification of migraine headache. *Biofeedback and Self-Regulation*, 1977, *2*, 241–254.
Finley, W. W. Effects of sham feedback following successful SMR training in an epileptic: Follow-up study. *Biofeedback and Self-Regulation*, 1976, *1*, 227–236.
Finley, W. W., Niman, C. A., Standley, J., & Wansley, R. A. Electrophysiologic behavior modification of frontal EMG in cerebral palsied children. *Biofeedback and Self-Regulation*, 1977, *2*, 59–79.
Ford, F. R. *Diseases of the nervous system in infancy, childhood, and adolescence*. Springfield, Ill.: Charles C Thomas, 1973.
Fordyce, E. *Behavioral methods for chronic pain and illness*. St. Louis: Mosby, 1976.
Forster, F. M. Conditioning in sensory evoked seizures. *Conditioned Reflex*, 1966, *1*, 224–234.
Forster, F. M. The classification and conditioning treatment of the reflex epilepsies. *International Journal of Neurology*, 1972, *9*, 73–83.
Freeman, J. *Comments on behavioral treatments of seizures*. Paper presented at the Johns Hopkins Conference on Behavioral Neurology, Baltimore, April 1980.
Gahn, N. H., Russman, B. S., & Cerciello, R. L., Fiorentino, M. R., & McGrath, D. M. Chronic cerebellar stimulation for cerebral palsy: A double-blind study. *Neurology*, 1981, *31*, 87–90.
Gardner, J. Behavior therapy treatment approach to a psychogenic seizure case. *Journal of Consulting Psychology*, 1967, *31*, 209–212.
Gardner, W. I. *Behavior modification in neutral retardation*. Chicago: Aldine, 1971.
Gastaut, H. Comments on "Biofeedback in epileptics: Equivocal relationship of reinforced EEG frequency to seizure reduction." *Epilepsia*, 1975, *16*, 487–490.
Gastaut, H., & Broughton, R. *Epileptic seizures*. Springfield, Ill.: Charles C Thomas, 1972.
Glaser, G. H. Treatment of intractable temporal lobe limbic epilepsy (complex partial seizures) by temporal lobectomy. *Annals of Neurology*, 1980, *8*, 455–459.
Gowers, W. The borderland of epilepsy. *Brain*, 1905, *41*, 68–79.
Haslam, R. H. A., Walcher, J. R., & Lietman, P. S. Dantrolene sodium in children with spasticity. *Archives of Physical Medicine and Rehabilitation*, 1974, *55*, 384–388.
Haynes, S. N., Griffin, P., Mooney, D., & Parise, M. Electromyographic biofeedback and relaxation instructions in the treatment of muscle contraction headaches. *Behavior Therapy*, 1975, *6*, 672–678.
Hersen, M., & Barlow, D. H. *Single-case experimental designs: Strategies for studying behavior change*. New York: Pergamon Press, 1976.
Hughes, H., Henry, D., & Hughes, A. The effect of frontal EMG biofeedback training on the behavior of children with activity level problems. *Biofeedback and Self-Regulation*, 1980, *5*, 207–219.
Ince, L. P. *Behavior modification in rehabilitative medicine*. Springfield, Ill.: Charles C Thomas, 1976.
Ince, L. P., Brucker, B. S., & Alba, A. Behavioral techniques applied to the care of patients with spinal cord injuries. *Behavioral Engineering*, 1976, *3*, 87–95.
Jackson, J. M. *Selected writings* (J. Taylor, Ed.). London: Hodder and Stoughton, 1931.
Jacobsen, E. *Progressive relaxation*. Chicago: University of Chicago Press, 1938.
Jacobsen, E. (Ed.). *Tension in medicine*. Springfield, Ill.: Charles C Thomas, 1967.
Jankel, W. R. Electromyographic biofeedback in Bell's palsy. *Archives of Physical Medicine and Rehabilitation*, 1978, *59*, 240–242.

Johnson, H. E., & Garton, W. H. Muscle re-education in hemiplegia by use of electromyographic device. *Archives of Physical Medicine and Rehabilitation*, 1973, *54*, 320–322.

Johnson, R. K., & Meyer, R. G. Phased biofeedback approach for epileptic seizure control. *Journal of Behavior Therapy and Experimental Psychiatry*, 1974, *5*, 185–187.

Johnson, R. T., & Herndon, R. M. Viologic studies of multiple sclerosis and other chronic and relapsing neurological diseases. *Progress in Medical Virology*, 1974, *18*, 214–228.

Joynt, R. L. Dantrolene sodium: Long-term effects in patients with muscle spasticity. *Archives of Physical Medicine and Rehabilitation*, 1976, *57*, 212–217.

Kaplan, B. J. Biofeedback in epileptics: Equivocal relationship of reinforced EEG frequencies to seizure reduction. *Epilepsia*, 1975, *16*, 447–458.

Keats, S. *Cerebral palsy*. Springfield, Ill.: Charles C Thomas, 1965.

Kohlenberg, R. J. Operant conditioning of human anal sphincter pressure. *Journal of Applied Behavior Analysis*, 1973, *6*, 201–208.

Kornhuber, H. H. Cerebral cortex, cerebellum, and basal ganglia: An introduction to their motor functions. In F. O. Schmitt & F. G. Worden (Eds.), *The neurosciences third study program*. Cambridge, Mass.: MIT Press, 1974.

Kuhlman, W. N. *EEG training in epileptic patients: Clinical and neurophysiological analysis*. Paper presented at the meeting of the Biofeedback Research Society, Colorado Springs, March 1976.

Lahey, B. B., Delamater, A., & Kupfer, D. Intervention strategies with hyperactive and learning disabled children. In S. M. Turner, K. S. Calhoun, & H. E. Adams (Eds.), *Handbook of clinical behavior therapy*. New York: Wiley, 1981.

Landau, W. M. Spasticity: The fable of a neurological demon and the Emperor's new therapy. (Editorial) *Archives of Neurology*, 1974, *31*, 217–218.

Liberman, R. P., & Davis, J. Drugs and behavior analysis. In M. Hersen (Ed.), *Progress in behavior modification (Vol. I)*. New York: Academic Press, 1975.

Lubar, J., & Bahler, W. W. Behavioral management of epileptic seizures following EEG biofeedback training of the sensorimotor rhythm. *Biofeedback and Self-Regulation*, 1976, *1*, 77–104.

Lubar, J., & Shouse, M. EEG and behavioral changes in a hyperkinetic child concurrent with training of the sensorimotor rhythm (SMR). *Biofeedback and Self-Regulation*, 1976, *1*, 293–298.

Luthe, W. *Autogenic therapy (Vols. 1–5)*. New York: Grune & Stratton, 1969.

Marinacci, A. *Applied electromyography*. Philadelphia: Lea & Febiger, 1968.

Marsden, C. D., & Harrison, M. J. G. Idiopathic torsion dystonia: A review of forty two patients. *Brain*, 1974, *97*, 793–810.

Martin, J., & Epstein, L. H. Evaluating treatment effectiveness in cerebral palsy: Single subject designs. *Physical Therapy*, 1976, *56*, 285–294.

Mason, J. *Neuroendocrine mechanisms of stress*. Paper presented at the first annual meeting of the Society for Behavioral Medicine, New York, December 1980.

McKhann, G. M. A cellular approach to neurological disease. *The Johns Hopkins Medical Journal*, 1978, *143*, 48–57.

Meichenbaum, D. Cognitive factors in biofeedback therapy. *Biofeedback and Self- Regulation*, 1976, *1*, 201–216.

Menkes, J. *Textbook of child neurology*. Philadelphia: Lea & Febiger, 1974.

Miller, N. E. Learning of visceral and glandular responses. *Science*, 1969, *163*, 434–445.

Miller, N. E. Biofeedback and visceral learning. *Annual Review of Psychology*, 1978, *29*, 373–404.

Molnar, G. E., & Alexander, J. Development of quantitive standards for muscle strength in children. *Archives of Physical Medicine and Rehabilitation*, 1974, *55*, 490–493.

Moser, H. W. Biochemical aspects of mental retardation. In D. B. Tower, (Ed.), *The nervous system (Vol. 2)*. New York: Raven Press, 1975.

Mostofsky, D. I. Recurrent paroxysmal disorders of the central nervous system. In S. Turner, K. Calhoun, & H. Adams (Eds.), *Handbook of clinical behavior therapy*. New York: Wiley, 1981.

Myers, R. The extrapyramidal system. *Neurology*, 1952, *24*, 627–655.

Nathan, P. W. Spasticity and its amelioration. In D. Williams (Ed.), *Modern trends in neurology (Vol. 5)*. London: Butterworths, 1970, pp. 41–59.

Neef, N. A., Iwata, B. A., & Page, T. J. Public transportation training in vivo vs. classroom instruction. *Journal of Applied Behavior Analysis*, 1978, *11*, 331–344.

Netsell, R., & Cleeland, C. S. Modification of lip hypertonia in dysarthria using EMG feedback. *Journal of Speech and Hearing Disorders*, 1973, *38*, 131–140.

Niedermeyer, E. *Compendium of the epilepsies*. Springfield, Ill.: Charles C Thomas, 1974.

Niedermeyer, E., Finyere, F., Rilley, T., & Bird, B. L. Myoclonus and the electroencephalogram: A review. *Clinical Electroencephalography*, 1979, *10*, 75–95.

Nordan, R. The psychological reaction of children with neurological problems. *Child Psychiatry and Human Development*, 1976, *6*, 214–223.

Nusselt, L., & Legewie, H. Biofeedback und systematisch Desensibilisierung bei Parkinson-Tremor: Eine Fallstudie. *Zeitschrift für Klinische Psychologie*, 1975, *4*, 112–123.

O'Leary, S. G., & Dubey, D. R. Applications of self-control procedures by children: A review. *Journal of Applied Behavior Analysis*, 1979, *12*, 449–465.

Ortega, D. F. Relaxation exercise with cerebral palsied adults showing spasticity. *Journal of Applied Behavior Analysis*, 1978, *11*, 447–454.

Peabody, F. W. The care of the patient. *Journal of the American Medical Association*, 1927, *88*, 877–882.

Phelps, W. M. Observations of a new drug in cerebral palsy athetoids. *Western Medicine*, 1963, Supplement 4.

Pippenger, C. E., Penry, J. K., & Kutt, H. (Eds.). *Antiepileptic drugs: Quantitative analysis and interpretation*. New York: Raven Press, 1978.

Pratt, K. L., Mattson, R. H., Weikers, N. J., & Williams, R. EEG activation of epileptics following sleep deprivation: A prospective analysis of 114 cases. *Electroencephalography and Clinical Neurophysiology*, 1968, *24*, 11–15.

President's Committee on Mental Retardation. Washington, D. C.: U. S. Government Printing Office, DHEW Publication Number (OHD) 76–21008, 1975.

Rafi, A. A. Learning theory and the treatment of tics. *Journal of Psychosomatic Research*, 1962, *6*, 71–76.

Research news: Regeneration in the central nervous system. *Science*, 1980, *209*, 378–380.

Rodenbaugh, J. E., Sato, S., Penry, J. K., Driefuss, F. E., & Kupferberg, H. J. Sodium valproate: pharmacokinetics and effectiveness in treating intractable seizures. *Neurology*, 1980, *30*, 1–6.

Romanczyk, R. G., & Goren, E. Severe self-injurious behavior: The problem of clinical control. *Journal of Consulting and Clinical Psychology*, 1975, *43*, 730–738.

Rosen, M., & Wesner, C. A behavioral approach to Tourette's syndrome. *Journal of Consulting and Clinical Psychology*, 1973, *41*, 308–312.

Roth, S. R., Sterman, M. B., & Clemente, C. D. Comparison of EEG correlates of reinforcement, internal inhibition, and sleep. *Electroencephalography and Clinical Neurophysiology*, 1967, *23*, 509–520.

Russo, D. C., Bird, B. L., & Masek, B. J. Assessment issues in behavioral medicine. *Behavioral Assessment*, 1980, *2*, 1–18.

Sachs, D. A., & Mayhall, B. Behavioral control of spasms using aversive conditioning with a cerebral palsied adult. *Journal of Nervous and Mental Disease*, 1971, *152*, 363.

Sachs, D. A., & Mayhall, B. The effects of reinforcement contingencies upon pursuit rotor performance by a cerebral palsied adult. *Journal of Nervous and Mental Disease*, 1972, *155*, 36–41.

Salam, M. Z., & Adams, R. D. Research in the clinical expressions and pathological basis of mental retardation. In D. B. Tower (Ed.), *The nervous system (Vol. 2)*. New York: Raven Press, 1975.

Schain, R. J. *Neurology of childhood learning disorders*. Baltimore: Williams and Wilkins, 1977.

Schwartz, G. E. Biofeedback and physiological patterning in human emotion and consciousness. In J. Beatty & L. Heiner (Eds.), *Biofeedback and behavior*. New York: Plenum Press, 1977.

Shapiro, A. K., Shapiro, E. S., Bruun, R. D., Sweet, R., Wayne, H., & Soloman, G. Gilles de la Tourette's syndrome: Summary of clinical experience with 250 patients and suggested nomenclature for tic syndromes. In R. Eldridge & S. Fahn (Eds.), *Advances in Neurology, Vol. 14: Dystonia*. New York: Raven Press, 1976, 277–283.

Smith, L. H. *Improving your child's behavior chemistry*. Englewood cliffs, N. J.: Prentice–Hall, 1976.

So, E. L., & Penry, J. K. Epilepsy in adults. *Annals of Neurology*, 1981, *9*, 3–16.

Sterman, M. B. Effects of sensorimotor EEG feedback training on sleep and clinical manifestations of epilepsy. In J. Beatty & H. Legewie (Eds.), *Biofeedback and behavior.* New York: Plenum Press, 1977.

Sterman, M. B., & Friar, L. Suppression of seizures in an epileptic following sensorimotor EEG feedback training. *Electroencephalography and Clinical Neurophysiology*, 1972, *33*, 89–95.

Sterman, M. B., & MacDonald, L. R. Effects of central cortical EEG feedback training on incidence of poorly controlled seizures. *Epilepsies*, 1978, *19*, 207–222.

Sterman, M. B., & Wyrwicka, W. EEG correlates of sleep: Evidence for separate forebrain substrates. *Brain Research*, 1967, *6*, 143–163.

Sterman, M. B., MacDonald, L. R., & Stone, R. K. Biofeedback training of the sensorimotor electroencephalogram rhythm in man: Effects in epilepsy. *Epilepsies*, 1974, *15*, 395–416.

Stevens, J. R., Milstein, V. M., & Dodds, S. A. Endogenous spike discharges as conditioned stimuli in man. *Electroencephalography and Clinical Neurophysiology*, 1967, *23*, 57–66.

Stoyva, J. Why should muscular relaxation be useful? In J. Beatty & L. Heiner (Eds.), *Biofeedback and behavior*. New York: Plenum Press, 1977.

Stroebel, C. *The quieting response*. Paper presented at the Johns Hopkins Conference on Behavioral Neurology, Baltimore, April 1980.

Swaan, D., van Wieringen, P. C., & Fokkema, S. D. Auditory electromyographic feedback to inhibit undesired motor activity. *Archives of Physical Medicine and Rehabilitation*, 1974, *55*, 251–254.

Tarnopol, L. *Learning disorders in children*. Boston: Little, Brown, 1971.

Taub, E., & Stroebel, C. F. Biofeedback in the treatment of vasoconstrictive syndromes. *Biofeedback and Self-Regulation*, 1978, *3*, 363–373.

Thomas, E. J., Abrams, K. S., & Johnson, J. B. Self-monitoring and reciprocal inhibition in the modification of multiple tics with Gilles de la Tourette's syndrome. *Journal of Behavior Therapy and Experimental Psychiatry*, 1971, *2*, 159–171.

Thoresen, C. E., & Mahoney, M. J. *Behavioral self-control*. New York: Holt, Rinehart, & Winston, 1974.

Varni, J. W., Boyd, E. F., & Cataldo, M. F. Self-monitoring, external reinforcement, and timeout procedures in the control of high rate tic behaviors in a hyperactive child. *Journal of Behavior Therapy and Experimental Psychiatry*, 1978, *9*, 353–358.

Vining, E. P. G., Accardo, P. J., Rubenstein, J. E., Farrell, S. E., & Roizen, N. J. Cerebral Palsy: A pediatric developmentalist's overview. *American Journal of Diseases of Children*, 1976, *130*, 643–649.

Wahler, R. G. Some structural aspects of deviant child behavior. *Journal of Applied Behavior Analysis*, 1975, *8*, 27–42.

Wannstedt, G. T., & Herman, R. M. Use of augmented sensory feedback to achieve symmetrical standing. *Physical Therapy*, 1978, *58*, 553–559.

Weiss, B., & Santelli, S. Dyskinesias evoked in monkeys by weekly administration of haloperidol. *Science*, 1978, *200*, 799–800.

Wender, P. H. *Minimal brain dysfunction in children*. New York: Wiley, 1971.

Wetherby, B., & Baumeister, A. A. Mental retardation. In S. M. Turner, K. S. Calhoun & H. E. Adams (Eds.), *Handbook of clinical behavior therapy*. New York: Wiley, 1981.

Whalen, C., Henker, B., Collins, B., Finck, D., & Dotemoto, S. A social ecology of hyperactive boys: Medication effects in structured classroom environments. *Journal of Applied Behavior Analysis*, 1979, *12*, 65–81.

Wolf, J. M. (Ed.). *The results of treatment in cerebral palsy*. Springfield, Ill.: Charles C Thomas, 1969.

Wooldridge, C. P., & Russell, G. Head position training with the cerebral palsied child: An application of biofeedback techniques. *Archives of Physical Medicine and Rehabilitation*, 1976, *57*, 407–414.

Wright, L. Aversive conditioning of self-induced seizures. *Behavior Therapy*, 1973, *10*, 712–713.

Wright, T., & Nicholson, J. Physiotherapy for the spastic child: An evaluation. *Developmental Medicine and Child Neurology*, 1973, *15*, 146–163.

Wyler, A. R., Fetz, E. E., & Ward, A. A. Effects of operantly conditioning epileptic unit activity on seizure frequencies and electrophysiology of neocortical experimental foci. *Experimental Neurology*, 1974, *44*, 113–125.

Wyler, A. R., Lockard, J. S., Ward, A. A., & Finch, C. A. Conditioned EEG desynchronization and seizures. *Electroencephalography and Clinical Neurophysiology*, 1976, *41*, 501–512.

Yu, J. Functional recovery with and without training following brain damage in experimental animals: A review. *Archives of Physical Medicine and Rehabilitation*, 1976, *57*, 38–41.

Zlutnick, S., Mayville, W. J., & Moffat, S. Modification of seizure disorders: The interruption of behavioral chains. *Journal of Applied Behavior Analysis*, 1975, *8*, 1–12.

5

Treatment of Urinary and Fecal Incontinence in Children

Lynn Parker and William Whitehead

INTRODUCTION

Bladder and bowel control typically occurs by age 3. Toilet training is usually accomplished by parents without the need for professional intervention. However, because of the use of ineffective training procedures or physical limitations, many children remain incontinent throughout the early school years. Lack of bladder or bowel control is one of the most prevalent childhood disorders. Although spontaneous remissions occur from early childhood through adolescence, the majority of incontinent children will require treatment.

Achievement of bladder and bowel control is one of the first developmental milestones of childhood, and lack of self-control over bodily functions may result in serious psychosocial problems. Children with incontinence have difficulty in developing peer relations and are often excluded from public school until continence is achieved. Social diffi-

---

LYNN PARKER • Children's Hospital of New Orleans, New Orleans, Louisiana 70118. WILLIAM WHITEHEAD • Gerontology Research Center; Johns Hopkins University School of Medicine, Baltimore, Maryland 21205. Preparation of this chapter was supported by Grant 1 R01 N515781 from the National Institute of Neurological and Communicative Disorders and Stroke and by Research Scientist Development Award 5 K01 MH 00133 from the National Institute of Mental Health.

culties related to incontinence are not restricted to children. For example, adults with incontinence secondary to spina bifida are far less likely to marry or to be employed than are people with identical neurological problems but without incontinence (Laurence & Beresford, 1975). Aside from the social problems produced by incontinence, elimination disorders have been reported to be the second most common reason for child abuse (Kempe & Helfer, 1972).

URINARY INCONTINENCE

Definition and Incidence

Urinary incontinence is typically defined as frequent urination in inappropriate settings after the age of 3. The criterion frequency has varied from 3 times per week to 7 times per week (Schaefer, 1979). However, it is generally accepted that an infrequent accident or bedwetting episode is not considered a problem requiring intervention.

Urinary incontinence may be due to organic pathology or refer to wetting in the absence of neurological or urological abnormalities. Approximately 95% of reported cases are of the latter type (i.e., functional) with no accompanying organic impairment (Campbell, 1951).

Functional incontinence or enuresis can be classified as either primary or secondary and diurnal (daytime) or nocturnal (nighttime). Primary enuresis refers to children who have never achieved dryness, whereas secondary enuresis implies that a child has had at least a 6-month period of daytime and nighttime continence. By far, the majority of enuretics are primary types, with bedwetting rather than daytime incontinence as the presenting complaint.

Daytime wetting is not as prevalent as nighttime wetting and occurs more frequently in girls than in boys. Bloomfield and Douglas (1956) reported that 2.9% of children at age 6 were incontinent by day with a male/female ratio of 1 to 2. Bedwetting is much more common than diurnal incontinence, occurring in 10–16% of 6-year-olds (Bloomfield & Douglas, 1956; Oppel, Harper, & Rider, 1968). Boys tend to be bedwetters more frequently than girls at age 4, but at this age the sex difference is not significant. This sex difference becomes highly significant by age 8 because of a more rapid decline in prevalence among girls (Blomfield & Douglas, 1956). Spontaneous remission of incontinence occurs for both sexes. Forsythe and Redmond (1974) found spontaneous remissions in 14% of 5- to 9-year-olds and in 16% of 10- to 14-year-olds who had failed to respond to surgical and pharmacological interventions.

Relapses are reported to occur in 25% of children who attain nighttime dryness by age 12 (Oppel *et al.*, 1968).

Prevalence of incontinence may be much higher for children with neurological disorders. For example, Lister, Zachary, and Brereton (1977) found that 21% of children with meningomyelocele (a type of spina bifida) were dry at age 6 and that only 25% were dry by age 10. Lorber (1971) reported an even higher incidence of urinary incontinence in a series of 134 children with meningomyelocele. Only 17% of these children were actually considered to have normal sphincter control. Twenty-seven percent of the children required surgical intervention to divert the urine to collection appliances. The remaining 56%, while not requiring surgery, were chronically incontinent of urine.

Physiological Mechanisms

Figure 1 shows the anatomy of the normal urinary tract. The urinary tract consists of the kidneys, the ureters, and the bladder. The kidneys are bean-shaped structures located below the diaphragm. They excrete

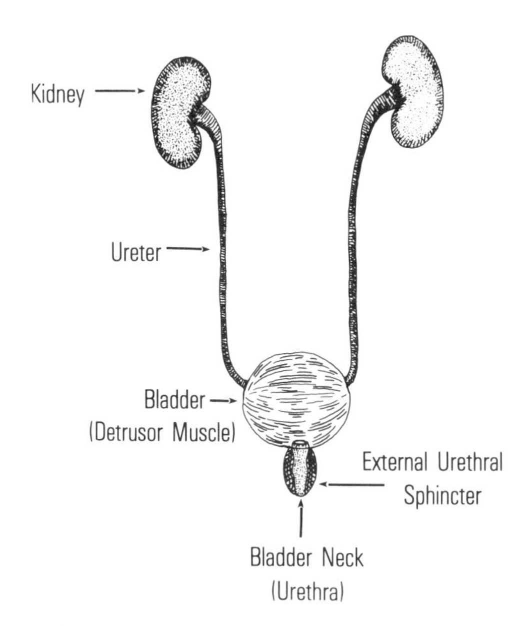

Figure 1. Anatomy of the urinary tract. From "Behavioral Treatment of Urinary and Fecal Incontinence" by W. Whitehead and L. Parker, in *Behavioral Medicine: A Practical Handbook*, edited by T. Coates, Champaign, Illinois: Research Press, 1982. Copyright 1982 by Research Press. Reprinted by permission.

excess water and the waste products of protein metabolism and regulate the acid–base balance, electrolyte concentration, and volume of the plasma. The kidneys filter and concentrate approximately 50 gallons of plasma every day to produce one quart of urine. Urine is collected at the renal pelvis and transported down the ureters to the bladder by peristalsis. The adult bladder is a large reservoir which holds approxiamtely 250–400 ml of fluid. The bladder wall consists of three layers of interwoven smooth muscles known collectively as the detrusor. These same muscle fibers are also present in the bladder neck and comprise the internal sphincter. The external sphincter and perineal muscles are striated and under voluntary control.

The known mechanisms of urination are as follows: urine accumulates in the bladder with little increase in intravesical pressure until a threshold volume (250–400 ml) is reached. At this threshold a massive contraction of the detrusor muscle (reflex bladder contraction) occurs unless it is voluntarily inhibited by the cerebral cortex. The ureterovesical junction joining the ureters and bladder prevents fluid from running into the ureters. The walls of the ureters also collapse during voiding so that the urine is expelled only through the bladder neck. Contraction of the detrusor causes a reflexive inhibition (relaxation) of the resting tone of the internal sphincter and also the external sphincter. If the external sphincter is not voluntarily contracted at this time, micturition will result. However, if the striated external sphincter and perineal muscles are contracted voluntarily, the smooth muscle of the bladder accommodates again to an increased volume of fluid for a short period of time.

The neurophysiology of micturition is poorly understood. What is clear is that passive collection and active expulsion are at least partially under voluntary control. However, the extent of this control is uncertain. Early studies on the role of striated muscle in urination have shown that micturition can be initiated and terminated on command without the use of striated muscle. Striated muscle appears to be necessary only for sudden, rapid inhibition of urination (Lapides, Sweet, & Lewis, 1957). Secondly, it is no longer assumed that bladder innervation is under parasympathetic dominance of the sacral cord (S2, S3, and S4). Histochemical studies have found adrenergic nerve endings (T10 and T11) concentrated in the base of the bladder and in the urethra (Marchant, 1977). It is thought that parasympathetic cholinergic receptors produce detrusor contractions. Control of the flow of urine through the bladder neck is likely to be mediated by adrenergic muscle receptors.

Attainment of continence in the young child is the result of a complex interaction of developmental and learning processes. Muellner (1960) described the transformation of the infantile bladder into the adult bladder as a multi-staged process. The child must first learn to sense (discriminate) bladder fullness. This occurs following the maturation of parasympathetic nerves which carry bladder sensations. Brief periods of urine retention occur by age 3 and are due to voluntary control of the pelvic floor muscles. By age 4½ most children can start a urine stream when the bladder is full by controlling the thoracic diaphragm and abdominal muscles. Initiation of urination during partial bladder fullness is achieved by age 6.

In addition to the development of nerve pathways and acquisition of skills necessary for urine control, the young child also experiences anatomical changes in bladder size. A retentive capacity of 294 to 355 ml is needed before the child can sleep through the night without an urge to void. Early studies have demonstrated that bladder capacity of enuretic children is apt to be less than that of nonenuretics and that enuretics are more likely to void frequently during the day (Bloomfield & Douglas, 1956).

In addition to problems of learning and developmental delay, incontinence may be due to organic pathology. Incontinence due to organic abnormalities is frequently differentiated into either congenital or acquired disorders. Congential causes of incontinence include extrophy of the bladder (eversion or turning inside out of the bladder), ureteral ectopy (ureteral orifice located somewhere other than the bladder, e.g., the vagina), and defects of the spinal cord (as occur in some types of spina bifida). Acquired types of incontinence are those associated with disease or injury of the spinal cord, detrusor dyssynergia (lack of muscular coordination of the detrusor with the sphincters), stress incontinence (loss of urine due to an impaired sphincter mechanism), urgency incontinence (inability to inhibit bladder contractions once the threshold for reflex contractions is reached), and incontinence due to cerebral trauma.

As is evident, the breakdown in one or more systems may result in inappropriate voiding. Some physiological abnormalities evidenced in organic incontinence include: (1) hyperactivity of the reflex bladder mechanism, (2) diminishing sensation of bladder fullness, and (3) weakness of the internal or external sphincter (Wilson, 1948; Marchant, 1977).

Medical Approaches

Stimulants, anticholinergics, tranquilizers, monoaminoxidase inhibitors, and tricyclic antidepressants have all been studied in the treatment of nocturnal enuresis (Blackwell & Currah, 1973; Breger, 1962; Harrington, 1960; McConaghy, 1969; MacLean, 1960; Wallace & Forsythe, 1969).

Tricyclics (imipramine, amitriptyline, nortriptyline, desipramine) are the only group of drugs shown to be more effective than placebos in double-blind studies (Agarwala & Heycock, 1968; Blackwell & Currah, 1973; Epstein & Guilfoyle, 1965; Poussainte & Ditman, 1965). In one such study on the effect of imipramine in 73 enuretics, over 50% of the children receiving the drug were markedly improved as compared with a 12% improvement in the placebo control group. These favorable results were attributed to increased internal sphincter tone due to sensitization of alpha adrenergic receptors to sympathetic innervation. Only four children maintained their continence following withdrawal of imipramine (Mahony, Laferte & Mahoney, 1973). Blackwell *et al.* (1973) also reported a 70–90% relapse rate in responders once drugs were stopped. Accordingly, although temporarily effective, tricyclics should not be considered a permanent cure (Safer, 1978).

Adverse side effects of tricyclics in doses up to 75 mg/day are dizziness and anorexia. These effects are reportedly temporary and of minor intensity (Safer, 1978). Werry and Aman (1975) also reported cardiovascular changes in males who received 50 mg of imipramine nightly. They found an increase in diastolic blood pressure and resting pulse rate as well as a minor but consistent weight loss across patients.

Surgical procedures are often used to remedy incontinence due to neurological bladder disorders (disorders of the nerve supply to the bladder and perineum). Urinary diversion procedures are used to divert the flow of urine from the bladder when the bladder is not functioning properly. The most common methods of diversion include ureterosigmoidostomy, cutaneous ureterostomy, and ileal conduit (Keuhnelian & Sanders, 1970).

In uretersigmoidostomy an opening is formed between the ureters and sigmoid colon such that urine flows freely into the bowel. Urine is passed through the rectum, and no external collection appliance is needed. However, renal infection and other complications are common after prolonged use. For this reason this procedure is seldom performed on children and young adults.

Cutaneous ureterostomy refers to the transplantation of the ureters onto the wall of the abdomen. Bilateral openings on the skin are created and two appliances are usually worn. This type of diversion has multiple disadvantages, one being the need for permanent indwelling catheters. Indwelling catheters frequently lead to infection.

The ileal conduit is constructed by suturing the ureters to a short segment of the small intestine called the terminal ileum. This segment of small bowel is dissected out and the bowel reconnected. One end of the new conduit is closed while the other end is brought out of the abdomen as a stoma. This method works quite efficiently to drain urine by means of peristalsis. Reflux into the kidneys is rare.

Other more conservative treatments of neurogenic bladder disorders include pharmacological agents (e.g., adrenergic blocking agents for lower motor neuron disorders) and intermittent catheterization. Catheterization, when applicable, is the method of choice for bladder drainage. This method is more socially acceptable than the use of external appliances and it leaves the urinary tract intact should future research development produce relevant new treatment approaches. Catheterization is done 3 or 4 times within 24 hours. The daily output of urine is maintained between 1500–2000 ml/day for adults (Rossier, 1978).

Recently, children with urinary incontinence secondary to myelodysplasia (a developmental defect of the spinal cord) have been taught self-catheterization using behavioral shaping procedures (Hannigan, 1979). Preschool children were taught components of catheterization using dolls equipped with male or female genitalia. All children were successful in learning catheterization skills within two days. These studies are ongoing at the Johns Hopkins Medical Institutions and are examples of needed collaborative work between behavioral psychologists and medical professionals.

Behavioral Treatments

Table I summarizes 29 studies which used behavioral procedures in the treatment of urinary incontinence. Six of these studies reported treatment of adults. They are included because the procedures discussed are potentially applicable to urinary incontinence in children. Behavioral interventions for urinary incontinence have been evaluated far more frequently than similar approaches to fecal incontinence. Generally, treatments were implemented across large groups of subjects as is evidenced by the 1,081 cases of incontinence reported in these 29 studies. The vast majority of articles focused on populations with nocturnal enuresis as a presenting problem. Only 3 studies reported the efficacy of behavioral treatments for incontinence due to organic pathology.

Although a large number of patients with different types of urinary incontinence have been studied, the literature is not free of experimental deficiencies. Objective dependent measures were not employed to evaluate treatment outcome in 32% of these articles. Baseline or pretreatment rates of incontinence were reported in only 53% of studies, while either group or single subject controls were employed in only 39%.

Long-term follow-ups have been conducted in the majority of studies. Follow-up periods ranged from 2 weeks to 5 years, with most in-

Table I. Behavioral Treatment of Urinary Incontinence

Author	No.	Age range	Diagnosis	Procedure	Objective dependent measure	Baseline	Controls	Length of follow-up	Number continent post-treatment	Number continent at follow-up
Lovibond, 1963	36	6–14	Nocturnal enuresis	Gr.1: bell & pad ($n=12$) Gr.2: Crosby electrical stimulus instrument ($n=12$) Gr.3: Twin signal instrument ($n=12$)	No	No	None	24 mos.	Gr.1: 11/12 Gr.2: 11/12 Gr.3: 12/12	Gr.1: 7/11[a] Gr.2: 7/11 Gr.3: 8/12
Forrester, Stein, & Susser, 1964	33	8–14	Nocturnal enuresis	Gr.1: amphetamine ($n=17$) Gr.2: bell & pad ($n=16$)	No	No	Amphetamine control group	None	Gr.1: 3/17 Gr.2: 10/16	—
Young & Turner, 1965	299	4–15	Nocturnal enuresis	Gr.1: bell & pad ($n=105$) Gr.2: bell & pad & dexedrine ($n=84$) Gr.3: bell & pad & methedrine ($n=110$)	No	No	Bell & pad control for addition of dex. & meth.	6–12 mo.	Gr.1: 68/105 Gr.2: 64/84 Gr.3: 90/110	Gr.1: 59/68 Gr.2: 45/64 Gr.3: 78/90
Turner & Young, 1966	142	4–15	Nocturnal enuresis	Gr.1: bell & pad ($n=41$) Gr.2: bell & pad & dexedrine ($n=41$) Gr.3: bell & pad & methedrine ($n=60$)	No	No	Bell & pad control for addition of dex. & meth.	9 mo. to 5 yr.	N/A (follow-up study of continent subjects)	Gr.1: 23/41 Gr.2: 9/41 Gr.3: 26/60
Browning, 1967	1	10	Nocturnal enuresis	1) Reinforcement of awaking to bell 2) Use of bell & pad	Yes	No	None	None	1/1	—
Kimmel & Kimmel, 1970	3	4–10	Nocturnal enuresis	Retention control training	No	No	None	12 mo.	3/3	3/3

Turner, Young, & Rachman, 1970	62	4–15	Nocturnal enuresis	Gr.1: Mowrer-type signal ($n = 15$) Gr.2: twin-signal ($n = 15$) Gr.3: intermittent ($n = 15$) Gr.4: placebo control ($n = 17$) Gr.5: arousal control	Yes	No	Placebo and arousal control groups	3 yr.	Gr.1: 12/15 Gr.2: 11/15 Gr.3: N/A (maintenance strategy only, subjects taken from Gr.1 & Gr.2) Gr.4 & 5: N/A (treated only 1 month)	Gr.1 & 2: 7/20 Gr.3: 8/11
Azrin & Foxx, 1971	9	20–62	Nocturnal enuresis, profound retardation	1) Increase fluid intake 2) Reinforcement 3) Delay for accidents 4) Use of signalling apparatus 5) Shaping toileting skills 6) Cleanliness training 7) Staff reinforcement	Yes	Yes	None	None	90% reduction in $\bar{x}$ number of accidents/day	—
Tough, Hawkins, McArthur, & Ravenswaay, 1971	2	4 & 8	Nocturnal enuresis (one child was retarded & multiply handicapped)	1) Punishment of bedwetting 2) DRO for nonbedwetting	Yes	Yes	None	18 mo.	1/2	1/2
Paschalis, Kimmel, & Kimmel, 1972	31	4–13	Nocturnal enuresis	Gr.1: Reinforcement for prolongation of urine retention Gr.2: delayed treatment control	Yes	Yes	Delayed treatment control group	3 mo.	15/31	15
Samaan, 1972	1	7	Enuresis	1) Bell & pad 2) Reinforcement of appropriate nighttime toiletings	No	No	None	24 mo.	1/1	1

Table I. (Continued)

Author	No.	Age range	Diagnosis	Procedure	Objective dependent measure	Baseline	Controls	Length of follow-up	Number continent post-treatment	Number continent at follow-up
Stedman, 1972	1	13	Nocturnal enuresis	Reinforcement for prolongation of urine retention	Yes	Yes	None	3 mo.	1/1	0
Young & Morgan, 1972	144	4–15	Nocturnal enuresis	Gr.1: bell & pad ($n = 77$) Gr.2: bell & pad & increased fluid intake ($n = 67$)	No	No	Bell & pad control group	24 mo.	Gr.1: 46/77 Gr.2: 55/67	Gr.1: 30/74 Gr.2: 48/67
Azrin, Sneed, & Foxx, 1973	12	$\bar{x} = 37$ years	Nocturnal enuresis, profound retardation	1) Reinforcement of appropriate urinations and dry bed 2) Overcorrection 3) Cleanliness training 4) Use of urine alarm	Yes	Yes	Urine alarm control condition	3 mo.	11/12	11/12
Miller, 1973	2	13–14	Nocturnal enuresis	Retention control training	Yes	Yes	Attention placebo	4–7 mo.	2/2	2/2
Azrin, Sneed, & Foxx, 1974	26	$\bar{x} = 6$ years	Nocturnal enuresis	Gr.1: urine alarm & training toileting & increased fluid & positive reinforcement ($n = 13$) (dry bed procedure) Gr.2: urine alarm ($n = 13$)	Yes	Yes	Urine alarm control group (later training provided controls)	None	24/24 (2 children continent with urine alarm, not trained with dry bed procedure)	—
Creer & Davis, 1975	9	Not given	Nocturnal enuresis	Staggered wakening	Yes	Yes	Multiple baseline design	2–6 wks.	0/9	4/9
Doleys & Wells, 1975	1	3	Nocturnal enuresis	Retention control training	Yes	Yes	None	14 wks.	1/1	1/1

Butler, 1976	49	Not given	Urinary & fecal incontinence	Parent training using Azrin & Foxx's *Toilet Training in Less than a Day*	Yes	Yes	None	2 mo.	46/49	47/49
Doleys, Ciminero, Tollison, Williams, & Wells, 1977	19	$\bar{x}=7.8$ for Gr. 1, 6.6 for Gr. 2	Nocturnal enuresis	Gr.1: dry bed training ($n=10$) Gr.2: retention control training ($n=9$)	Yes	Yes	None	6–11 mos.	Gr.1: approx. 1.5 $\bar{x}$ wets/weeks Gr.2: approx. 6.5 $\bar{x}$ wets/week	Gr.1: 5/11 Gr.2: no follow-up
Pearne, Zigelbaum, & Peyser, 1977	1	27	Chronic urinary retention & incontinence	EMG-assisted relaxation (frontalis)	Yes	No	None	4 mo.	1/1	1/1
Azrin & Thienes, 1978	55	3–14	Nocturnal enuresis	Gr.1: 8 component operant approach ($n=28$) Gr.2: bell & pad ($n=27$)	Yes	Yes	Bell & pad control group	12 mo.	Unclear—presents $\bar{x}$% wet nights for each group	51/51 (after all subjects, less 4 dropouts, received operant procedure)
Cardozo, Stanton, Hafner, & Allan, 1978	6	22–47	Detrusor instability	Biofeedback of detrusor pressure	Yes	Yes (pre-test)	None	3 mo.	Not reported	3/6 (bladder stability during cystometry)
Cardozo, Abrams, Stanton, & Feneley, 1978	27	18–64	Idiopathic bladder instability	Biofeedback of detrusor pressure	Yes	No	None	3 mo.	11/27	11/27
Wear, Wear, & Cleeland, 1979	8	19–72	Urinary retention (5) urinary incontinence (5), pelvic pain (1)	Periurethral EMG feedback & intraurethral or intravesical pressure feedback	Yes	Yes (pre-test)	None	None	4/8 reported to be moderately to markedly improved (state of continence unclear)	—

[a]Denominator reflects number of total subjects available at follow-up.

vestigators following patients for at least 6 months after the termination of treatment.

Enuresis

Three types of procedures have emerged in the treatment of functional urinary incontinence: (1) the Mowrer and Mowrer (1938) bell-and-pad method; (2) behavioral procedures employing reinforcement, punishment, and extinction to modify the frequency of appropriate toiletings and accidents; and (3) urine retention training.

The bell-and-pad (urine alarm) apparatus has been the most widely studied treatment for nocturnal enuresis. It has been used alone and in combination with a number of other procedures. As developed by Mowrer and Mowrer, it involves having the child sleep on a pad which detects urine. The urine closes an electrical circuit and causes an alarm to sound, awakening the child and his parents. The child is then taken to the bathroom where urination is completed. The bell-and-pad method is based on the classical conditioning paradigm. Bladder fullness and the initiation of detrusor contraction is the unconditioned stimulus (UCS) giving rise to the unconditioned response (UCR), micturition. By pairing bladder fullness (UCS) with the alarm, the UCS becomes a conditioned stimulus (CS) for the conditioned response (CR) of awakening and contracting the urethral sphincters. Lovibond (1963) suggested that the conditioning treatment employed by Mowrer and Mowrer follows an avoidance paradigm rather than one of classical conditioning. This explanation is forwarded to account for the lack of extinction of the conditioned response following removal of the bell stimulus. A classical conditioning model would predict a weakening of the conditioned response in these circumstances.

Early research on the effectiveness of the urine alarm (Forrester, Stein & Susser, 1964; Lovibond, 1963; Turner & Young, 1966; Young & Turner, 1965) showed remission of enuresis in 56%–91% of patients. However, relapse rates during follow-up for these studies were as high as 30%. Relapse rates were even higher in groups receiving bell-and-pad training plus certain pharmacological agents, such as dexedrine (Turner & Young, 1966). This high relapse rate may be due to extinction following the removal of the alarm or to lack of generalization to levels of bladder fullness not experienced during training. In an effort to increase maintenance of treatment effects, Turner, Young, and Rachman (1970) employed an intermittent signal procedure so as to establish greater resistance to extinction. The group receiving intermittent signal training had a lower relapse rate at follow-up than the group for whom the maintenance strategy was not employed.

Young and Morgan (1972) used a training procedure designed to increase maintenance by continuing treatment with levels of bladder fullness not experienced during the original conditioning. This "overlearning" procedure required the child to consume 295 to 946 ml (depending on body size) prior to retiring. The same urine alarm contingencies were in effect during this phase of training. Following a 2-year follow-up the overlearning group showed only a 10% relapse as compared to a 22% relapse in the group that did not receive the overlearning component.

Operant procedures involving principles of reinforcement, shaping, punishment, and extinction have been used in toilet training and in the treatment of nocturnal enuresis (Doleys, 1977). Foxx and Azrin (1973) published a detailed manual for toilet training the retarded and later revised these procedures for use in training children of normal intelligence (Azrin & Foxx, 1974). Their program for daytime continence training involved four stages: (1) training in lowering and raising clothes and in sitting on the toilet; (2) training to urinate in the toilet when instructed to do so; (3) training in the self-initiation of urinating in the toilet; and (4) maintenance, during which the individual is rewarded for being dry. These procedures have been very effective, especially when applied by trained staff (Azrin & Foxx, 1973). The utility of this approach was further extended by Butler (1976), who demonstrated its effective use by parents in toilet training their children. Parents of normal children were given three lectures describing the procedure and phone calls following the lectures. Of the 49 treated children, 46 were successfully toilet trained by their parents. Results were maintained at the end of a 2-month follow-up.

Dry-bed training (DBT), a program similar to the daytime program, has been developed to eliminate bedwetting in the retarded. Components of the program are: (1) reinforcement of appropriate urinations and a dry bed, (2) increased fluid intake, (3) use of a urine alarm for urine detection, (4) hourly awakenings, (5) overcorrection for accidents, and (6) cleanliness training. This program has been successful in eliminating nocturnal enuresis in approximately 90% of reported cases (Azrin, Sneed, & Foxx, 1973).

Similar procedures, with and without the addition of a urine alarm apparatus, have been reported in the treatment of enuretic children of normal intelligence. Operant approaches with and without the inclusion of the urine alarm have been shown to be superior to the use of the urine alarm alone. (Azrin, Sneed, & Foxx, 1974; Azrin & Thienes, 1978). Especially noteworthy is the Azrin and Thienes (1978) study, in which all 51 patients treated using an intensive operant procedure were still continent at a one-year follow-up. These findings are in contrast to the high relapse rates reported when using the bell-and-pad method (Lovibond, 1963; Turner *et al.*, 1970).

A variety of less complex procedures such as punishment for wet nights, differential reinforcement for the absence of bedwetting (DRO), and systematic desensitization have also been reported in the treatment of urinary incontinence (Creer & Davis, 1975; Taylor, 1972; Tough, Hawkins, McArthur & Ravenswaay, 1971). These procedures were implemented in a single case studies or in unreplicated reports of small numbers of patients. Further investigation will be needed to demonstrate their effectiveness relative to other more widely researched methods.

Another type of behavioral intervention, retention control training (RCT), was introduced by Kimmel and Kimmel (1970) as an alternative to the urine alarm in the treatment of nocturnal enuresis. This procedure was developed so that training could be conducted during the day rather than disrupting the sleep of the child and parent at night. A second impetus for its development was the finding that enuretics had smaller bladder capacities than nonenuretics (Muellner, 1960). Training of detrusor inhibition should result in the accommodation of the bladder to larger volumes of urine and allow the child to sleep through the night without accidents.

Kimmel and Kimmel (1970) reported the first use of RCT for the treatment of enuresis in three patients. RCT involved increasing each child's fluid intake and reinforcing successively longer periods of urine retention. All three patients were continent after a maximum of two weeks of training and maintained treatment gains throughout a one-year follow-up. A systematic replication of these procedures was conducted using a larger number of patients and employing a delayed treatment control group (Paschalis, Kimmel, & Kimmel, 1972). Approximately 50% of 31 patients were continent following training and at a three-month follow-up.

Doleys, Ciminero, Tollison, Williams, and Wells (1977) compared dry-bed training (DBT) with retention control training (RCT) in the treatment of 19 enuretics. DBT was superior to RCT in decreasing the number of wets per week. These data were not supportive of earlier data reported by Kimmel and Kimmel (1970) and Paschalis *et al.* (1972). This study, as well as a case study by Doleys and Wells (1975), also suggested that for some patients increased functional bladder capacity was not a necessary prerequisite to nocturnal continence. Certain subjects in the DBT group actually showed a decrease in capacity while showing a reduction in the number of nightly wettings.

Dyssynergia and Neurogenic Bladder

Biofeedback procedures have recently been reported effective in the treatment of urinary incontinence associated with organic pathology in adults (Cardozo, Abrams, Stanton, & Feneley, 1978; Cardozo, Stanton, Hafner, & Allan, 1978; Wear, Wear, & Cleeland, 1979). Dyssynergia is typically a lack of muscle coordination of the urinary sphincter. Neurogenic bladder refers to involuntary, reflex contraction of the bladder upon filling. These studies are presented in Table I because of their potential application to work with children.

Cardozo and her colleagues (in both 1978 studies) reported the use of manometric feedback in the treatment of idiopathic bladder instability (a type of neurogenic bladder). Bladder instability is the most common cause of urgency incontinence and frequent micturition. The biofeedback training procedure consisted of inserting a catheter into the bladder and recording pressure during bladder filling. Auditory feedback of changes in bladder pressure was provided to the patient. The patient was instructed to inhibit bladder contraction, thereby increasing retention intervals. Following treatment, 12 of 27 patients no longer exhibited detrusor contractions, high amplitude pressure urges or urine leakage during cystometry.

Wear, Wear, and Cleeland (1979) also used biofeedback in the treatment of urinary retention and incontinence. Four of the eight patients in this study showed evidence of dyssynergia. Periurethral EMG feedback and intraurethral pressure feedback provided information about the contraction and relaxation of the external urethral sphincter. Patients with overly contracted sphincters and subsequent urine retention were taught relaxation of the sphincter. Patients with hypotonic sphincters associated with incontinence were taught to contract the sphincter. Four of the eight patients were reported to be moderately to markedly improved.

These few studies suggest that biofeedback procedures previously shown to be effective in conditioning other types of physiological responses may also be extended to the treatment of urological problems. Urinary incontinence in children with neurological impairments is common and is currently treated conservatively with surgery, medications, or catheterization. The development of less intrusive procedures based on a learning model would offer a significant contribution to the treatment of these populations. Further investigations of experimental treatments in this area are needed.

FECAL INCONTINENCE

Definition and Incidence

Fecal incontinence refers to defecation in inappropriate places at least once per week occurring after age 4 and enduring for at least one month (Doleys, 1979; Schaefer, 1979). Incontinence which has no clear physiological cause is referred to as *encopresis* and may be due to improper diet, toileting skills deficits, motivational problems, or other nonorganic factors. Fecal incontinence is also prevalent in neurologically impaired children in whom soiling is a result of a physiological abnormality such as evidenced in Hirschsprung's disease or spina bifida.

As in urinary incontinence, fecal incontinence can be classified as either primary or secondary. In primary fecal incontinence, toilet training has never been achieved, while in secondary incontinence bowel control has been attained and a relapse to soiling has occurred.

The incidence of encopresis is reported to be 1.5% in 7- to 8-year-old children (Schaefer, 1979). It occurs much less frequently than enuresis. Boys are more likely to be encopretic than are girls. The prognosis for untreated encopresis is good since this problem tends to disappear by adolescence. The incidence of fecal incontinence in neurological disorders is higher than that in the general population. For example, approximately 40% of children with meningomyelocele are fecally incontinent (Lorber, 1971).

Physiological Mechanisms

The anatomy of the anorectal area is shown in Figure 2. The major structures reponsible for continence and defecation are the rectum, the internal sphincter, and the external sphincter. The rectum is a reservoir approximately 12–15 cm in length. Fecal material is transported to the rectum through the intestinal tract. There it is stored until the rectal wall is distended with stool causing the subjective sensation of the urgency to defecate. This sensation is mediated by stretch receptors in the wall of the rectum which provide afferent information about rectal fullness. The rectum consists of smooth muscle. Its transmural thickness is greatest at its most distal end where it forms the concentrated ring of muscle known as the internal sphincter. In addition to the subjective sensation of urgency, rectal distension elicits a reflexive inhibition of the internal

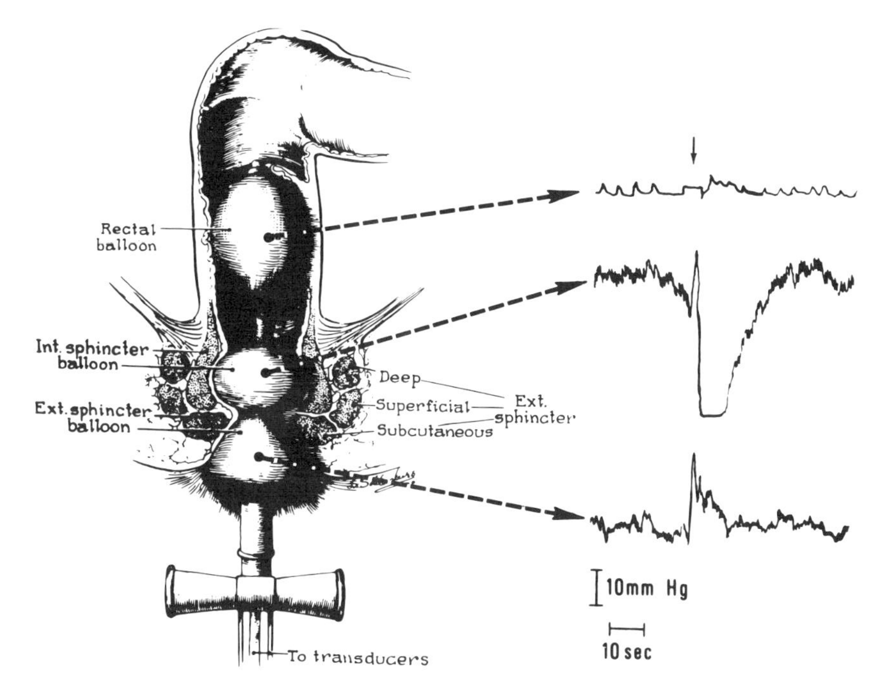

Figure 2. A diagram of the anorectal area. A measurement device, consisting of three pressure-sensitive balloons, is shown in relation to the sphincter muscles. A representative tracing of internal sphincter relaxation and external sphincter contraction following rectal distension (at the downward pointing arrow above the top tracing) is shown at right. From Engel (1978).

sphincter allowing the passage of fecal material into the anal canal. Inhibition of the internal sphincter is a myenteric reflex and thus not dependent on sensation of rectal distension (Schuster, 1968).

The external anal sphincter consists of striated muscle and is normally under voluntary control. The deeper layers of the external sphincter encircle the internal sphincter while the surface layer surrounds the terminal portion of the anal canal. The function of the external sphincter is to maintain continence by voluntary contraction during periods of reflexive inhibition of the internal sphincter (see Figure 2). Once the external sphincter is relaxed, defecation occurs.

The perianal skin and anal canal derive their sensory innervation from the pudendal nerves (S2, S3, S4), and the afferent pathways for the rectum are via the nervi erigentes also passing through the sacral portion of the spinal cord (S2, S3, S4). The external anal sphincter derives its efferent innervation from the pudendal nerve (S2, S3, S4) and the coccygeal plexus (S4, S5). The hypogastric nerves maintain high internal sphincter tone through sympathetic innervation from the lumbar cord (L4, L5). Parasympathetic innervation, supplied through the nervi erigentes (S1, S2, S3), serves to inhibit internal sphincter tone (Schuster, 1968).

As described above, the normal continence mechanism requires the discrimination of rectal distension and an ability to contract the external sphincter strongly for the duration of the relaxation of the internal sphincter. Either structural, sensory, or motor impairments in this system may result in fecal incontinence. The absence of these normal rectosphincteric responses is common in some neurological disorders (e.g., Hirschsprung's disease, spina bifida), and new treatment approaches for these disorders will be described later in this chapter.

Medical Approaches

Enemas, suppositories, laxatives, diet changes, and exercise have all been used to regulate bowel habits or relieve constipation in an effort to decrease the incidence of fecal soiling. These aids are used to eliminate impactions which may cause overflow incontinence. Impactions result in a loss of the normal urge to defecate due to the constant state of distension of the rectum, and a decrease in the muscle tone of the colon. Overflow incontinence refers to the escape of soft or liquid stool from the rectum through a distended and weakened internal sphincter.

The safest and most effective means of emptying the rectum is the use of enemas. This type of evacuation procedure works by distending the rectum and producing the stimulus for defecation. Water enemas are not irritating to the bowel and have the advantage of working immediately.

Suppositories are widely used and often considered by parents to be easier to administer than enemas. Glycerine suppositories act by softening and lubricating to stimulate evacuation. Unfortunately, the results are less reliable than those provided by enemas because this type of suppository may not work with children who have hard fecal impactions. Another frequently used suppository is bisacodyl (Ducolax). This suppository produces a more reliable effect than glycerine suppositories. However, it stimulates evacuation by irritating the bowel. A common side effect of frequent use is mucous production in the anal canal as a result of the chemical irritants in this product.

Laxatives include mineral oils, chemical irritants, and saline products. They are taken orally and are slow acting, requiring up to 12 hours to produce effects. Aside from their delayed action, laxatives have the disadvantage of relieving constipation at the expense of irritating the bowel lining. The child's system may become dependent on laxatives and loose its natural ability to evacuate waste via colonic motility. (Schaefer, 1979).

Proper diet can help in the prevention of constipation and has been used in lieu of medical regimens. Some foods, especially those high in fiber, have a natural laxative effect whereas others promote constipation due to their binding properties. Exercise has also been found to increase peristalsis and thus result in rapid elimination of waste (Schaefer, 1979).

Medication regimens, although frequently used, have not been extensively studied and reported in the literature. Levine and Bakow (1976) report an exceptionally thorough study of a pediatric intervention in which 51% of 127 encopretic children were continent following a multicomponent program which used enemas, suppositories, and laxatives in conjunction with counseling and toileting skills training. Other single case studies on the use of pharmacological agents have been reported (Halpern, 1977; Musicco, 1977). However, the comparative efficacy of each of these evacuation procedures has yet to be systematically evaluated.

Surgical procedures have been used as a treatment for incontinence in people with congenital abnormalities of the spinal cord or anorectal area and in patients who suffer from traumatic or operative injury of the external anal sphincter (Gross, Holcomb, & Swan, 1953). Current treatments in use include (1) repair of the existing sphincter, (2) plication procedures (explained below), (3) creation of functional slings, (4) muscle transplant, and (5) colostomies (Castro & Pittman, 1978).

The repair of the existing sphincter is done by suturing the separated ends of the sphincter so as to make it continuous. This is a difficult operation which often results in poor sphincter control. Plication procedures seek to achieve the same result but involve making a tuck in the muscle rather than rejoining the injured ends. This procedure is preferred by many surgeons and tends to have a more favorable outcome. The third type of repair, creation of functional slings, serves to control defecation by passing two sections of tissue (typically the fascia) around the anus and anchoring each end to the gluteus maximus. This allows for closure of the anus by contraction of the gluteal muscle. The fourth type of procedure, muscle transplant, allows for the reconstruction of a new sphincter using gracilis muscle tissue. Transplants were originally conceived in Russia for the treatment of neurogenic incontinence but are also applicable to other disorders (Hakami, Enamy, & Radi, 1976; Hakelius, Gierup, Grolte, & Jorulf, 1978). Colostomies, the final type of surgical intervention, are typically reserved as a last resort when less intrusive training procedures or less radical surgical methods have failed.

Behavioral Treatments

Several studies employing behavioral procedures in the treatment of fecal incontinence are presented in Table II. Fecal incontinence in children has received little attention in the research literature despite its high incidence and impact on social adjustment. The majority of these studies are single-case reports or reports involving fewer than 10 patients. Behavioral treatments have primarily been implemented in cases of encopresis. However, four studies reporting the successful treatment of fecal incontinence due to neurological impairment are also summarized in Table II.

Methodological rigor is generally lacking in treatment studies of fecal incontinence. None of the above studies reported the use of adequate single-subject or control group controls. Spontaneous cure rates, although not reported in the literature, are likely to be high considering the paucity of adolescent encopretics described in research studies. If encopresis can be conceptualized as a developmental disorder the incidence of which declines with age, delayed treatment control groups are especially needed to improve experimental design. Also absent are comparative studies examining the relative efficacy of behavioral treatments and medical management. It is not currently known which medical regimen is best nor how each compares with different types of behavioral intervention.

Objective dependent measures have been obtained in two thirds of the 18 summarized articles. Unfortunately, in many cases these data do not allow for empirical evaluation of treatment outcome because pretreatment levels of soiling and appropriate evacuations were not documented in 56% of the studies. The most commonly used dependent measure is soilings per unit of time. Although this measure is essential, a decrease in accidents does not necessarily imply that appropriate bowel movements are occurring. Fecal retention can occur without an increase in the frequency of evacuations. The use of multiple measures, including a measure of soiling, appropriate toiletings, and physiological responses (in the case of neurologically based incontinence) are needed for evaluation of treatment effectiveness.

Follow-up data are available for all but two studies in Table II. The length of the follow-up period has typically been adequate, from 2 months to 5 years, with the majority of studies monitoring the maintenance of treatment effects for at least 6 months.

Table II. Behavioral Treatment of Fecal Incontinence

Author	No.	Age range	Diagnosis	Procedure	Objective dependent measure	Baseline	Controls	Length of follow-up	Number continent post-treatment	Number continent at follow-up
Neale, 1963	4	7–10	Encopresis	Reinforcement of appropriate bowel movements	Yes	No	None	3–6 mo.	3/4	3/4
Gelber & Meyer, 1965	1	13	Encopresis	1) Reinforcement of appropriate bowel movements 2) Response cost	No	No	None	6 mo.	1/1	1/1
Keehn, 1965	1	5	Encopresis	Reinforcement of appropriate bowel movements	No	No	None	2 mo.	1/1	1/1
Lal & Lindsley, 1968	1	3	Encopresis	Contingent attention for appropriate bowel movements	Yes	Yes	None	8 mo.	1/1	1/1
Conger, 1970	1	9	Encopresis	Withdrawal of attention following soiling	Yes	Yes	None	3 mo.	1/1	1/1
Wagner & Paul, 1970	19	25–66	Encopresis (psychiatric populations)	Punishment of soilings	Yes	Yes	None	13 mo.	6/19 totally continent 8/19 daytime continent	13/13 totally continent 11/13 daytime continent
Edelman, 1971	1	12	Encopresis	Punishment & avoidance	Yes	Yes	None	3 mo.	0/1	0/1
Kohlenberg, 1973	1	13	Encopresis, decreased anal sphincter tone	Reinforcement of increased pressure recorded in anal sphincter area	Yes	No	None	12 mo.	0/1	0/1

Table II. (Continued)

Author	No.	Age range	Diagnosis	Procedure	Objective dependent measure	Baseline	Controls	Length of follow-up	Number continent post-treatment	Number continent at follow-up
Young, 1973	24	4–10	Encopresis	Conditioned gastroileal reflex training	No	No	None	6 mo.–5 yr.	22/24	18/24
Engel, Nikoomanesh, & Schuster, 1974	7	6–54	Mixed organic incontinence	Manometric biofeedback	Yes (sample)	Yes (pretest)	None	6 mo.–5 yr.	4/7	4/7
Ayllon, Simon, & Wildman, 1975	1	7	Encopresis	Reinforcement for accident-free periods	Yes	Yes	None	11 mo.	1/1	1/1
Ashkenazi, 1975	18	3–12	Encopresis (1 child with Hirschsprung's)	Glycerine suppositories and positive reinforcement	No	No	None	6 mo.	16/18	16/18
Bach & Moylin, 1975	1	6	Encopresis & urinary incontinence	1) Reinforcement of appropriate expulsion 2) Extinction of accidents	Yes	Yes	None	24 mo.	1/1	1/1
Doleys & Arnold, 1975	1	16	Encopresis	Full cleanliness training & positive reinforcement	Yes	Yes	None	6 mo.	1/1	0/1

Christophersen & Rainey, 1976	6	5–12	Encopresis	1) Positive reinforcement for appropriate toiletings 2) Use of suppositories & enemas 3) Cleanliness training 4) Fading medications	Yes	Yes	None	2–12 mo.	3/6 (3 still in treatment	3/6
Epstein & McCoy, 1977	1	3	Hirschsprung's disease (postsurgery)	Modified Foxx & Azrin (1973) procedure	Yes	Yes	None	4.5 mo.	1/1	1/1
Cerulli, Nikoomanesh, & Shuster, 1979	50	6–97	Postanal surgery (15) Postrectal surgery (9) Postspinal surgery (11)	Manometric biofeedback	No	No	None	4 mo.–9 yr.	20/50	unclear
Olness, McParland, & Piper, 1980	50	4–18	Imperforate anus ($n = 10$) chronic constipation ($n = 40$)	Manometric biofeedback	No	No	None	None	6/10 with imperforate anus 24/40 with functional constipation	—
Whitehead, Parker, Masek, Cataldo, & Freeman (1981)	8	5–15	Meningomyelocele	Manometric biofeedback	Yes	Yes	None	13–24 mo.	5/8	3/8

Encopresis

Treatments for encopresis have involved two types of approaches. The first type of intervention has used positive reinforcement or punishment to alter the frequency of accidents or appropriate toiletings. The second type of intervention is a more comprehensive approach involving a combination of behavioral procedures usually with the addition of evacuation aids, such as enemas, laxatives, or suppositories.

Ayllon, Simon, and Wildman (1975), Bach and Moylin (1975), Conger (1970), Edelman (1971), Gelber and Meyer (1965), Keehn (1965), Lal and Lindsley (1968), and Neale (1963) have all employed the first type of intervention either to increase appropriate bowel movements or to decrease soilings. Positive reinforcement for appropriate bowel movements has been the most widely used procedure in this group of studies (Bach & Moylin, 1975; Gelber & Meyer, 1965; Keehn, 1965; Lal & Lindsley, 1968; Neale, 1963). Unlike the above studies, Ayllon *et al.* (1975) reinforced periods without accidents by performing periodic pants checks to determine whether the pants were clean or soiled. One study (Conger, 1970) used extinction of soiling as the primary treatment, whereas two others relied either partially or totally on punishment procedures (Gelber & Meyer, 1965; Edelman, 1971). Edelman (1971) employed both a punishment and avoidance contingency to decrease soiling in a 12-year-old child. Punishment consisted of a 30-minute confinement to her bedroom following accidents. Avoidance of dishwashing was contingent on accident-free periods.

The second type of procedure involving a more comprehensive approach has been used by Ashkenazi (1975), Christophersen and Rainey (1976), Doleys and Arnold (1975), and Young (1973). One such program, full cleanliness training, is described by Doleys and Arnold (1975) in the successful treatment of an 8-year-old encopretic. This procedure involved shaping toileting skills, the use of a mild laxative to increase the frequency of bowel movements, reinforcement of "clean" pants checks, reinforcement of appropriate bowel movements, and discrimination training to develop subjective sensation of rectal fullness.

Another type of comprehensive training procedure used to treat constipation employs enemas or suppositories to establish a conditioned evacuation response (Young, 1973; Christophersen & Rainey 1976). The bowel is first emptied by means of enemas, suppositories or laxatives. The child is then placed on a bowel training regimen. He is instructed to sit on the toilet immediately preceeding or following some specified meal. If the child has a bowel movement, he is reinforced and given no enema or suppository. All unsuccessful attempts to evacuate are followed immediately by an enema or suppository which then results in a stimulated bowel movement. By pairing the meal and evacuation, the meal becomes a conditioned stimulus which sets the occasion for a bowel movement. Young's (1973) procedure, which requires the child to attempt a bowel movement following a meal, also takes advantage of the rectal motility stimulated by a meal, thereby increasing the probability of defecation. The use of suppositories or enemas also decreases the likelihood of accidents during the day in addition to increasing the sensitivity of the rectum by keeping the bowel relatively empty. Christophersen and Rainey (1976) suggest the use of the full cleanliness training procedure in conjunction with suppository or enema use. Daily suppositories or enemas can be faded after continence is demonstrated.

Young (1973) reported a variation of this procedure in a sample of 24 patients. Nineteen of these patients were successfully treated in an average of 5 months. Three other patients responded but required more than a year of treatment. Only four patients had relapsed after approximately $1\frac{1}{2}$ years of follow-up.

Fecal Incontinence Due to Neurological Impairment

Over the past 6 years, behavioral approaches have also been reported in the treatment of fecal incontinence due to neurological impairment. A high incidence of fecal incontinence occurs in injuries and birth defects including sphincterotomy in Hirschsprung's disease, meningomyelocele, diabetic neuropathy, and repair of imperforate anus. Preliminary studies using operant conditioning (Kohlenberg, 1973; Epstein & McCoy, 1977) and biofeedback training (Cerulli, Nikoomanesh, & Schuster, 1979; Engel, Nikoomanesh, & Schuster, 1974; Olness, McParland, & Piper, 1980; Whitehead, Parker, Masek, Cataldo, & Freeman, 1981) have reported the successful treatment of adults and children with incontinence secondary to these disorders.

Kohlenberg (1973) reported the modification of external anal sphincter tone in a 13-year-old child following surgical treatment for Hirschsprung's disease. Pressure changes from a fluid-filled balloon were displayed to the child using a water column. Increases in pressure were reinforced with money. Implementation of the reinforcement contingency resulted in an increase in anal sphincter pressure but did not produce continence.

Esptein and McCoy (1977) treated a child after sphincterotomy for Hirschsprung's disease using a standard Foxx and Azrin (1973) training procedure. This child gained both bowel and bladder control following 19 weeks of treatment. Although contingencies were placed only on urine accidents, soilings decreased as well. The authors imply that covariation between these two physiological responses may result in an economy of training time in cases of double incontinence.

Biofeedback training of patients with fecal incontinence due to neurological dysfunction was first reported by Engel *et al.* (1974). A hollow tube surrounded by two balloons was inserted into the anal canal to measure pressure changes in the internal and external anal sphincters. A third balloon was inserted through the tube into the rectum. Training was designed to teach patients sensation of bowel fullness. This was done by distending the rectum with large volumes of air and teaching patients to contract the external anal sphincter to the stimulus of rectal distension. Pressure feedback of both anal sphincters was provided to the patient by means of an ongoing polygraph record. After the patient had learned to make voluntary contractions of the external sphincter following a 50-ml rectal distension, the volume of the distension was gradually reduced. Four of seven treated patients were continent following this procedure.

Cerulli *et al.* (1979), using this same procedure, treated a larger number of fecally incontinent adults following anal, rectal or spinal surgery. Seventy-two percent of these patients achieved at least a 90% reduction in the amount of incontinence. Relapses occurred in two trained patients. However, both were successfully reconditioned. Olness *et al.* (1980) have also applied the Engel *et al.* procedure to children with functional encopresis and repair of imperforate anus. Of primary interest is that 6 of the 10 children with imperforate anus achieved continence.

Although biofeedback procedures have been used to treat a variety of fecally incontinent patients with neurological disorders, cases of incontinence due to meningomyelocele have not been successfully treated in the above studies. Meningomyelocele is a subtype of spina bifida and refers to a developmental defect of the spinal column due to a failure of fusion between the arches of the vertebrae. This condition results in spinal anomalies and neurological signs including bowel and bladder dysfunction (Henderson & Synhorst, 1977). In approximately 40% of these patients, external anal sphincter tone is either absent or markedly impaired, resulting in varying degrees of fecal incontinence. In a preliminary study by Whitehead *et al.* (1981), the Engel *et al.* (1974) procedure was altered to provide response-shaping rather than stimulus-discrimination training. Eight fecally incontinent children with meningomyelocele were shown a polygraph tracing of external anal sphincter contraction and reinforced for contractions when the rectum was distended with progressively larger volumes of air in the rectal balloon.

Despite evidence of spinal cord dysfunction and in contrast to adult-onset fecal incontinence, these patients showed normal sensation for rectal distension. Following 6 biofeedback training sessions, 5 patients were continent and a sixth patient showed an 80% reduction in the frequency of incontinence. At a one- to two-year follow-up, four children were having accidents once per month or less often.

SUMMARY AND CONCLUSION

Several behavioral approaches have been effective in the treatment of incontinence. Urinary incontinence occurs more frequently in children than fecal incontinence and has received more attention in the behavioral literature despite the severe social disability which accompanies fecal soiling.

Enuresis, which is generally not associated with organic impairment, has been the most commonly studied type of urinary incontinence. Mowrer and Mowrer's (1938) bell-and-pad procedure for nocturnal enuresis has resulted in a 75%–90% remission with a 30% relapse rate. This procedure is still the only treatment for nocturnal enuresis which has been systematically studied in a number of comparative group designs. Results of drug treatments have not been encouraging, most providing temporary relief with wetting reoccurring once drugs are withdrawn. The dry-bed procedure (Foxx & Azrin, 1973) has been effective in preliminary studies but requires further documentation.

Contrary to clinical lore, our current knowledge base for the treatment of encopresis is extremely limited. Simple behavioral procedures have been implemented with single cases, but there is a need for well-controlled group studies. Comprehensive approaches, like the one used by Young (1973), are more likely to deal effectively with the complexities of fecal incontinence. These approaches typicaly consolidate operant procedures, principles of gastrointestinal physiology, and medical management techniques into a multicomponent treatment package.

Bladder and bowel dysfunctions secondary to neurological disorders have been successfully treated using biofeedback. Cardozo *et al.* (Cardozo, Abrams, Stanton, & Feneley, 1978; Cardozo, Stanton, Hafner, & Allan, 1978) and Wear *et al.* (1979) demonstrated the efficacy of pressure feedback in the treatment of neurogenic bladder and dyssynergia. Engel *et al.* (1974), Cerulli *et al.* (1979), Olness *et al.* (1980), and Whitehead *et al.* (1981) effectively treated fecally incontinent children and adults with a wide variety of structural and neurological disorders. Development of biofeedback procedures are important in that they may offer a less intrusive therapeutic alternative to traditional surgical and pharmacological treatments now in use with these patients.

Both medical and behavioral interventions have been used with varying degrees of success in the treatment of incontinence. Behaviorists have contributed expertise in the objective assessment of environmental variables contributing to these disorders. Operant conditioning approaches have been successful in modifying inappropriate responses and correcting skills deficits by the application of learning principles. Medical professionals have provided knowledge of bladder and bowel physiology and have developed pharmalogical and surgical procedures to keep children free of accidents.

Future studies could profit from the integration of the independent contributions of behavioral and medical research. Past investigations have often been constricted to a set of methods which were thought to be applicable to all incontinent children. Collaboration between physicians and psychologists may result in greater familiarity with different types of incontinence and lead to concentrated study of specific disorders.

REFERENCES

Agarwala, S., & Heycock, J.B. A controlled trial of imipramine (Tofranil) in the treatment of childhood enuresis. *British Journal of Clinical Practice*, 1968, *22*, 296–298.

Ashkenazi, Z. The treatment of encopresis using a discriminative stimulus and positive reinforcement. *Journal of Behavior Therapy and Experimental Psychiatry*, 1975, *6*, 155–157.

Ayllon, T., Simon, S. J., & Wildman, R. W. Instructions and reinforcement in the elimination of encopresis: A case study. *Journal of Behavior Therapy and Experimental Psychiatry*, 1975, *6*, 235–238.

Azrin, N. H., & Foxx, R. M. Dry pants: A rapid method of toilet training children. *Behaviour Research and Therapy*, 1973, *4*, 435–442.

Azrin, N. H., & Foxx, R. M. *Toilet training in less than a day*. New York: Simon & Schuster, 1974.

Azrin, N. H., & Thienes, P. M. Rapid elimination of enuresis by intensive learning without a conditioning apparatus. *Behavior Therapy*, 1978, *9*, 342–354.

Azrin, N. H., Sneed, T. J., & Foxx, R. M. Dry bed: A rapid method of eliminating bedwetting (enuresis) of the retarded. *Behaviour Research and Therapy*, 1973, *11*, 427–434.

Azrin, N. H., Sneed, T. J., & Foxx, R. M. Dry-bed training: Rapid elimination of childhood enuresis. *Behaviour Research and Therapy*, 1974, *12*, 147–156.

Bach, R., & Moyland, J. J. Parents administer behavior therapy for inappropriate urination and encopresis: A case study. *Journal of Behavior Therapy and Experimental Psychiatry*, 1975, *6*, 239–241.

Blackwell, B., & Currah, J. The psychoparmacology of nocturnal enuresis. *Clinical Developmental Medicine*, 1973, *48/49*, 231–257.

Bloomfield, J. M., & Douglas, J. Enuresis: Prevalence among children aged 4–7 years. *Lancet*, 1956, *1*, 850–852.

Breger, E. Hydroxyzine hydrochloride and methylphenidate in enuresis, *Journal of Pediatrics*, 1962, *61*, 443–447.

Browning, R. M. Operantly strengthening UCR (awakening) as a prerequisite to the treatment of persistent enuresis. *Behaviour Research and Therapy*, 1967, *5*, 371–372.

Butler, J. F. The toilet training success of parents after reading *Toilet Training in Less than a Day*. *Behavior Therapy*, 1976, *7*, 185–191.

Campbell, M. F. *Clinical pediatric urology*. Philadephia: Saunders, 1951.

Cardozo, L. D., Abrams, P. D., Stanton, S. L., & Feneley, R. C. L. Idiopathic bladder stability treated by biofeedback. *British Journal of Urology*, 1978, *50*, 521–523.

Cardozo, L., Stanton, S. L., Hafner, J., & Allan, V. Biofeedback in the treatment of detrusor instability. *British Journal of Urology*, 1978, *50*, 250–254.

Castro, A. F., & Pittman, R. E. Repair of the incontinent sphincter. *Disorders of the Colon and Rectum*, 1978, *21*, 183–187.

Cerulli, M. A., Nikoomanesh, P., & Schuster, M. M. Progress in biofeedback conditioning for fecal incontinence. *Gastroenterology*, 1979, *76*, 742–746.

Christophersen, E. R., & Rainey, S. K. Management of encopresis through a pediatric outpatient clinic. *Journal of Pediatric Psychology*, 1976, *4*(1), 38–41.

Conger, J. C. The treatment of encopresis by the management of social consequences. *Behavior Therapy*, 1970, *1*, 386–390.

Creer, T. L., & Davis, M. H. Using a staggered-wakening procedure with enuretic children in an institutional setting. *Journal of Behavior Therapy and Experimental Psychiatry*, 1975, *6*, 23–25.

Doleys, D. M. Behavioral treatments for nocturnal enuresis in children: A review of the recent literature. *Psychological Bulletin*, 1977, *84*, 30–54.

Doleys, D. M. Assessment and treatment of childhood enuresis. *In* A. J. Finch & D. C. Kendall (Eds.), *Clinical treatment and research in child psychopathology*. New York: SP Medical & Scientific Books, 1979.

Doleys, D. M., & Arnold, S Treatment of childhood encopresis: Full cleanliness training. *Mental Retardation*, 1975, *13*, 14–16.

Doleys, D. M., & Wells, K. C. Changes in functional bladder capacity and bed-wetting during and after retention control training: A case study. *Behavior Therapy*, 1975, *6*, 685–688.

Doleys, D. M., Ciminero, A. R., Tollison, J. W., Williams, C. L., & Wells, K. C. Dry-bed training and retention control training: A comparison. *Behavior Therapy*, 1977, *8*, 541–548.

Edelman, R. I. Operant conditioning treatment of encopresis. *Journal of Behavior Therapy and Experimental Psychiatry*, 1971, *2*, 71–73.

Engel, B. T. The treatment of fecal incontinence by operant conditioning. *Automedica*, 1978, *2*, 101–108.

Engel, B. T., Nikoomanesh, P., & Schuster, M. M. Operant conditioning of rectosphincteric responses in the treatment of fecal incontinence. *New England Journal of Medicine*, 1974, *290*, 646–649.

Epstein, S. J., & Guilfoyle, F. M. Imipramine (Tofranil) in the control of enuresis. *American Journal of Diseases of Children*, 1965, *109*, 412–415.

Epstein, L. H., & McCoy, J. F. Bladder and bowel control in Hirschsprung's disease. *Journal of Behavior Therapy and Experimental Psychiatry*, 1977, *8*, 97–99.

Forrester, R. M., Stein, Z., & Susser, M. W. A trial of conditioning therapy in nocturnal enuresis. *Developmental Medicine and Child Neurology*, 1964, *6*, 158–166.

Forsythe, W. I., & Redmond, A. Enuresis and spontaneous cure rate: Study of 1129 enuretics. *Archives of Disease in Childhood*, 1974, *49*, 259–276.

Foxx, R. M., & Azrin, N. H. *Toilet training the retarded.* Champaign, Ill.: Research Press, 1973.

Gelber, H., & Meyer, V. Behavior therapy and encopresis: The complexities involved in treatment. *Behaviour Research and Therapy,* 1965, *2,* 227–231.

Gross, R. E., Holcomb, G. W., & Swan, H. Treatment of neurogenic urinary and fecal incontinence in children. *Archives of Surgery,* 1953, *66,* 143,–154.

Hakami, M., Enamy, H., & Radi, H. A. Vastus internus muscle transplant for the correction of anal incontinence. *The American Journal of Proctology,* August 1976, 50–52.

Hakelius, L., Gierup, J., Grotte, G., & Jorulf, H. A new treatment of anal incontinence in children: Free autogenous muscle transplantation. *Journal of Pediatric Surgery,* 1978, *13,* 77–82.

Halpern, W. I. The treatment of encopretic children. *Journal of Child Psychiatry,* 1977, *16,* 478–499.

Hannigan, K. F. Teaching intermittent self-catheterization to young children with myelodysplasia. *Developmental Medicine and Child Neurology,* 1979, *21,* 365–368.

Harrington, M. Phenometrazine in the treatment of nocturnal enuresis. *Practitioner,* 1960, *185,* 343–346.

Henderson, M. L., & Synhorst, D. M. Bladder and bowel management in the child with myelomeningocele. *Pediatric Nursing,* September/October 1977, 24–31.

Keehn, J. D. Brief case-report: Reinforcement therapy of incontinence. *Behaviour Research and Therapy,* 1965, *2,* 239–240.

Kempe, C. H., & Helfer, R. E. *Helping the battered child and his family.* Oxford, England: Lippincott, 1972.

Keuhnelian, J. C., & Sanders, V. E. *Urologic nursing.* London: Macmillan, 1970.

Kimmel, H. D., & Kimmel, E. An instrumental conditioning method for the treatment of enuresis. *Journal of Behavior Therapy and Experimental Psychiatry,* 1970, *1,* 121–123.

Kohlenberg, R. J. Operant conditioning of human anal sphincter pressure. *Journal of Applied Behavior Analysis,* 1973, *6,* 201–208.

Lal, H., & Lindsley, O. R. Therapy of chronic constipation in a young child by rearranging social contingencies. *Behaviour Research and Therapy,* 1968, *6,* 484–485.

Lapides, J., Sweet, R. B., & Lewis, L. W. Role of striated muscle in urination. *The Journal of Urology,* 1957, *77,* 247–250.

Laurence, K. M., & Beresford, A. Continence, friends, marriage, and children in 51 adults with spina bifida. *Developmental Medicine and Child Neurology,* 1975, *17* (35), 123–128.

Levine, M. D., & Bakow, H. Children with encopresis: A study of treatment outcome. *Pediatrics,* 1976, *58,* 845–852.

Lister, J., Zachary, A. B., & Brereton, R. Open myelomeningocele: A ten year review of 200 consecutive cases. *Progress in Pediatric Surgery,* 1977, *10,* 161–176.

Lorber, J. Results of treatment of myelomeningocele. *Developmental Medicine and Child Neurology,* 1971, *13,* 279–303.

Lovibond, S. H. The mechanism of conditioning treatment of enuresis. *Behavior Research and Therapy,* 1963, *1,* 17–21.

MacLean, R. E. Imipramine hydrochloride (Tofranil) and enuresis. *American Journal of Psychiatry,* 1960, *117,* 551.

Mahony, D. T., Laferte, R. O., & Mahoney, J. E. Observations on sphincter augmenting effect of imipramine in children with urinary incontinence (Part VI). *Urology,* 1973, *1,* 317–323.

Marchant, D. J. Urinary incontinence in the female. *Hospital Medicine,* March 1977, 60–91.

McConaghy, N. A controlled trial in imipramine, amphetamine, pad-and-bell conditioning and random awakening in the treatment of nocturnal enuresis. *Medicine Journal of Australia,* 1969, *2,* 237–239.

Miller, P. M. An experimental analysis of retention control training in the treatment of nocturnal enuresis in two institutionalized adolescents. *Behavior Therapy*, 1973, *4*, 288–294.

Mowrer, O. H., & Mowrer, W. M. Enuresis: A method for its study and treatment. *The American Journal of Orthopsychiatry*, 1938, *8*, 436–459.

Muellner, S. R. Development of urinary control in children: A new concept in cause, prevention, and treatment of primary enuresis. *The Journal of Urology*, 1969, *84*, 714–716.

Musicco, N. Encopresis: A good result in a boy with UTP (uridine–5–triphosphate). *American Journal of Proctology*, 1977, *28*, 43–46.

Neale, D. H. Behavior therapy and encopresis in children. *Behaviour Research and Therapy*, 1963, *1*, 139–149.

Olness, K., McParland, M. D., & Piper, J. Biofeedback: A new modality in the management of children with fecal soiling. *Pediatrics*, 1980, *96*, 505–509.

Oppel, W. C., Harper, P. A., & Rider, R. V. The age of attaining bladder control. *Pediatrics*, 1968, *42*, 614–628.

Paschalis, A., Kimmel, H. D., & Kimmel, E. Further study of diurnal instrumental conditioning in treatment of enuresis nocturna. *Journal of Behavior Therapy and Experimental Psychiatry*, 1972, *3*, 253–256.

Pearne, D. H., Zigelbaum, S. D., & Peyser, W. P. Biofeedback-assisted EMG relaxation for urinary retention and incontinence: A case report. *Biofeedback and Self-Regulation*, 1977, 213–217.

Poussaint, A. F., & Ditman, K. S. A controlled study of imipramine (Tofranil) in the treatment of childhood enuresis. *The Journal of Pediatrics*, 1965, *67* (2), 283–290.

Rossier, A. B. Classification and management of the neurogenic bladder. *Resident & Staff Physician*, 1978, *24* 105–112.

Safer, D. Drug treatment in child psychiatry. In L. L. Iversen, S. D. Iversen, & S. H. Snyder (Eds.), *Handbook of psychopharmacology*. New York: Plenum Press, 1978,

Samaan, M. The control of nocturnal enuresis by operant conditioning. *Journal of Behavior Therapy and Experimental Psychiatry*, 1972, *3*, 103–105.

Schaefer, C. E. *Childhood encopresis and causes and therapy*. New York: Van Nostrand Reinhold, 1979.

Schuster, M. M. Motor action of rectum and anal sphincters in continence and defecation. In C. F. Code (Ed.), *Handbook of physiology: Alimentary canal IV*. Washington, D. C.: American Physiological Society, 1968.

Stedman, J. M. An extension of the Kimmel treatment method for enuresis to an adolescent: A case report. *Journal of Behavior Therapy and Experimental Psychiatry*, 1972, *3*, 307–309.

Taylor, D. W. Treatment of excessive frequency of urination by desensitization. *Journal of Behavior Therapy and Experimental Psychiatry*, 1972, *3*, 311–313.

Tough, J. H., Hawkins, R. P., McArthur, M. M., & Ravenswaay, S. Modification of enuretic behavior by punishment: A new use for an old device. *Behavior Therapy*, 1971, *2*, 567–574.

Turner, R. K., & Young, G. C. CNS stimulant drugs and conditioning treatment of nocturnal enuresis: A long-term follow-up study. *Behaviour Research and Therapy*, 1966, *4*, 225–228.

Turner, R. K., Young, G. C., & Rachman, S. Treatment of nocturnal enuresis by conditioning techniques. *Behaviour Research and Therapy*, 1970, *8*, 367–381.

Wagner, B. R., & Paul, G. L. Reduction of incontinence in chronic mental patients: A pilot project. *Journal of Behavior Therapy and Experimental Psychiatry*, 1970, *1*, 29–38.

Wallace, I. R., & Forsythe, W. I. The treatment of eneuresis: A controlled clinical trial of propanthelien, propanthelien-and-phenobarbitone, and a placebo. *British Journal of Clinical Practices*, 1969, *23*, 207–215.

Wear, J. B., Wear, R. B., & Cleeland, C. Biofeedback in urology using urodynamics: Preliminary observations. *The Journal of Urology,* 1979, *121,* 464–468.

Werry, J. S., & Aman, M. G. Methylphenidate and haloperidol in children: Effects on attention, memory, and activity. *Archives of General Psychiatry,* 1975, *32,* 790–795.

Whitehead, W. E., Parker, L. H., Masek, B. J., Cataldo, M. F., & Freeman, J. M. Biofeedback treatment of fecal incontinence in patients with myelomeningocele. *Developmental Medicine and Child Neurology,* 1981, *23,* 313–322.

Wilson, T. S. Incontinence of urine in the aged. *The Lancet,* September 1948, 374–377.

Young, C. G. The treatment of childhood encopresis by conditioned gastro-ileal reflex training. *Behaviour Research and Therapy,* 1973, *11,* 499–503.

Young, C. G., & Morgan, R. T. T. Overlearning in the conditioning treatment of enuresis. *Behaviour Research and Therapy,* 1972, *10,* 147–151.

Young, C. G., & Turner, R. K. CNS stimulant drugs and conditioning treatment of nocturnal enuresis. *Behaviour Research and Therapy,* 1965, *3,* 93–101.

III

PAIN, DISCOMFORT, AND STRESS SECONDARY TO DISEASE AND MEDICAL PROCEDURES

6

Behavioral and Neurochemical Aspects of Pediatric Pain

James W. Varni, Ernest R. Katz, and Jerry Dash

INTRODUCTION

Pain represents a complex psychophysiological phenomenon involving cognitive, neurochemical, sensory, affective, emotional, and motivational components which act in synergistic fashion to produce varying degrees of intensity perception and reactions differing across individuals and socioenvironmental conditions (Beecher, 1959; Melzack, 1973; Merskey, 1970; Varni, 1981a, 1981b). Pain may be described along the dimensions of quality (*dull, sharp, burning*), anatomical location and duration and within a threshold/tolerance perspective (*unbearable*). The verbal expression of pain along these linguistic dimensions represents a developmental learning process, whereby the child learns to label nociceptive impulses in the language of his/her particular cultural environment.

Fabrega and Tyma (1976) have delineated what constitutes the structure and content of verbal descriptions of the pain experience, tracing the historical roots of primary pain terms (e.g., *ache, hurt, sore*), which

---

JAMES W. VARNI • Orthopaedic Hospital; University of Southern California School of Medicine, Los Angeles, California 90007. ERNEST R. KATZ • Childrens Hospital of Los Angeles; University of Southern California School of Medicine, Los Angeles, California 90027. JERRY DASH • Childrens Hospital of Los Angeles; University of Southern California School of Medicine, Los Angeles, California 90027.

are basic terms generally restricted to the pain experience; secondary pain terms (e.g., *sharp, burning, jabbing*), whose principal function is to qualify the primary pain terms; and tertiary pain terms (e.g., *deep, intense, mild, depressing*), which further elaborate on the pain experience but do not bear a special connection to pain *per se,* but which are often used to qualify any subjective experience. The authors point out that the term *pain* represents an abstract concept which economically describes a multiplicity of sensations of various etiologies influenced by cultural and social factors.

From a philosophical point of view, Degenaar (1979) notes that the term *pain* is derived from the Latin *poena,* which means punishment. In the philosophical context, the significance of pain as suffering or as punishment for a moral transgression differentiates it from the neurophysiologic detection of tissue damage (nociception). Thus, the philosophical meaning of the pain experience represents a further dimension in the individual's detection and expression of nociceptive impulses (Degenaar, 1979). For the young child in the preabstract stage of cognitive development, the term *pain* may be meaningless. Rather, nociceptive impulses may be viewed as temporally noncontingent punishment for a perceived transgression from parental standards, further adding to the complexity of the pain experience and the subsequent communicative intent of verbal pain descriptors.

ACUTE AND CHRONIC PAIN DIMENSIONS

Although there are a number of pediatric conditions in which a differentiation between acute and chronic pain may be simply a matter of semantics rather than a distinct delineation (e.g., frequent and recurrent painful hemorrhages in the hemophiliac), there are a number of dimensions which can facilitate a differential assessment and subsequent intervention in the pediatric population.

Acute pain serves as an adaptive biological warning signal, directing attention to an injury or disease, acting as a deterrent against harmful stimuli, and signaling the necessity for immobilization and protection of an injured area. The various reactions of a child's explorations and subsequent burns from a hot stove or match illustrate these adaptive functions. From a disease perspective, acute pain signals the need for a diagnosis of the underlying pathological process causing the pain, such as acute hemorrhage, which may result from a number of diseases and require immediate and appropriate treatment. In these cases, pain represents an abstract term serving as a symptom of a particular organic pathology. However, the pain sensation may be considerably greater than necessary; that is, the intensity of the pain may supercede or be disproportionate to its functional intent as a signaling stimulus. Although neurophysiological mechanisms may differentiate acute and chronic pain (Bonica, 1977; Dennis & Melzack, 1977), it is precisely the severe intensity of pain and its associated anxiety which may most parsimoniously distinguish acute and chronic pain (Varni, 1981b). Thus, the experience of pain contains two components, the original sensation and the reaction to this sensation. This reactive component represents the fearful response which can modulate the pain sensation and in acute pain often serves to intensify the reaction to noxious stimulation.

In chronic pain, the fearful component is lacking or greatly diminished, as the patient evidences an adaptive response to the initial acute experience, with the distinguishing features characterized by chronic pain behaviors, depressed mood, or inactivity. These chronic pain reactions can become reinforced independently of the original nociceptive impulses and tissue damage by socioenvironmental influences (Varni, Bessman, Russo, & Cataldo, 1980). The possibilities for narcotic dependence become greater because of this chronicity, which only further maintains the pain reaction (Fordyce, 1976; Varni & Gilbert, in press). Eventually, chronic pain reactions may become completely independent of the underlying organic pathology and persist even after the pathogenic factor has resolved (Bonica, 1977). This stands in marked contrast to the acute pain reaction, which appears to be more closely associated with the underlying pathology. These dimensions will become clearer in the sections on acute and chronic pain mangement.

PREVALENCE

Four categories of pain may be delineated: pain associated with an observable physical injury or trauma (burns, lacerations, fractures); pain associated with a disease state (hemophilia, sickle cell anemia, arthritis); pain not associated with a disease state or identifiable physical trauma (recurrent abdominal pain syndrome, tension headaches); and pain associated with medical procedures (lumbar punctures, surgery, bone marrow aspirations). Whereas the prevalence of pain secondary to physical injury, disease processes, or medical procedures is directly linked to and is studied in relationship to these particular conditions, the prevalence of pediatric pain of unknown etiology (idiopathic pain) has also been systematically studied.

Øster (1972) studied the prevalence rates of recurrent abdominal pain, headache, and limb pains in an 8-year longitudinal study of 18,162 school-age children and adolescents between 6 and 19 years of age. The prevalence of recurrent headaches was 20.6%; of recurrent abdominal pain, 14.4%; and of recurrent limb or "growing"pain, 15.5%, with a higher prevalence in all three classes of pain for girls than for boys. From a prognostic perspective, Christensen and Mortensen (1975) conducted a 28-year follow-up investigation of 34 patients who were initially diagnosed during childhood as evidencing recurrent abdominal pain (RAP) in comparison to a control group without childhood RAP. The average age of the clinical and control populations at follow-up was 35.9 and 36.4 years, respectively. At follow-up, 53% of the clinical group reported gastrointestinal pain consistent with a diagnosis of irritable colon syndrome, peptic ulcer/gastritis, and duodenal ulcer, as well as milder symptoms such as diarrhea, constipation, and meteorism (abdominal or intestinal distention by gas). Of considerable interest from a behavioral-medical perspective, 89% reported that these symptoms were provoked by stress. In the control group, 29% reported gastrointestinal pain as adults. Nongastrointestinal pain symptoms were also more frequent among the clinical patients (32%) than among the controls (13%), with such complaints as tension and migraine headaches, back pain, and gynecological pain. There were no differences between the clinical and control groups in the incidence of abdominal pain in their children (average age, 8.5 and 9.3 years, respectively). However, in combining the two groups, 28% of those children whose parents were complaining of abdominal pain at the time of the follow-up evidenced RAP, whereas only 7% of those children whose parents had no abdominal pain at follow-up reported RAP.

These findings are consistent with other investigations (Øster, 1972; Stone & Barbero, 1970), which have shown that the children of adults with abdominal pain are more likely to suffer from abdominal pain themselves than are children whose parents do not demonstrate such symptoms, even though only an estimated 5% of pediatric recurrent abdominal pain has an organic etiology (Maddison, 1977). When taken together, these findings suggest the potential role of socioenvironmental learning and modeling of parental pain behavior on the occurrence of pediatric abdominal pain complaints. In further support, Øster (1972) obtained data indicating that parent's frequent manifestations of pain may be a precipitating factor or discriminative stimulus for their children's painful experiences, delineating recurrent headache and abdominal and limb pains as the most common pediatric complaints secondary to observed parental pain behavior.

PAIN ASSESSMENT

Assessment of pediatric pain involves not only verbal reports of pain intensity and location, but also such nonverbal expressions as facial grimaces, compensatory posturing, restricted movement, limping, and the absence of developmentally appropriate behaviors. Additionally, the potential socioenvironmental factors which may influence the frequency of verbal and nonverbal pain behaviors (Varni *et al.*, 1980), as well as the delineation of the relative contribution of physiological and medical parameters to the pain experience, represent essential components of the assessment process.

Fordyce, Fowler, Lehmann, and DeLateur (1968) initially suggested the objective assessment of verbal and nonverbal pain behaviors in the management of adult chronic pain patients. This initial work has also been extended to the assessment of pain behaviors in pediatric populations (Miller & Kratochwill, 1979; Sank & Biglan, 1974; Varni *et al.*, 1980). In this model, verbal pain behaviors include the objective assessment of the frequency of such statements as "It hurts," "Ouch," "I have a stomachache." Although such an objective assessment allows for the determination of potential verbal pain behavior interactions with socioenvironmental factors, it does not include the subjective evaluation of the pain experience. Pain questionnaires and rating scales are designed for this purpose.

Melzack (1975a) developed the McGill Pain Questionnaire to assess empirically the discriminating characteristics of three major classes of verbal pain descriptors: 1) sensory (temporal, spatial, pressure, thermal); 2) affective (tension, fear, autonomic properties); and 3) evaluative (subjective intensity ratings). A primary purpose of the questionnaire was to provide quantitative measures of clinical pain which could then be statistically analyzed to be sensitive to differences among pain relief treatment modalities. Gracely, McGrath, and Dubner (1978) subsequently developed ratio scales for sensory (from extremely weak to extremely intense) and affective (from bearable to excruciating) verbal pain descriptors, reporting specific functional relationships between measurable psychophysical stimuli and matched ratio scales of verbal stimuli. In contrast, Fordyce and his associates (1978) conducted an investigation to assess whether there existed a relationship between how patients describe their pain problems using a fixed set of bipolar adjectives and physician diagnostic judgments, activity measures through patient-recorded diary forms, and the MMPI. The major conclusion by Fordyce *et al.* (1978) was that chronic pain patients evidence an interrelated group of problems too complex to be reliably discriminated by a single measure such as word sets.

As previously indicated, verbal and nonverbal pain patterns may be disproportionately greater in some patients than is deemed appropriate for the assumed severity of the underlying pathophysiological condition, or they may even exist without obvious concurrent pathology or trauma (Fordyce, 1976; Varni *et al.*, 1980). Given this multidimensionality of pain etiology and expression, then the assessment of the various behavioral-medical parameters is most ethically conducted within the context of an interdisciplinary approach.

Pediatric headache represents an exemplary area demonstrating the necessity for an interdisciplinary assessment in determining the etiology and maintaining conditions of pain complaints. For example, headache report may increase once the child discovers that this strategy successfully allows him or her to avoid aversive or negative school situations, such as an impending exam. What may develop is a self-perpetuating cycle whereby the child progressively falls further and further behind in school, subsequently setting the occasion for more negative school situations and resultant headache-avoidance behaviors. This does not imply that the child originally did not experience a headache or that his headaches do not continue to reoccur, but rather that the intensity and frequency of headache perception and complaint may be influenced by stress and other socioenvironmental factors. On the other hand, the headache complaint may be a symptom of a brain tumor or intracranial hemorrhage (Curless & Corrigan, 1976; Tomasi, 1979). Evidence of antecedent brain damage or dysfunction of possible etiological significance has ranged as high as 69% in one clinic sample (Millichap, 1978), whereas other authors investigating migraine and tension headaches point to socioenvironmental stress factors as the most frequent precipitating event (Brown, 1977; Moe, 1978).

Additional pediatric pain syndromes illustrate the need for a multidimensional assessment. In pediatric recurrent abdominal pain, an organic etiology is estimated at only 5% (Maddison, 1977), with a high frequency of such potential socioenvironmental factors as parental abdominal pain complaints and recurrent school absences (Berger, Honig, & Liebman, 1977). The pathogenesis of recurrent leg aches has consisted of such diverse etiologies as rapid growth ("growing pains"), rheumatism or rheumatic conditions, overexertion, emotional stress, and orthopedic defects (Peterson, 1977), with a psychophysiological etiology proposed for the temporomandibular joint and myofacial pain-dysfunction syndrome (Sheppard & Sheppard, 1977). Significantly, in a long-term follow-up study of 161 children with recurrent abdominal pain, approximately 20% underwent surgical or medical treatments of doubtful necessity; that is, no organic cause was evident (Stickler & Murphy, 1979). Only with a truly interdisciplinary assessment can the various behavioral-medical factors which may influence pediatric pain perception and report be differentiated. Such an assessment is essential and mandatory prior to any intervention program.

TREATMENT

The primary modalities utilized in the treatment of pediatric pain may be delineated into *pain perception regulation* through such self-regulatory processes as imagery techniques and meditation, and *pain behavior regulation* which focuses on the socioenvironmental factors which may influence pain expression and rehabilitation. Although a number of investigations are reported on the use of transcutaneous nerve stimulation (TNS) for adult pain patients (cf. Eriksson, Sjölund, & Nielzen, 1979; Melzack, 1975b), only a case report exists in the literature using this procedure in pediatric pain management (Stilz, Carron, & Sanders, 1977). Thus, the following treatment sections will emphasize the self-regulation of pain perception and pain behavior regulation because of their greater empirical base in pediatric pain control.

As will become evident during the course of this section, self-regulatory processes are the primary modality in the management of pediatric acute pain, whereas self-regulation of pain and pain behavior regulation are utilized in chronic pain syndromes depending on the particular disorder and existing socioenvironmental influences. The self-regulation techniques share common features with hypnosis, autogenic therapy, meditation, progressive muscle relaxation exercises, and biofeedback training (Varni, 1981b). Laboratory research on experimental pain has indicated the role of distraction, dissociation or refocusing of attention from thoughts concerned with pain, anxiety reduction, suggestions of pain relief, and the imagination of past experiences that were incompatible with pain as potent cognitive variables in the reduction of pain perception (cf. Hilgard, 1975). On the other hand, the pain behavior regulation modality follows the operant approach initially developed for adult chronic pain patients (cf. Fordyce, 1976). Although similar mechanisms may be operating in both pain perception regulation and pain behavior regulation (e.g., distraction from pain perception as the patient concentrates on emitting well behaviors or increases in mobility and sleep as pain perception decreases), it will be instructive to present these techniques separately in the following sections on pain secondary to a chronic disorder (hemophilia and sickle cell anemia), pain secondary to an initial physical injury, and pain secondary to medical procedures.

Pain Perception Regulation: Acute and Chronic Pain in Hemophilia

Medical Aspects

Hemophilia represents a congenital hereditary disorder of blood coagulation, transmitted as an X-linked trait by a female carrier to her male offspring. This life-long chronic disorder is characterized by recurrent, unpredictable internal hemorrhaging affecting any body part, especially the joints and the extremities. The severity of the disorder varies among individuals with the failure to produce one of the plasma proteins required to form a blood clot being the identifying characteristic (Hilgartner, 1976). A major breakthrough in the medical management of hemophilia came less than 15 years ago with the discovery and separation of the essential clotting factors from the plasma of normal blood. The intravenous infusion of the factor replacement concentrate temporarily replaces the missing clotting factor and converts the clotting status to normal, allowing a functional blood clot to form (Kasper, 1976). Even though muscle and soft tissue hemorrhages (hematomas) may be quite serious, the most frequent problem and major cause of disability in severe hemophilia is repeated bleeding into the joint areas (hemarthroses) (Arnold & Hilgartner, 1977; Dietrich, 1976a). With repeated hemarthroses, a pathophysiological condition similar to osteoarthritis develops, marked by articular cartilage destruction, pathological bone formation, and impaired function (Kisker, Perlman, & Benton, 1971; Sokoloff, 1975). An estimated 75% of hemophilic adolescents and adults are affected by degenerative hemophilic arthropathy, with chronic arthritic pain representing the most frequent problem in the care of the adolescent and adult hemophiliac (Dietrich, 1976a, 1976b).

Arthritic Pain Management

Whereas acute pain in the hemophiliac is associated with a specific bleeding episode, chronic arthritic pain represents a sustained condition over an extended period of time (Dietrich, Luck, & Martinson, in press). Thus, pain perception represents a complex psychophysiological event, further complicated in the hemophiliac by the existence of both acute bleeding and chronic arthritic pain (Varni, 1981a). More specifically, acute pain of hemorrhage provides a functional signal, indicating the necessity of intravenous infusion of factor replacement to allow a functional blood clot to form (Sergis-Deavenport & Varni, 1981). Arthritic pain, on the other hand, represents a potentially debilitating chronic condition which may result in impaired life functioning and analgesic dependence (Varni & Gilbert, in press). A substantial number of analgesics and anti-inflammatory medications are of limited usefulness since they inhibit platelet aggregation and prolong the bleeding time in hemophiliacs (Arnold & Hilgartner, 1977). Consequently, the challenge has been to develop an effective intervention in the reduction of perceived arthritic pain, while not interfering with the essential functional signal of acute bleeding pain.

Medical observations have suggested the possible therapeutic value of warming and heat application in the management of arthritic joints (Lehman, Warren, & Scham, 1974). White (1973) reported data from a questionnaire administered to 30 osteoarthritic and rheumatoid arthritic patients, with 27 associating pain relief with past experiences of warmth and massage, with the application of a counterirritant (10% menthol and 15% methyl salicylate) producing active tissue hyperemia and a sensation of heat accompanied by decreased pain perception. Citing earlier findings by Wasserman, Oester, Oryschkevich, Montgomery, Poske, and Ruksha (1968) which demonstrated abnormal electromyographic readings in muscles adjacent to arthritic joints, White (1973) proposed that the counterirritant-produced hyperemia reduced the ischemia in the contracted muscles surrounding the arthritic joint. Swezey (1978) reviewed the evidence on the relative efficacy of superficial heating in articular disorders and suggested that the threshold for pain can be raised by superficial heating, but he cautioned that the increases in superficial circulation in relationship to pain relief and muscle relaxation must still be experimentally determined. Hilgard's (1975) laboratory research on hypnosis-controlled pain has indicated the role of distraction and refocusing of attention, anxiety reductions, suggestions of pain relief, and the imagination of past experiences that were incompatible with pain as potential cognitive variables in the reduction of pain perception.

Varni (1981a, 1981b) reported on the successful self-regulation of arthritic pain perception by hemophiliacs involving a treatment strategy consisting of progressive muscle relaxation exercises, meditative breathing, and guided imagery techniques. In these studies, the patient was instructed to imagine himself actually in a scene previously identified or associated with past experiences of warmth and arthritic pain relief. The scene was initially evoked by a detailed multisensory description by the therapist, and then further details of the scene were described aloud by the patient. Once the scene was clearly described and visualized by the patient, the therapist's suggestions included imagining the gentle flow of blood from the forehead down all the body parts to the targeted arthritic joint, images of warm colors such as red and orange, and the sensations of warm sand and sun on the joint in the context of a beach scene. Further suggestions consisted of statements indicating reduction of pain as the joint progressively felt warmer and more comfortable. Each patient had previously identified one joint as the site of greatest arthritic pain, and this joint was targeted for active warming during clinic and home practice sessions.

In the initial investigation, preliminary findings with two hemophiliacs suggested the potential utility of these self-regulation techniques in the reduction of arthritic pain (Varni, 1981a). The next investigation was designed to extend these earlier findings systematically to three additional hemophiliacs under improved methodological conditions, including longer baseline and follow-up assessments, and measurement of multiple subjective and medical parameters in relationship to arthritic pain (Varni, 1981b). Each patient was instructed in the use of a daily self-monitoring multidimensional pain assessment instrument. In addition, each patient's subjective evaluation of the overall impact of the intervention was determined by a comparative assessment inventory administered at the final follow-up session. An analysis of each patient's medical chart and pharmacy records provided data on the number of analgesics ordered, factor replacement units for bleeding episodes, and the number of hemorrhages. These records were analyzed retrospectively at the completion of the investigation to represent the patient's status 6 months prior to the self-regulation intervention and during the follow-up period (averaged monthly for both pretreatment and posttreatment assessment). Finally, a thermal biofeedback unit served as a physiological assessment device rather than as a training technique, with the patient instructed to actively attempt to increase the temperature at the joint site during the self-regulation conditions without the benefit of incremental feedback.

An analysis of Figure 1 demonstrates that the self-regulation techniques were effective in significantly reducing the number of days of perceived arthritic pain per week for all three patients, maintained over an extended follow-up assessment. Further analysis of the daily pain assessment instrument showed that on the 10-point scale (1 = mild pain, 10 = most severe pain), arthritic pain perception decreased from an average of 5.1 during baseline to 2.2 on those follow-up days in which arthritic pain occurred (Table I). Average bleeding pain was essentially unchanged from baseline to follow-up conditions (6.9 to 6.8). Each patient's evaluation at the final follow-up session on the comparative assessment inventory indicated substantial positive changes in arthritic

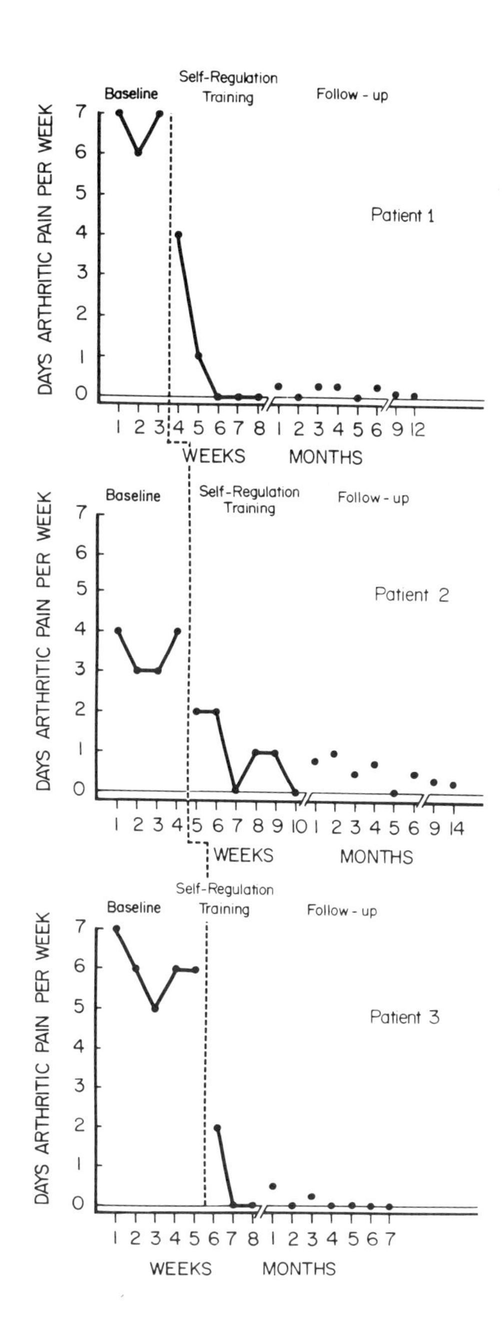

Figure 1. Days of total arthritic pain perception per week during baseline, self-regulation training, and follow-up for three patients. From "Self-Regulation Techniques in the Management of Chronic Arthritic Pain in Hemophilia" by J. W. Varni, *Behavior Therapy,* 1981, *12,* 185–194. Copyright 1981 by the Association for Advancement of Behavior Therapy. Reprinted by permission.

Table I. Subjective Ratings on the 10-Point Scales[a]

Patient	Arthritic pain intensity		Bleeding pain intensity	Daily tension rating
1	Pre	5.3	6.8	5.6
	Post	1.8	6.9	4.9
2	Pre	5.2	5.9	4.3
	Post	3.3	5.7	4.4
3	Pre	4.8	7.9	3.6
	Post	1.4	7.7	2.9
$\bar{x}$	Pre	5.1	6.9	4.5
	Post	2.2	6.8	4.1

[a]From "Self-Regulation Techniques in the Management of Chronic Arthritic Pain in Hemophilia" by J. W. Varni, *Behavior Therapy*, 1981, *12*, 185–194. Copyright 1981 by the Association for Advancement of Behavior Therapy. Reprinted by permission.

pain, mobility, sleep, and general overall functioning, with no reported changes in bleeding pain perception (Table II). Assessment of the surface skin temperature over the targeted arthritic joint showed increased thermal readings from baseline to self-regulation conditions, averaging an increase of 4.1°F. These findings were consistent across patients and were maintained over the follow-up period (Table III). The analysis of the medical charts and pharmacy records demonstrated that the units of factor replacement prescribed covaried with the number of bleeding episodes for all three patients, with the average number of bleeding episodes and units of factor replacement remaining essentially constant across pretreatment and posttreatment assessment periods. The number of analgesic tablets showed a consistent decrease across patients (Table III).

These data on both the medical parameters of bleeding frequency and factor replacement units, as well as the subjective pain ratings,

Table II. Subjective Evaluations of Improvement on the ± 3 Scales[a]

Patient	Comparative arthritic pain	Comparative bleeding pain	Comparative mobility	Comparative sleep	Overall improvement
1	+3	0	+3	+3	+3
2	+2	0	+2	+2	+2
3	+3	0	+2	+2	+3
$\bar{x}$	+2.7	0	+2.3	+2.3	+2.7

[a]From "Self-Regulation Techniques in the Management of Chronic Arthritic Pain in Hemophilia" by J. W. Varni, *Behavior Therapy*, 1981, *12*, 185–194. Copyright 1981 by the Association for Advancement of Behavior Therapy. Reprinted by permission.

Table III. Physiological and Medical Parameters[a]

Patient		Arthritic joint skin temperature (°F)[b]	Analgesic tablets (#)[b]	Factor replacement (units)[b]	Bleeding frequency[b]
1	Pre	84.4	4.3	10667	5.1
	Post	87.9	0	9893	4.8
2	Pre	83.1	15.8	6264	3.7
	Post	88.3	7.9	8919	4.6
3	Pre	85.8	2.7	1166	.33
	Post	89.3	0	858	.29
x̄	Pre	84.4	7.6	6032	3.0
	Post	88.5	2.6	6557	3.2

[a]From "Self-Regulation Techniques in the Management of Chronic Arthritic Pain in Hemophilia" by J. W. Varni, *Behavior Therapy*, 1981, *12*, 185–194. Copyright 1981 by the Association for Advancement of Behavior Therapy. Reprinted by permission.
[b]For skin temperature, premeasures were recorded during the baseline condition, with postmeasures recorded during the follow-up condition. For the other parameters, the medical charts and pharmacy records were analyzed retrospectively to yield a monthly average 6 months prior to the baseline condition and during the follow-up condition.

support the proposal that these self-regulation techniques did not affect bleeding pain perception. Substantial decreases in reported bleeding frequency and factor replacement units would have been expected had bleeding pain perception been affected. It is essential to emphasize that these techniques are not meant to supercede proper and correct medical care. Nevertheless, in the case of chronic degenerative arthropathy, the arthritic pain, in contrast to the acute bleeding pain, exceeds its functional intent, potentially resulting in limitations in normal activities, with the possibility of analgesic dependence always of concern (Varni & Gilbert, in press). Thus, within the context of a detailed consideration of each patient's physical and psychological status, these techniques represent a noninvasive and supportive therapeutic intervention which when applied within an interdisciplinary hemophilia center, provides an additional treatment modality in the comprehensive management of hemophilia.

Pain and Analgesia Management Complicated by Factor VIII Inhibitor

The treatment of children with hemophilia has evidenced significant advances in recent years as a result of the development of antihemophilic concentrate, home infusion programs, and the opportunity for comprehensive care through regional hemophilia centers (Dietrich, 1976b). Unfortunately, approximately 10% of hemophilic children develop an inhibitor to Factor VIII, presenting a serious problem in the management of bleeding episodes (Buchanan & Kevy, 1978). Although the frequency of hemorrhage is no different, the neutralization of Factor VIII replacement by an inhibitor (antibody) makes the control of bleeding ineffective. A recent advance in the treatment of hemophilic patients with inhibitors has been the use of prothromin-complex concentrates (PCC), effecting some level of hemostasis through apparent activated by-products present in the reconstituted preparation. However, recent clinical reports have indicated decreased effectiveness of these products (Parry & Bloom, 1978; Scranton, Hasiba, & Gorenc, 1979). The pain associated with uncontrolled hemorrhage can be extremely severe, with narcotic analgesics traditionally prescribed (Dietrich, 1976b). Thus, although the acute pain of hemorrhage provides a functional signal indicating the necessity of factor replacement therapy, in the hemophilic child with Factor VIII inhibitor the intensity of the pain supercedes its functional intent, with analgesic dependence of constant concern. Consequently, an effective alternative to analgesic dependence in the reduction of perceived pain in the patient with an inhibitor has been greatly needed.

Varni, Gilbert, and Dietrich (1981) recently reported on a case study involving a 9-year-old hemophilic child with Factor VIII inhibitor. At 4 years of age, when the inhibitor developed and subsequent Factor VIII replacement therapy became impossible, the patient began to require narcotics in order to tolerate the pain of each hemorrhage. Progressively, the need for pain medication increased for both bleeding pain and for arthritic pain in his left knee secondary to degenerative arthropathy. Since the arthritic pain eventually occurred almost daily, the requests for analgesics further increased so that the acute pain of hemorrhage required ever larger doses for pain relief, even though home PCC therapy and joint immobilization continued for the management of bleeding episodes. As a consequence of bleeding and arthritic pain in the lower extremities, the patient was wheelchair-bound nearly 50% of the time, had been hospitalized 16 times in the $4\frac{1}{2}$-year period prior to the study for a total of 80 days after the development of the inhibitor, with analgesic medication also kept at his school for pain control. The final precipitating event in this steadily worsening cycle occurred during an evening visit to the emergency room because of a very painful and severe left knee hemorrhage which had not responded to home PCC therapy, with the administration of an adult dose of meperidine hydrochloride and IV diazepam providing no pain relief.

Training in the self-regulation of pain perception consisted of techniques developed earlier by Varni (1981a, 1981b), with modification in the guided imagery techniques required for the bleeding pain intensity.

The patient recorded the severity of his pain on the 10-point scale for a 2½-week baseline prior to self-regulation training. The average score for both arthritic and bleeding pain during this period was 7, indicating rather intense pain. At 1-year follow-up after the self-regulation training, the patient reported that both arthritic and bleeding pain were reduced to 2 on the scale when he engaged in the self-regulation techniques. In addition to this measure of pain perception, the patient's evaluation at the 1-year follow-up session on the comparative assessment inventory indicated substantial positive changes in arthritic and bleeding pain, mobility, sleep, and general overall functioning. As seen in Table IV, once the patient began using the self-regulation techniques for pain management, there were no further requests for meperidine during the 1-year posttreatment assessment, with substantially decreased amounts of acetaminophen with codeine required. Table IV also shows that significant improvements in other areas of functioning, including improved mobility as evidenced by the physical therapy measures on his arthritic left knee in comparison with his normal right knee on the dimensions

Table IV. Parameters Associated with Pain Intensity[a]

Parameters	1 year before initiation of self-regulation training	1 year after initiation of self-regulation training
Pain intensity (1 = mild; 10 = severe)	7[b]	2[c]
Meperidine	74 tablets (50mg/ea.)	0 tablets
Acetaminophen/codeine elixir	438 doses (24mg codeine/dose)	78 doses (24mg codeine/dose)
Physical therapy measures		
Range of motion	Normal right knee 0°–150° Arthritic left knee 15°–105°	Right knee 0°–150° Left knee 0°–140°
Quadricep strength (0–5 scale)	Normal right knee 4– Arthritic left knee 3+	Right knee 4+ Left knee 4
Girth (knee joint circumference)	Not available	Right knee 26 cm Left knee 25.8 cm
Ambulation on stairs	2–3 maximum	No limitation
School days missed	33	6
Hospitalizations		
Total days	11	0
Number of admissions	3	0

[a]From "Behavioral Medicine in Pain and Analgesia Management for the Hemophilic Child with Factor VIII Inhibitor" by J. W. Varni, A. Gilbert, and S. L. Dietrich, *Pain*, 1981, *11*, 121–126. Copyright 1981 by Elsevier/North-Holland Biomedical Press B. V. Reprinted by permission.
[b]2½-week preassessment during pain episodes just prior to self-regulation training.
[c]1-year average rating during pain episodes when using self-regulation techniques.

of range of motion (0–150° = normal) and quadricep strength (1 = no joint motion, 5 = complete range of motion against gravity with full resistance). Normalization of psychosocial activities is suggested by increased school attendance, decreased hospitalizations, and parental report, noting a distinct elevation of the child's overall mood, that he was considerably less depressed during pain episodes now that he had the skills actively to reduce his pain perception without having to depend on pain medication.

The analysis of the various parameters assessed in this study suggests a significant improvement across a number of areas. As envisioned by the authors, a deteriorating cycle was evident prior to the intervention, schematically represented as: hemorrhage → pain → analgesics/joint immobilization → atrophy of muscles adjacent to the joints/joint deterioration → hemorrhage. Thus, as has been previously suggested (Dietrich, 1976 a,b), pain-induced immobilization results in muscle weakness surrounding the joints, setting the occasion for future hemorrhaging. By breaking this deteriorating cycle at the point of pain severity, the patient was offered the opportunity to decrease immobilization and increase therapeutic activities such as swimming, subsequently improving the strength and range of motion in the left knee, and with this improved ambulatory status, school attendance and his general activity level were consequently increased. The possibility that this early intervention may have prevented or reduced the likelihood of later drug abuse must also be considered (Varni & Gilbert, in press). Finally, it is important to reiterate that these procedures were used for a child with an inhibitor. For the hemophiliac without an inhibitor, bleeding pain serves as a functional signal and is best managed with factor replacement therapy. However, in the present case, no effective medical procedure was available to control severe bleeding pain other than powerful analgesics, clearly an undesirable therapy modality.

Pain Behavior Regulation: Chronic Pain Secondary to an Initial Trauma

Behavioral regulation of pediatric chronic pain focuses on the external accompaniments of an assumed internal pain state. External manifestations include the absence of well behavior and the manifestation of pain behavior, that is, verbal reports of pain and such nonverbal expressions as facial grimaces, compensatory posturing, restricted movement, and limping (Fordyce, 1976). Well behavior deficits include decreased activity levels and restricted interpersonal contacts periodically associated with depression. Although pediatric chronic pain often appears to have a clear history of a precipitating trauma involving surgery or physical injury (Epstein & Harris, 1978), socioenvironmental factors may influence the continuation of the pain behaviors. Assessment of the potential socioenvironmental influences on pain behavior and well behavior can often delineate which stimulus-consequent events are candidates for change. Fordyce (1976) has developed a behavioral treatment package successful in the regulation of chronic pain behavior with adult patients. Since Fordyce's original publication in this area (Fordyce, Fowler, Lehmann, & Delateur, 1968), an increasing amount of attention has been paid to the behavioral management of adult chronic pain (Swanson, Marita, & Swenson, 1979). However, to date, little empirical evidence exists on the behavioral regulation of pediatric chronic pain behavior. The following case investigation represents an illustration of such an approach.

Varni, Bessman, Russo, and Cataldo (1980) reported on an extensive assessment and treatment program for a 3-year-old child who had been hospitalized for 10 months for the treatment of second- and third-degree burns to her buttocks, legs, and peroneum as the result of immersion in hot water. Circumstances surrounding the burn incident indicated the possibility of child abuse. Scar contractures and subsequent decreased range of motion in both knees necessitated the wearing of Jobst stockings and knee extension splints. At the time of the referral for pain regulation, the patient was exhibiting an array of chronic pain behaviors which interfered significantly with her rehabilitation and constructive patient–caregiver interactions. Furthermore, these pain responses appeared to increase in both intensity and frequency during attention-seeking and demand avoidance situations. Data were obtained in three different settings: (1) the clinic room, where the patient wore the knee extension splints in a contrived setting, (2) the bedroom, where the patient wore the splints in the hospital environment, and (3) a physical therapy situation, during which the physical therapist focused on improved range of motion and independent ambulation.

Three categories of pain behavior were recorded: crying, which ranged in intensity from sobbing to screaming; verbal pain behavior, which consisted of such statements as "My leg/ankle/foot/stomach hurts," "Ouch," or "I can't stand up"; and nonverbal pain behavior consisting of any gestural response expressing pain or discomfort such as facial grimaces, rubbing her legs or buttocks, or not standing. In addition, an activity measure, number of steps descended, was measured in physical therapy since it was important for the child's range of

SPLINTS IN CLINIC

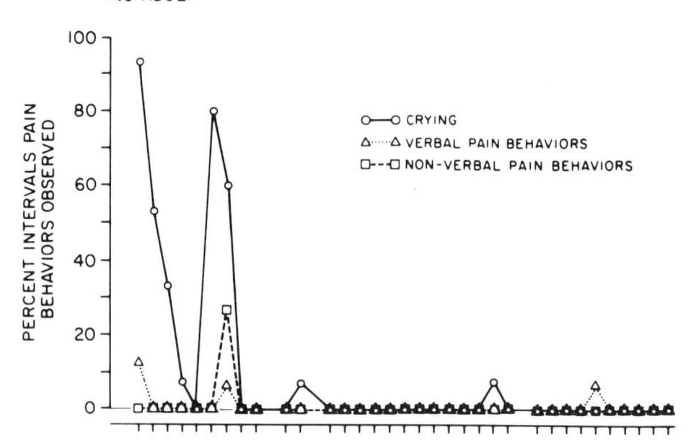

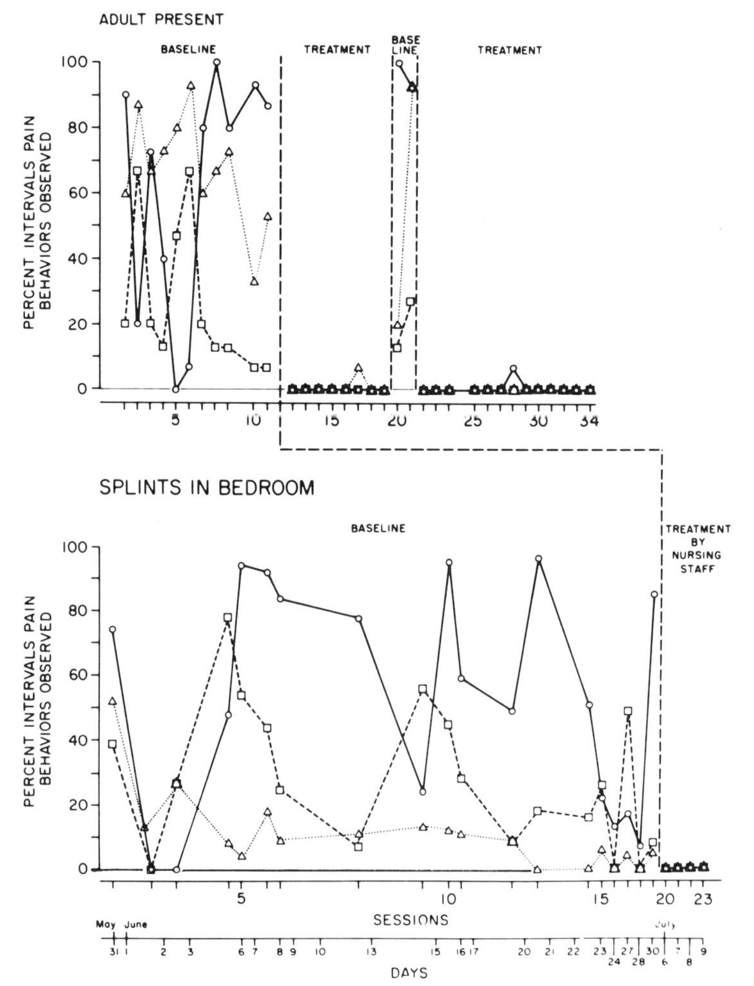

motion and independent ambulation. During the baseline assessment, it became evident that the child's pain behavior was a function of adult attention and demand situations (Figures 2 and 3). In the absence of adults, chronic pain behavior was noticeably infrequent. Perhaps more important, it was observed that when the child was engaged in interesting activities with accompanying staff attention for this appropriate behavior, pain complaints were reciprocally low.

Since the baseline assessment suggested that the pain behavior was influenced by socioenvironmental factors, treatment consisted of rearranging the existing contingencies. Thus, in both the clinic and bedroom settings where the child wore her knee extension splints, social praise and a treat were contingent upon such well behavior as helping to put on the splints, positive verbalizations, and smiling, with all pain responses ignored. Once the techniques were shown to be effective in decreasing pain patterns in the clinic setting, both the rationale and a demonstration were presented to other staff caregivers, along with a review of data collection procedures. The nursing staff were then instructed on how to carry out the procedures in the hospital bedroom setting. In the physical therapy environment, since the child was required to descend a number of steps as part of her rehabilitation exercises, treatment consisted of the therapist rewarding the child for descending the steps and ignoring pain responses. Reliable behavioral data on pain and appropriate well or rehabilitative behaviors obtained throughout baseline and treatment conditions indicated a successful program (Figures 2 and 3), demonstrating through a functional analysis that the child's chronic pain behaviors were under socioenvironmental influences.

In addition to the objective measures, other clinically significant changes were noted. Whereas the child initially resisted splinting attempts and actively attempted to remove the the splints, she subsequently began asking to assist, saying, for example, "I'll do it," or "I want to help you." She began to make positive statements about her accomplishments instead of statements of pain and resistance to rehabilitation. Rather than seeking attention from her caregivers for pain behavior, there was a shift to the utilization of well behavior to attract

Figure 2. Percent of observation intervals per session in which pain behaviors were noted in each of three situations. From "Behavioral Management of Chronic Pain in Children: Case Study" by J. W. Varni, C. A. Bessman, D. C. Russo, and M. F. Cataldo, *Archives of Physical Medicine and Rehabilitation*, 1980, *61*, 375–379. Copyright 1980 by *Archives of Physical Medicine and Rehabilitation*. Reprinted by permission.

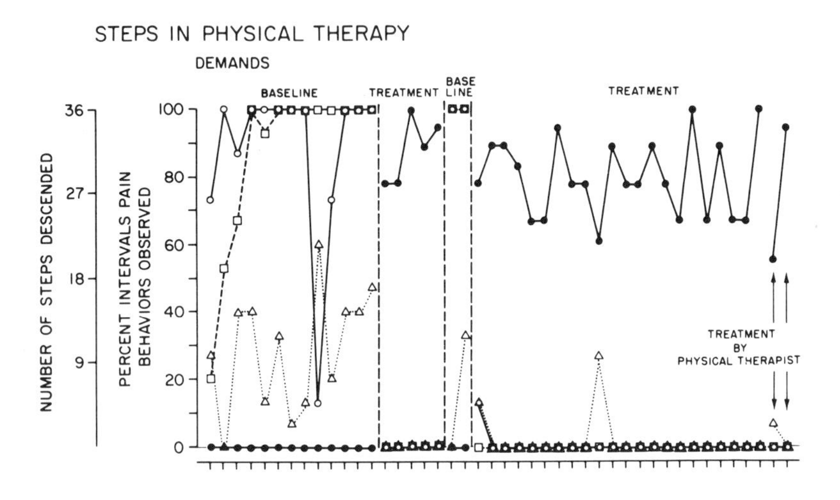

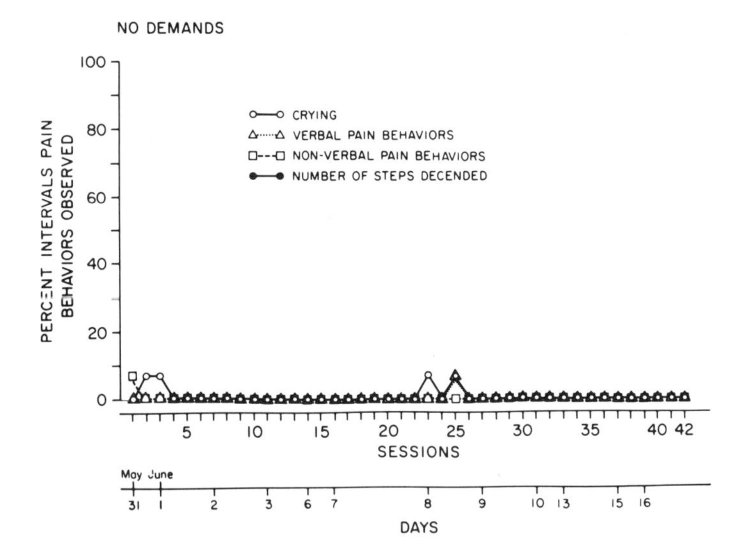

Figure 3. Number of steps descended and percent of observation intervals per session in which pain behaviors were noted. From "Behavioral Management of Chronic Pain in Children: Case Study" by J. W. Varni, C. A. Bessman, D. C. Russo, and M. F. Cataldo, *Archives of Physical Medicine and Rehabilitation*, 1980, *61*, 375–379. Copyright 1980 by *Archives of Physical Medicine and Rehabilitation*. Reprinted by permission.

social attention and praise. She no longer cried during naptime and at night as she had done prior to treatment and therefore did not need to be moved to a separate room to avoid disturbing the other children in her bedroom. Although it was not possible to determine whether the patient actually felt pain or simply displayed the associated behaviors, in this case no further pain displays were observed with the onset of the intervention. Fordyce (1976) has suggested that during periods of initial trauma and its resultant pain, the patient has many opportunities for pairing environmental stimuli to feelings of pain. Through learning, the patient may actually come to experience pain in certain circumstances in excess of the accompanying physical basis for the pain, or even in the absence of a physical basis for the perceived pain. Whether the subjective experience of pain abates over the course of time may be independent of the pain behavior which the patient displays. In such cases, pain behavior may serve multiple functions, such as to avoid rehabilitative or demand situations and/or to gain the attention of others. Pain expression may often become situation-specific, occurring only under circumstances in which it is likely to result in attention or therapy avoidance, significantly hindering long-term rehabilitation (Varni *et al.*, 1980).

In acute cases, the pain is an important symptom of underlying physical distress, requiring sympathy, empathy, and a detailed search for the sources of the acute pain and appropriate therapies. Continued expression of pain in chronic cases, however, may serve as a signal for the therapist to look more closely into the conditions of the child's daily care and therapeutic environment. Consideration should be given to secondary gains of the pain behaviors, particularly if the result is to increase staff attention and to avoid therapy. The important aspects to consider are the extent of empathetic attention provided to the child and its contingent relationship to pain expression. The care staff must delineate when and what consequences for pain behavior are in the therapeutic interests of the patient. Particular situations in which pain expression occurs should result in empathy, whereas others should result in alternative consequences including ignoring the pain behavior and providing attention or other reinforcement, if necessary, for adhering to rehabilitative programs. Thus, the therapeutic goal should not be the total suppression of pain expression, since such communication often serves an essential function, but rather the relatively greater expression of well behavior in those situations in which pain behavior interfers with the child's long-term rehabilitation. Such differential expression of emotion represents both a normal and adaptive developmental process.

Pain Perception Regulation: Acute Pain Secondary to Medical Procedures

Conditioned anxiety associated with recurrent diagnostic and treatment procedures represents a major source of distress in children with chronic or life-threatening disease (Varni, Kellerman, & Dash, in press). Such anxiety is often anticipatory, occurring prior to the administration of a potentially painful medical procedure, and manifests itself in terms of a variety of symptoms including nausea and emesis, insomina, anorexia, nightmares, withdrawal, and depression (Katz, Kellerman, & Siegel, 1980). Though the medical procedures themselves may be the primary stimuli that evoke anxiety, substantial stimulus generalization can occur, so that children may exhibit anxiety in response to hospital locations and personnel and to the temporal and sensory stimuli of sights, sounds, or smells associated with medical treatments (Katz *et al.*, 1980). Since the area of pediatric hematology/oncology provides an exemplary model for investigating the management of acute pain secondary to medical procedures (Kellerman & Varni, in press), it will be discussed in detail.

Medical Aspects

Although chronic, debilitating pain is often associated with adult cancer, it is less commonly encountered in children with malignant disease. This is primarily because differing types of cancers are found in adult and pediatric populations. In the former, the most common malignancies are carcinomas of various organ systems (e.g., lung, breast, stomach, ovaries). These diseases are known to cause prolonged, often severe pain. In children, the modal malignancy is acute leukemia, comprising approximately 40% of all pediatric cancers (Kellerman, 1980). Acute leukemia is not an inevitably painful disease, though it may be accompanied by bone pain, fatigue, and influenza-like symptoms during states of relapse. However, children with leukemia do experience a high rate of painful and discomforting medical procedures associated with diagnosis (bone marrow aspirations and biopsies, lumbar punctures, blood tests) and with treatment (intravenous and intramuscular injections of chemotherapeutic agents, blood transfusions, radiotherapy). Katz, Kellerman, and Siegel (1980) conducted behavioral observations of 115 children with acute leukemia undergoing bone marrow aspirations, finding that anxiety and discomfort were virtually ubiquitous, with behavioral distress tending to decrease with advancing age and with females demonstrating higher levels of distress than males. Subsequent follow-up studies refining the observational methodology and incorporating the use of structured self-reports of general affect, fear, and pain confirmed these initial findings, as well as delineating a similar pattern associated with lumbar punctures, demonstrating no habituation to the procedures (Katz, Kellerman, & Siegel, 1981b).

Typically, procedures that are considered minor (venipunctures, blood transfusions, intramuscular and intravenous injections) are administered without accompanying chemoanesthesia. Exceptionally anxious children may be presedated with agents such as diazepam. For procedures that are regarded as somewhat more intrusive, such as bone marrow aspirations and lumbar punctures, it is common medical practice to utilize intramuscular injections of local anesthetic agents, resulting in eventual pain reduction. However, the degree of perceived anesthesia varies widely from child to child, with some children reporting that the pain and anxiety associated with anesthetic injections are equal to or greater than the distress associated with the procedures themselves (Katz *et al.*, 1980). All of these factors combine to create a situation in which these children repeatedly undergo medical procedures which are anxiety-provoking and painful and to which habituation does not routinely occur, indicating the need for clinical intervention aimed at reducing the pain/anxiety pattern (Katz *et al.*, 1980).

Pain Management

Crasilneck and Hall (1973) described the case of a 4-year-old child with inoperable brain cancer who was in continual pain, was anorexic, and cried constantly. Induction techniques were used to elicit a state of relaxation, and suggestions for reduced pain, increased appetite and sleep, and general well-being were made. La Baw and his associates (1975) reported on group training in self-hypnosis with 27 patients ranging in age from 4 to 20 years. Progressive muscle relaxation was the induction used, followed by guided imagery of restful scenes. Finally, Gardner (1976) described the use of hypnotherapy in helping reduce distress in a terminally ill child with acute lymphoblastic leukemia. Although these studies are of clinical significance, the recent case study by Ellenberg, Kellerman, Dash, Higgins, and Zeltzer (1980) represents one of the first published studies collecting pretreatment baseline information. In the study, a 16-year-old girl with chronic myelogenous leukemia typically tolerated medical procedures without complaint but evidenced acute pain and anxiety related to bone marrow aspirations (BMAs). The patient recorded on a self-report form her subjective ratings of pain and anxiety on a 1 (no symptoms) to 5 (severe symptoms) scale

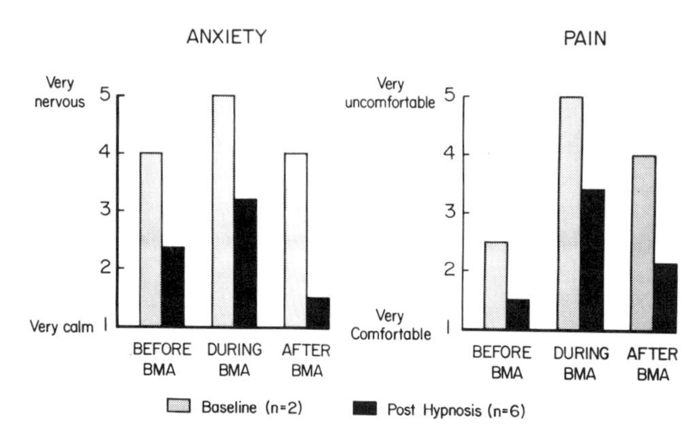

Figure 4. Mean self-report ratings of anxiety and pain before, during, and after bone marrow aspirations for baseline and post-hypnosis procedures. From "Use of Hypnosis for Multiple Symptoms in an Adolescent Girl with Leukemia" by L. Ellenberg, J. Kellerman, J. Dash, G. Higgins, and L. Zeltzer, *Journal of Adolescent Health Care*, 1980, *1*, 132–136. Copyright 1980 by the *Journal of Adolescent Health Care*. Reprinted by permission.

for intervals prior to, during, and following a BMA procedure. These subjective ratings were recorded during two BMAs to establish a pre-hypnosis baseline level of pain and anxiety, as well as during six BMAs following hypnosis. The patient also reported severe chronic headaches and backaches following the first course of chemotherapy, with numerous episodes of nausea and emesis. Subsequent to the hypnosis intervention, there were decreases in both pain and anxiety before, during, and after BMAs (Figure 4), with substantial reductions in headache and backache pain as well as nausea and emesis.

Pain Perception Regulation: Acute Pain in Sickle Cell Anemia

Although in pediatric leukemia the acute pain associated with medical procedures represents a primary pain pattern, acute pain episodes secondary to hematological disease represent an additional area of substantial importance. This section on pain management in sickle cell anemia provides another model for pediatric acute pain regulation.

Medical Aspects

In general, patients with sickle cell anemia manifest a relatively healthy pattern, with periodic sickling crises which may be of sudden onset and with occasional fatal outcome (Song, 1971). Infarctive or painful crises occur most frequently in chest, abdomen, and bones. Environmental pathogenic factors which may induce vasoconstriction associated with sickling include cold climatic conditions, consumption of large quantities of ice water, and swimming (Diggs & Flowers, 1971; Redwood, Williams, Desai, & Sergeant, 1976). Environmental changes which impair adequate oxygenation of arterial blood are deleterious and interfere with the need for efficient cardiopulmonary function and efficient utilization of inspired air (Lehmann, Huntsman, Casey, Lang, Lorkin, & Comings, 1977). Within the sickle cell pain crisis, the course of events includes sickling, increased blood viscosity resulting in further vascular stasis, more sickling, with vascular obstruction possible, and subsequent infarction leading to tissue anoxia and tissue death, clinically manifested as pain (Lehmann *et al.*, 1977). Thus, the painful crisis, which complicates a chronic hemolytic anemia and is the hallmark of sickle cell anemia, results from the vasoocclusion associated with rigid, tangled, sickled, or crescent-shaped erythrocytes. Leichtman and Brewer (1978) have suggested a role for intravascular coagulation in the pathogenesis of sickle cell pain crisis, observing elevations of fibrinopeptide during pain crisis as contrasted to pain-free periods and to normal controls. They propose that tests of such factors as fibrinopeptide may provide an objective assessment of pain crisis and provide an evaluation of new therapeutic interventions. Finally, as discussed by Lehmann *et al.* (1977), prophylaxis of infarctive crises and management during crises require keeping the patient warm and administering oxygen and, at times, transfusions of blood. The authors further point out that the recurrent nature of the acute, painful vasoocclusive crises may result in analgesic dependence.

Pain Management

If vasoconstriction represents a major component of the sickling crisis, then therapy aimed at specifically producing vasodilation and associated warmth may have a therapeutic impact on the painful nature of the crisis. Such was the rationale for a recent study by Zeltzer, Dash, and Holland (1979), teaching two adolescents with sickle cell anemia self-hypnosis techniques. Specifically, the authors employed eye fixation and progressive relaxation as the induction, leading to guided imagery techniques centering around a pleasant, pain-free scene, with suggestions of increased body warmth and vasodilation. A thermal biofeedback device was used periodically to monitor any physiological changes in peripheral temperature. Each patient was instructed to practice these techniques at the perceived onset of a sickling crisis. The authors re-

Table V. Comparison of Pain-Related Hospital Contacts before and after Hypnosis Training[a]

Pain-related contacts	Patient	12 months before hypnosis	4 months after hypnosis	8 months after hypnosis
Emergency/outpatient department	J.S.	17	2	—
	M.B.	37	0	1
Hospitalizations	J.S.	6	0	—
	M.B.	10	0	0
Total days hospitalization	J.S.	48	0	—
	M.B.	47	0	0

[a]From "Hypnotically Induced Pain Control in Sickle Cell Anemia" by L. K. Zeltzer, J. Dash, and J. P. Holland, *Pediatrics*, 1979, *64*, 533–536. Copyright 1979 by the American Academy of Pediatrics. Reprinted by permission.

ported significant reductions over an 8-month follow-up on the frequency and intensity of pain crises as well as analgesic need, with increased peripheral skin temperature observed as measured by the thermal biofeedback device. Of particular note, a comparison of the hospital records 12 months preintervention and 8 months postintervention demonstrated marked reductions in pain-related outpatient visits and in the number and total days of hospitalizations (Table V).

The sum of the findings in these sections indicates significant progress in the development of treatment techniques. The rest of the chapter will now be devoted to the potential neurochemical mechanisms which may further our understanding of pediatric pain perception and treatment.

NEUROCHEMICAL DIMENSIONS: ENDOGENOUS OPIATES

In the early 1970s, several laboratories were actively investigating the mode of action of morphine, focusing specifically on the mechanisms of addiction (Benedetti, 1979). Specific receptors for morphine and related opiates were postulated to exist because of the high degree of steric and structural specificity inherent in many observed actions of these compounds (Goldstein, 1974; Simon, Hiller, & Edelman, 1973). In 1973, three independent laboratories almost simultaneously identified highly specific radioactively labeled opiate agonist and antagonist binding to brain tissue, effectively identifying the opiate receptors (Pert, Pasternak, & Snyder, 1973; Simon, Hiller, & Edelman, 1973; Terenius, 1973).

Characteristic of this early work, Pert *et al.* (1973) isolated opiate receptors from the synaptosomal fraction of rat brain homogenate. Opiate receptors have since been determined to exist in all vertebrates studied, but not invertebrates (Benedetti, 1979; Simon & Hiller, 1978), and have been found to be distributed in various body regions, including the brain, spinal cord, and gastrointestinal tract. In humans as well as in lower primates, the brain contains the highest concentration of receptors (Kuhar, Pert, & Snyder, 1973; Simon & Hiller, 1978). Areas of the brain and nervous system heavily endowed with receptor sites include the amygdala, hypothalamus, corpus striatum, caudate nucleus, periaqueductal gray, thalamus, raphe nuclei, and gray matter of the spinal cord, most specifically in the substantia gelatinosa (Benedetti, 1979; Simon & Hiller, 1978; Terenius, 1978).

The existence of receptor sites which would bind with exogenous substances found in the opium poppy (i.e., morphine) was immediately noted to be a peculiar phenomenon (Benedetti, 1979; Simon & Hiller, 1978). Researchers hypothesized that exogenous opiates must activate a neuromodulation system normally mobilized by endogenous substances, pharmacologically similar to morphine. In 1975, Terenius (1975) and Hughes (1975) independently succeeded in isolating a substance from animal brain with opiate-like characteristics, the effects of which were reversible by naloxone, an opiate antagonist.

The first endogenous opiates isolated were pentapeptides, termed enkephalins, meaning "in the head" (Hughes, Smith, Kosterlitz, Fothergill, Morgan, & Morris, 1975). Two nearly identical enkephalins were identified differing only in their terminal amino acid being either methionine or leucine. These are referred to as methionine and leucine enkephalins (Benedetti, 1978; Snyder & Childers, 1979).

The amino acid sequence of methionine-enkephalin was found to be identical with residues 61–65 of the pituitary hormone beta-lipotropin (β-LPH) (Hughes *et al.*, 1975). This very significant finding gave rise to the isolation of other peptides from extracts of hypothalami and pituituary glands. The amino acid sequences of all these substances were also found to be present in β-LPH (Simon & Hiller, 1978). The proliferation in the number of endogenous opiate-like peptides led to the adoption of the generic term *endorphin* (endogenous morphine-like material) to refer to all of these compounds (Simon & Hiller, 1978).

Guillemin, Ling, and Burgus (1976) isolated two peptides from pig brain and pituitary extracts: alpha-endorphin (β-LPH sequence 61–76) and gamma-endorphin (β-LPH sequence 61–77). Cox, Goldstein, and Li (1976) and Bradbury, Smyth, Snell, Birdsall, and Hulme (1976) independently found that the C-terminal fragment of β-LPH (sequence

61–91) was the most potent endogenous opioid peptide so far isolated, now referred to as β-EP (Simon & Hiller, 1978).

Because all identified endorphins appeared to be fragments of the C-terminal of the pituitary hormone β-LPH, it has been suggested that β-LPH may serve as the prohormone, or precursor for all endorphins (Bradbury, Smyth, & Snell, 1976; Lazarus, Ling, & Guillemin, 1976). Although this may be true for endorphins of pituitary-hypothalamic origin (alpha, beta, and gamma endorphin), it is not likely for the enkephalins (Terenius, 1978). Clement-Jones, Lowry, Rees, and Besser (1980) cite current evidence suggesting that precursors for methionine and leucine enkephalin derive from extracts of the adrenal medulla; this evidence is supported by findings of Kimura, Lewis, Stern, Rossier, Stein, and Udenfriend (1980).

Terenius (1978) has posited the existence of two major, separate, endorphin-releasing systems: one appears localized, mainly in the pituitary-hypothalamic area with β-EP as the principal active agent, the other localized in brain and neural tissues where enkephalins are abundant. Support for this hypothesis is derived from the difference in sustained action between β-EP and the enkephalins. Where β-EP has been noted to exert a sustained, hormonal effect, the enkephalins are rapidly degradable and appear to function as classic neurotransmitters (Snyder & Childers, 1979; Terenius, 1978; Terenius & Wahlstrom, 1979). This difference in duration of action could explain why β-EP is more potent than all other endorphin compounds (Snyder & Childers, 1979), excluding recently discovered dynorphin (Goldstein, Tachibana, Lowney, Hunkapiller, & Hood, 1979).

Endorphins and Endogenous Pain Modulation

Basbaum and Fields (1978) cite the earliest evidence for the existence of an endogenous pain modulatory system as deriving from the observed phenomenon known as stimulation-produced analgesia (SPA). It was first noted that electrical stimulation of certain discrete brainstem sites in animals produced profound analgesia without general behavioral depression (Mayer & Liebeskind, 1974; Mayer, Wolf, & Akil, 1971). Numerous brain sites have since been found that produce analgesia when stimulated, but electrode placement in the ventro-lateral periaqueductal gray (PAG) of the mesencephalon (midbrain) appears to be most consistently effective (Basbaum & Fields, 1978; Benedetti, 1979; Mayer & Liebeskind, 1974).

SPA has been demonstrated to be a very effective analgesic for severe clinical pain in humans. Midbrain and diencephalic stimulation at brain sites analogous to those effective in laboratory animals have produced relief of pain of various causes (Basbaum & Fields, 1978; Cannon, Liebeskind, & Frenk, 1978; Hosobuchi, Adams, & Linchitz, 1977). The duration of SPA appears to be much greater in human patients than in experimental animals and may persist for hours after cessation of stimulation (Basbaum, Clanton, & Fields, 1976).

Concomitant with the discovery and development of SPA, neuropharmacological studies alluded to above demonstrated receptors in the brain to which opiate compounds bind stereospecifically. The isolation of endorphins led various investigators to hypothesize that SPA is effective because of its stimulation of endorphins (Basbaum & Fields, 1978).

Opiate analgesia (OA) and SPA both activate an efferent, descending brain stem system that suppresses and modulates pain transmissions (Basbaum *et al.*, 1976; Mayer & Liebeskind, 1974; Mayer & Price, 1976). The identification and historical mapping of opiate receptors and endorphins in various parts of the nervous system directly involved in pain suppression and modulation indicate that OA and SPA are, in fact, stimulating an endogenous system (Chung & Dickenson, 1980).

Endogenous pain modulation operates in a complex feedback loop. Highly simplified, a nociceptor transmits a pain signal to the spinal dorsal horn, where it is transmitted up the spine along ascending nerve tracts to the brainstem. The brainstem communicates with higher cortical areas, and a physiological response to the stimulus is transmitted down from the brainstem along descending spinal tracts that impact finally with the spinal dorsal horn. Endorphins stimulated by the brainstem are hypothetically capable of modulating the pain experience at each of the schematic levels outlined (Basbaum & Fields, 1978; Chung & Dickenson, 1980; Sherman & Liebeskind, 1980).

Dorsal Horn and Ascending Spinal Tracts

The dorsal horn neurons of the spinal cord are the central terminal of the A-lambda and C fibers from nociceptors in the skin, viscera, and muscles (Zimmerman, 1979). There are two distinct groups of spinal neurons which respond to painful stimuli, both located within the substantia gelatinosa of the dorsal horn. The first is in the superficial layer (lamina I), involved mainly in transmitting sharp pain. The other is located in deeper layers of the dorsal horn (laminae IV and V), responding to intense pain (Chung & Dickenson, 1980). Through a dense meshwork of short and long axons, these cells transmit messages further centrally. Some of these axons terminate within the spinal cord, while others ascend to the brainstem and above (Benedetti, 1979; Chung & Dickenson, 1980).

In the spinal cord, methionine-enkephalin is found in the interneurons of the substantia gelatinosa and the proprioceptive afferents (Hökfelt, Ljundahl, Terenius, Elde, & Nilsson, 1977). Immunohistochemical analyses indicate a strict relationship between methionine-enkephalin and substance P (SP), an excitatory neurotransmitter that is necessary for transmission of noxious impulses from the periphery to higher brain centers (Konishi & Ótsuka, 1974). The demonstration of opiate receptors on the superficial laminae of the dorsal horn, and their marked decrease after dorsal rhizotomy (surgical cutting of spinal nerve roots), suggests that they are localized on the presynaptic junction of the primary nociceptive afferents (Lamotte, Pert, & Snyder, 1976). The binding of endogenous or exogenous opiates to these receptors appears to inhibit the release of SP, preventing the synaptic transmission of noxious impulses (Iverson, 1979; Jessel & Iverson, 1977; Marx, 1979).

Brainstem and Descending Spinal Tracts

It is well established that in all sensory systems, including pain, the ascending information is modulated by centrifugal influences from the brain (Zimmerman, 1979). In the somatosensory system, such backward modification is exerted via pathways descending in the white matter of the spinal cord that originate in the cortex, diencephalon, and brainstem.

In the context of pain, midbrain and brainstem systems appear to be crucial to descending modulation (Basbaum & Fields, 1978; Zimmerman, 1979). Cells in the periaqueductal gray (PAG) project to and activate cells in the nucleus raphe magnus (NRM) and adjacent structures. These medullary cells project to the spinal dorsal horn via the dorsal lateral funiculus, where they inhibit pain-transmission neurons in laminae I and V.

Neurons connecting the PAG and NRM have been found to have high concentrations of methionine-enkephalin, SP, and opiate receptors (Hökfelt *et al.*, 1977). From the NRM, these neurons project down the dorsolateral funiculus and release serotonin at their synaptic terminals (Basbaum et al., 1976). Neural inhibition in the dorsal horn appears to result from this serotonergic system exerting axo-axonic influence: serotonin-containing terminals of NRM axons synapse on enkephalin-containing interneurons of the dorsal horn, causing secretion of enkephalin and the blocking or inhibition of SP-mediated pain transmissions (Fields, Basbaum, Clanton, & Anderson, 1977).

Higher Brain Mediation

Pert and Yaksh (1974) have suggested that opiates may induce analgesia by specifically interfering with the processing of pain information in terminal brain regions of the spinal pain pathways. By preventing or modulating the access of pain information from brainstem to limbic structures, opiates may influence the affective and motivational components of pain experience. Neurons of raphe medianus and dorsalis nuclei are known to send axons to higher brain regions of the limbic system and cortex, providing potential pathways for pain sensation (Messing & Lytle, 1978). These pathways contain dopamine and norepinephrine as neurotransmitters, both of which are known to modulate pain (Akil & Liebeskind, 1975).

Herkenham and Pert (1980) performed *in vitro* autoradiography of opiate receptors in rat brains in order to map nerve tracts that utilize endorphins as neurotransmitters. They found that sensory input areas from nearly all modalities contained opiate receptors. The densest regions of opiate receptors in the cortex were found within the limbic system, a group of brain structures which regulate emotional and motivational behaviors. Lewis and his colleagues (1981), in their work on monkey brains, found that opiate receptors increased in a gradient along hierarchically organized cortical systems that sequentially process modality-specific sensory information of a progressively more complex nature. The authors suggest that specific opiate receptors may play a role in affective filtering of sensory stimuli at the cortical level. Their proposal offers a neural mechanism whereby the limbic-mediated emotional states essential for individual and species survival could influence selective attention to specific sensory stimuli.

Endorphins, Pain, and Stress: Experimental and Clinical Data on Humans

Following the discovery of endorphin peptides, active research has attempted to elucidate their physiological, neurochemical, and behavioral implications in animals and man (Bunney, Pert, Klee, Costa, Pert, & Davis, 1979). Most of the original hypotheses generated and tested derive from decades of previous research on morphine and other exogenous opiates. Major areas of functional inquiry, not necessarily independent, may be categorized as follows:

1. Pain perception, response, and endogenous modulation
2. Stress and sensory reactivity

3. Schizophrenia and psychotic disturbances
4. Addiction
5. General endocrinological and metabolic interactions

Bunney *et al.* (1979) delineated four research strategies employed in the functional analysis of endorphins. Modified versions of these strategies will be presented, along with major research findings on human subjects related to pain and stress. The dual phenomena of pain and stress are the focus of this review because of the fundamental interaction between these psychophysiological processes (Chapman, 1977; Bonica, 1977; Katz, Kellerman, & Siegel, 1981a).

Use of Narcotic Antagonists

The first strategy used, and still in use today, involves the introduction of narcotic antagonists that displace exogenous opiates from the opiate receptors, such as naloxone or naltrexene. This strategy rests on the premise that if a specific physiological or behavioral response is endorphin-mediated, then the administration of an opiate antagonist should create a measurable alteration in that response.

Studies on the effects of naloxone on experimentally induced pain in normal subjects generally have not found any interaction (El-Sobky, Dostrovsky, & Wall, 1976; Grevert & Goldstein, 1978). However, Buchsbaum, Davis, and Bunney (1977) found naloxone to alter pain perception in normal subjects identified as either sensitive or insensitive to pain. In pain-sensitive subjects, naloxone induced mild analgesia, whereas pain-insensitive subjects experienced hyperalgesia.

In clinical pain experiments, naloxone was found to increase reported pain intensity following a dental extraction more than placebo did (Levine, Gordon, Jones, & Fields, 1978). In another study, Levine, Gordon, and Fields (1979) found that dental patients experienced analgesia to low doses of naloxone and hyperalgesia to high doses. Placebo nonresponders, however, failed to demonstrate a significant response to naloxone at all. Chronic pain patients were found to experience no changes in spontaneous heat and pain thresholds after naloxone (Lindbloom & Tegner, 1979).

Naloxone has also been used to evaluate various analgesic interventions for their potential interaction or modulation of the endorphin system. Dental analgesia induced by electrical stimulation is partially or totally impaired by naloxone (Chapman & Benedetti, 1977), with a similar finding noted for acupuncture-induced analgesia of tooth pulp stimulation (Mayer, Price, Barber, & Rafi, 1976). Sjölund and Eriksson (1976) found that analgesia induced by electroacupuncture in chronic pain patients became ineffective after naloxone administration.

Several attempts have been made to evaluate the potential involvement of the endorphin system in hypnotic analgesia. Goldstein and Hilgard (1975) found naloxone to have no impact on hypnotic analgesia. This finding is supported by data collected by Mayer *et al.* (1976) and Barber and Mayer (1977), who found hypnosis to be a very potent analgesic resistant to naloxone interruption. These findings suggest that hypnosis operates independent of the endorphin system. However, Frid and Singer (1979) found that under conditions of stress, hypnotic analgesia was reduced by naloxone administration. Their results support the notion that the endorphin system is active during stress and appears to contribute to hypnotic analgesia under such a condition.

Support for endorphin activity's being related to stress derives from a study by Lewis, Cannon, and Liebeskind (1980). Rats exposed to prolonged intermittent foot shock were found to have naloxone-reversible analgesia for a subsequent pain threshold test. No such stress-analgesia was found for nonstressed controls. Brief continuous shock also led to stress-analgesia, but this was not naloxone-reversible. The authors concluded that there are both opioid and nonopioid mechanisms for stress analgesia.

Studies have been performed utilizing naloxone that implicate the endorphin system in physiological shock reactions in animals and humans. Faden and Holaday (1979) reported data on the potential role of endorphins in mediating hypotension resulting from both endotoxic and hypovolemic-induced shock in rats. In actual clinical cases with humans, Dirksen and his colleagues (1980) found naloxone to help alleviate symptoms of cardiogenic and septicemic shock. The authors suggest that some forms of stress may act as an input or trigger to the endorphin stress system, which may then serve an important mediating role in the organism's response (i.e., slowing down specific neuromuscular phenomenon). Carr (1981) hypothesized that endorphins may mediate a variety of physiological and psychological phenomena that have been described by people in near-death and terminal conditions, such as euphoria, dissociation, and sensory hallucinations.

As a group, these studies suggest that the endorphin system is not tonically active but rather is reactive to real, acute stress (Snyder & Childers, 1979).

Direct Administration of Endorphin

Rather than inferring the involvement of the endorphin system by naloxone reversal, this method focuses on direct administration of synthetic endorphins while observing alterations in behavior. In humans, β-EP has been found to be a very potent analgesic for clinical pain when introduced into the cerebrospinal fluid (CSF) (Foley, Kourides, Inturrisi, Kaiko, Zaroulis, Posner, Huude, & Li, 1979; Oyama, Jin, & Yamaya, 1980). Administered intravenously (IV), β-EP has been analgesic for some patients (Catlin, Hui, Loh, & Li, 1979) but not for others (Foley *et al.*, 1980). Although synthetic endorphins have generated great interest for their pain management potential, they have since been demonstrated to be as addictive as morphine (Snyder & Childers, 1979).

Direct Assay in Biological Fluids

The development of sensitive radioimmunoassay procedures (RIA) that can identify minute concentrations of endorphins in biological fluids and tissue samples provided an important tool for the study of these neuropeptides (Snyder & Childers, 1979). It is now possible to measure and compare endorphin concentrations in various sample populations, such as normals, chronic pain patients, pregnant women, or neonates. Studies of this sort are a much more direct means of implicating the endorphin system than are naloxone studies. RIA, however, is subject to many potential sources of variance and one must be careful in interpreting results (Jaffe & Behrman, 1978; Tietz, 1976).

In general, results of an RIA for endorphins are conservatively interpreted as endoprhin-like immunoreactivity. One cannot state conclusively that a specific endorphin concentration has been measured without first establishing the specificity of the RIA. As many RIAs are also sensitive to certain related compounds (i.e., cross-reactive), chromatography and/or other techniques are necessary to ensure that the assay is valid and to derive cross-reactivity ratios.

For the purpose of this general review, endorphin values will be reported as direct measures. The reader is cautioned that techniques and metrics vary between laboratories, making direct comparison of results difficult. It should be stressed, however, that RIA represents the most advanced technology for study of endorphins and other endocrinological substances, and results of the same lab are often highly consistent (Sharp, Pekary, Meyer, & Hershman, 1980).

In humans, the major biological fluids studied include blood plasma and cerebrospinal fluid (CSF). Wardlaw and Frantz (1979) found measurable quantities of β-EP in normal human plasma that averaged 21 $\pm$ 7.3 pg/ml (pg = picogram = 10^{-12}gr). Following administration of an ACTH-releasing drug (Metyrapone), mean plasma concentration increased to 55.4 $\pm$ 10.1 pg/ml. These data are consistent with previous findings on rats that β-EP and ACTH are secreted concomitantly by the pituitary in response to stress (Guillemin, Vargo, Rossier, Minick, Ling, Rivier, Bloom, & Vale, 1977). Suda, Liotta, and Krieger (1978) failed to detect β-EP in normal human plasma but found significant quantities in patients having endocrinological diseases associated with increased ACTH (mean concentration of 148 fmole/ml). Wilkes, Stewart, Bruni, Quigley, Yen, Ling, and Chretien (1980) found plasma concentrations of β-EP in normal men and women to be 115 ± 9 pg/ml, respectively. Metyrapone produced a rise in β-EP from 37–155%.

Akil, Watson, Barchas, and Li (1979) found pregnant women to have significantly higher β-EP plasma concentrations than nonpregnant women (62.5 ± 10 fmoles/ml versus 12 ± 3.9 fmoles/ml). Wardlaw, Stark, Baxi, and Frantz (1979) evaluated plasma endorphin concentrations relevant to the trauma of birth in humans. They measured β-EP in umbilical cord plasma of 45 term infants and found mean concentrations of 91 ± 16 pg/ml. These values were significantly higher than the normal adult level of 30.7 ± 2.7 pg/ml determined by the same assay. Highly significant negative correlations were found between umbilical arterial pH, oxygen content, and β-EP. No interaction was found between mode of delivery and endorphin concentrations, although the data suggested that neonates delivered vaginally had higher endorphin levels than those delivered by cesarean section. The authors concluded that the stress of childbirth on the fetus is likely to contribute to the higher levels of endorphin found, with secondary acidosis and hypoxia as potential stimuli for endorphin release.

In CSF, Almay, Johansson, Von Knorring, Terenius, and Wahlstrom (1978) found significant differences in endorphin concentrations between chronic pain patients with organic and psychogenic pain syndromes. Using a radioreceptor assay to measure methionine-enkephalin, patients with predominantly organic pain syndrome had mean concentrations in lumbar CSF of .60 ± .14 pmole/ml, whereas psychogenic pain patients had 1.25 ± .14 pmole/ml. Almay *et al.* also found a positive correlation between depressive symptomatology and CSF enkephalins in their total sample, as well as in each subgroup (organic and psychogenic). The authors suggest that their methodology could have potential diagnostic value for distinguising psychogenic and organic etiologies of chronic pain.

Evaluating lumbar CSF of normal subjects with an RIA, Akil, Watson, Sullivan, and Barchas (1978) found a mean methionine-enkephalin concentration of 3.12 ± 1.23 pmoles/ml, and a median of 1.7 pmoles/ml. Their findings demonstrate significant variability between subjects. Using the same assay, CSF withdrawn from the third ventricle of patients with intractable pain undergoing neurosurgical pain alleviation was also evaluated for methionine-enkephalin concentrations. The levels in these samples were all lower than the lowest normal (mean = .69 ± .24 pmole/ml). Differences in extraction sites and methods were noted as possible contaminants contributing to the different sample means, although the authors' unpublished data suggested this not to be significant. Wilkes *et al.* (1980) detected no β-EP in normal human lumbar CSF.

Jeffcoate, Rees, McLoughlin, Ratter, Hope, Lowry, and Besser (1978) evaluated β-EP concentrations in plasma and CSF of 20 patients with various disorders. Using a highly specific RIA, they found concentrations in CSF to range from 50 to 145 pmoles/l, consistently higher than in plasma (5–19 pmoles/l). No other relationship between CSF and plasma concentrations was found. Consistent findings were demonstrated by Nakao, Nakai, Oki, Matsubara, Konishi, Nishitani, and Imura (1980). They found β-EP concentrations in lumbar CSF of nonendocrine patients to have a mean of 91.6 ± 11.3 pg/ml. This mean, lower than that found by Jeffcoate *et al.* (1978), was attributed to the different RIA procedures used.

Intervention-Induced Alterations in Endorphin Concentrations

The fourth strategy involves the direct measurement of endorphins both before and after introduction of an intervention hypothesized to affect pain. Many investigators have evaluated the impact of direct electrical stimulation of the brain on pain experience and endorphin levels. Hosobuchi, Rossier, Bloom, and Guillemin (1979) found electrical stimulation of the periaqueductal gray matter in patients with pain of peripheral origin resulted in significant analgesia, with concomitant increases in ventricular β-EP of 50% to 300%. Akil, Richardson, Barchas, and Li (1978) found similar results with medial thalamic stimulation in patients with intractable pain.

Sjölund, Terenius, and Eriksson (1977) found increased endorphin in lumbar CSF after successful electroacupuncture in patients with lumbar pain, but not pain in other sites. Their findings suggest that local release of endorphins may contribute to analgesia, and evidence of endorphin activity may not always be evident in spinal fluid.

Clement-Jones, McLoughlin, Lowry, Besser, and Rees (1979) evaluated the analgesic potential of electroacupuncture on heroin addicts during withdrawal, while monitoring endorphin concentrations in blood and CSF. In addicts showing behavioral symptoms of withdrawal, β-EP levels were found to be elevated in blood and CSF. After successful electroacupuncture (i.e., reduction of symptoms), no alterations in β-EP were found. Methionine-enkephalin levels were not elevated prior to treatment but rose significantly in CSF after successful treatment. No comparable rise was detected in blood.

Cohen and his colleagues (1981) evaluated the acute stress of surgery on the concentration of β-EP measurable in plasma by RIA. Plasma samples were collected before, during, and after surgery. Five of the six patients showed marked rises in β-EP during surgery (mean peak response = +140%), accompanied by rises in cortisol. The authors concluded that surgical stress increases plasma β-EP levels, and they suggest that future research should evaluate postoperative pain and behavior as they relate to these differential levels.

Pediatric Endorphin Research Relevant to Pain and Stress

In a recent study (Katz, 1980; Katz, Sharp, Kellerman, Marston, Hershman, & Siegel, 1982), β-EP immunoreactivity was measured by RIA in CSF of 75 children with leukemia undergoing routine lumbar puncture (LP). These data were related to four behavioral measures of distress collected during the procedures: objective distress (PBRS), nurse ratings of anxiety, self-report of pain, and self-report of fear (Katz, Kellerman, & Siegel, 1980; Katz, 1979; Katz, 1980). Children ranged in age from 8 months to 18 years, 4 months, and all had experienced at least 3 prior LPs.

Mean β-EP for the entire sample was 111.27 pg/ml (SD = 42, range = 37–253). As one means of evaluating age effects on β-EP, the age range was divided into 6 age groups of roughly equal size and interval. Mean β-EP for each age group is presented in Figure 5. One-way analysis of variance indicated that β-EP significantly varied by age group ($p \leq .01$). Visual inspection of Figure 5 suggests a general decline in β-EP with increasing age, if the youngest children (under 4 years) are excluded. This is supported by a significant negative correlation between age and β-EP for these subjects ($r = -.31$, $p \leq .05$). The inconsistent mean for the youngest children appears to be related to antineoplastic chemotherapeutic variables. Females had a significantly lower mean β-EP than males (94 pg/ml, $SD = 41$; vs. 119, $SD = 40$: $p \leq .01$), with no age-by-sex interactions noted.

Data for the behavioral measures are presented by age group in Figure 6. No self-report data were available for children under 4 years of age. All behavioral measures showed general declines with advancing age, as evidenced by significant inverse correlations between age and

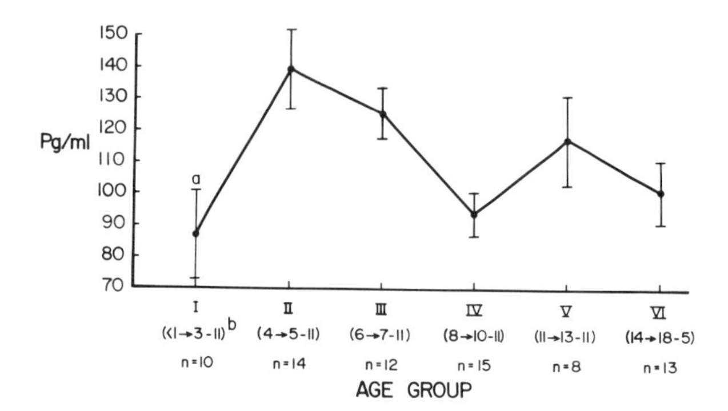

Figure 5. Mean beta-endorphin immunoreactivity for each age group. (a) Standard error of measurement. (b) Age ranges in years–months. From "Beta-Endorphin Immunoreactivity and Acute Behavioral Distress in Children with Leukemia" by E. R. Katz, B. Sharp, J. Kellerman, A. R. Marston, J. N. Hershman, and S. E. Siegel, *Journal of Nervous and Mental Disease*, 1982, *170*, 72–77. Copyright 1982 by Williams & Wilkins. Reprinted by permission.

each measure (PBRS: $r = -.67$, $p \leq .001$; nurse rating: $r = -.67$, $p \leq .001$; Fear Self-Report: $r = -.40$, $p \leq .01$; Pain Self-Report: $r = -.26$, $p \leq .05$). Females had higher distress scores than males on all measures, but these differences were not statistically significant.

After controlling for age, sex, and chemotherapy, β-EP and nurse ratings of anxiety were significantly positively correlated (partial correlation coefficient $= .31$, $p \leq .05$). Correlations between β-EP and the

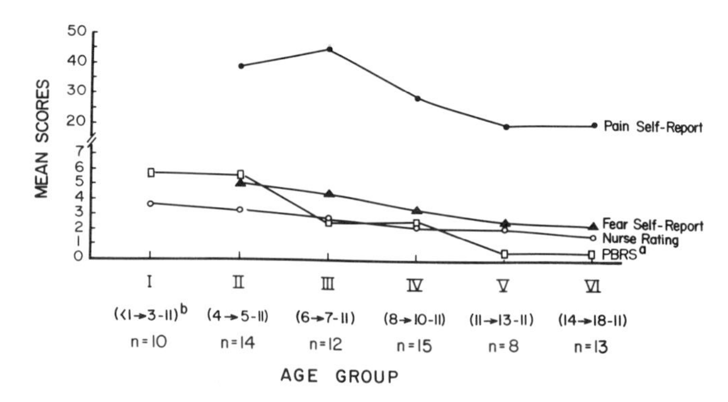

Figure 6. Mean scores on behavioral distress measures for each age group. (a) Procedure Behavioral Rating Scale. (b) Age ranges in years–months. From "Beta-Endorphin Immunoreactivity and Acute Behavioral Distress in Children with Leukemia" by E. R. Katz, B. Sharp, J. Kellerman, A. R. Marston, J. N. Hershman, and S. E. Siegel, *Journal of Nervous and Mental Disease*, 1982, *170*, 72–77. Copyright 1982 by Williams & Wilkins. Reprinted by permission.

other behavioral measures demonstrated positive trends. Nurse ratings appeared to be most sensitive to behavioral changes associated with alterations in β-EP immunoreactivity.

Results of this study support the reactive nature of β-EP in CSF to acute distress. It was hypothesized that increased endorphin activity in response to pain and stress is part of the organism's attempt to control or modulate the effects of perceived, noxious stimulation. Younger children tended to experience greater distress and generally had correspondingly higher β-EP. These findings are consistent with previous investigators who have viewed endorphins as having an important role in psychophysiological homeostasis (Stein & Belluzzi, 1978; Vereby, Volavka, & Clouet, 1978).

Since females generally have higher levels of distress than males under the same conditions (Katz *et al.*, 1980; Maccoby & Jacklin, 1974; Woodrow, 1972), one might have expected females to have higher β-EP immunoreactivity than males also. That females were found to have lower β-EP than males suggests the existence of sex differences in the reactive synthesis/secretion of β-EP in CSF to acute distress. If β-EP is a homeostatic facilitator, males may be biologically more adept at minimizing pain and stress because of their enhanced ability to synthesize/secrete β-EP. An alternative hypothesis suggests that females have less of a need for enhanced endorphin, perhaps because of better expressive abilities and less internal stress. These hypotheses are highly speculative at present and will require further, carefully controlled investigation.

Results of this investigative study on children with leukemia must be interpreted cautiously. Even after careful statistical evaluation, the potential complexities of illness, combination chemotherapy, and psychosocial adjustment may have impacted the data in unforseen ways. These findings should be considered preliminary, requiring further controlled research.

Summary of Endorphin Research and Future Directions for Pediatric Study

Endorphins are an important class of neuropeptides with demonstrated involvement in various aspects of sensory experience, including pain and stress. It appears that a primary trigger for the synthesis/secretion of endorphins is the physiological and/or cognitive perception of stress. The reactive increase in endorphin concentrations under such conditions probably serves many homeostatic-engendering functions, one of which is the endogenous modulation of pain through neural inhibition.

Because endorphin activity is associated with alterations in various other neurochemical compounds, future pediatric research should include concurrent study of other stress-related endocrine substances, such as cortisol and catecholamines. By developing comprehensive neuroendocrinological patient profiles (Mason, 1975), coupled with behavioral profiles under controlled circumstances, it may be possible to tailor more successfully psychological and pharmacological pain- and stress-reducing strategies to individual patients. Such research would also help clarify the interface between cognitive experiences and neurobiochemical responses and determine how each potentially effects the other.

CONCLUSION

Even though the systematic behavioral and neurochemical study of pediatric pain represents a relatively recent area of investigation when compared to the extensive literature on adult pain, the pediatric field has developed a solid empirical foundation from which to proceed. Although this research has primarily been limited to a small population of children and adolescents, this is both the most parsimonious and ethical approach in the development of a new field. The cumulative positive results of these diverse studies suggest that it is now appropriate to begin a series of investigations with larger groups of pediatric patients.

The application of these behavioral pediatrics interventions may have a profound impact across a number of pediatric disorders, where pain is a primary or secondary symptom of acute and chronic diseases. Additionally, the neurochemical investigations hold the promise of advancing an understanding of pain mechanisms which may ultimately result in enhancing treatment effectiveness. Thus, the potential of the behavioral pediatrics model in substantially adding to our understanding of the pediatric pain experience, as well as a preventive approach to later adult pain, signals the ultimate significance of pediatric pain investigations as a major contribution to the comprehensive study of acute and chronic pain.

REFERENCES

Akil, H., & Liebeskind, J. C. Monoaminergic mechanisms of stimulation-produced analgesia. *Brain Research*, 1975, *94*, 279–296.

Akil, H., Richardson, D. E., Barchas, J. D., & Li, C. H. Appearance of β-endorphin-like immunoreactivity in human ventricular cerebrospinal fluid upon analgesic electrical stimulation. *Proceedings of the National Academy of Sciences, U.S.A.*, 1978, *75*, 5170–5172.

Akil, H., Watson, S. J., Sullivan, S., & Barchas, J. D. Enkephalin-like material in normal human CSF: Measurement and levels. *Life Sciences*, 1978, *23*, 121–126.

Akil, H., Watson, S. J., Barchas, J. D., & Li, C. H. β-endorphin immunoreactivity in rat and human blood: Radioimmunoassay, comparative levels, and physiological alterations. *Life Sciences*, 1979, *24*, 1659–1665.

Almay, B. G. L., Johansson, F., Von Knorring, L., Terenius, L., & Wahlstrom, A. Endorphins in chronic pain: I. Differences in CSF endorphin levels between organic and psychogenic pain syndrome. *Pain*, 1978, *5*, 153–162.

Arnold, W. D., & Hilgartner, M. W. Hemophilic arthropathy: Current concepts of pathogenesis and management. *Journal of Bone and Joint Surgery*, 1977, *59*, 287–305.

Barber, J., & Mayer, D. Evaluation of the efficacy and neural mechanism of a hypnotic analgesia procedure in experimental and clinical dental pain. *Pain*, 1977, *4*, 41–48.

Basbaum, A. I., & Fields, H. L. Endogenous pain control mechanisms: Review and hypothesis. *Annals of Neurology*, 1978, *4*, 451–462.

Basbaum, A. I., Clanton, C. H., & Fields, H. L. Opiate and stimulus-produced analgesia: Functional anatomy at a medullospinal pathway. *Proceedings of the National Academy of Sciences, U.S.A.*, 1976, *73*, 4685–4688.

Beecher, H. K. *Measurement of subjective responses: Qualitative effects of drugs*. New York: Oxford, 1959.

Benedetti, C. Neuroanatomy and biochemistry of antinociception. In J. J. Bonica (Ed.), *Advances in pain research and therapy (Vol. 2)*. New York: Raven Press, 1979.

Berger, H. G., Honig, P. J., & Liebman, R. Recurrent abdominal pain: Gaining control of the symptom. *American Journal of Diseases of Children*, 1977, *131*, 1340–1344.

Bonica, J. J. Neurophysiologic and pathologic aspects of acute and chronic pain. *Archives of Surgery*, 1977, *112*, 750–761.

Bradbury, A. F., Smyth, D. G., & Snell, C. R. Liptropin: Precursor to two biologically active peptides. *Biochemical and Biophysical Research Communication*, 1976, *69*, 950–956.

Bradbury, A. F., Smyth, D. G., Snell, C. R., Birdsall, N. J. M., & Hulme, E. C. C fragment of lipotropin has a high affinity for brain opiate receptors. *Nature*, 1976, *260*, 793–795.

Brown, J. K. Migraine and migraine equivalents in children. *Developmental Medicine and Child Neurology*, 1977, *19*, 683–692.

Buchanan, G. R., & Levy, S. V. Use of prothrombin complex concentrates in hemophiliacs with inhibitors: Clinical and laboratory studies: *Pediatrics*, 1978, *62*, 767–774.

Buchsbaum, M. S., Davis, G. C., & Bunney, W. E. Naloxone alters pain perception and somatosensory evoked potentials in normal subjects. *Nature*, 1977, *270*, 620–622.

Bunney, W. E., Pert, C. B., Klee, W., Costa, E., Pert, A., & Davis, G. C. Basic and clinical studies of endorphins. *Annals of Internal Medicine*, 1979, *91*, 239–250.

Cannon, J. T., Liebeskind, J. C., & Frenk, H. Neural and neurochemical mechanisms of pain inhibition. In R. A. Sternbach (Ed.), *The psychology of pain*. New York: Raven Press, 1978,

Carr, D. B. Endorphins at the approach of death. *Lancet*, 1981, *I*, 390.

Catlin, D. H., Hui, K. K., Loh, H. H., & Li, C. H. β-endorphin: Initial clinical studies. In E. Usdin, W. E. Bunney, & N. S. Kline (Eds.), *Endorphins in mental health research*. New York: Oxford, 1979.

Chapman, C. R. Psychological aspects of pain patient treatment. *Archives of Surgery*, 1977, *112*, 767–772.

Chapman, C. R., & Benedetti, C. Analgesia following transcutaneous electrical stimulation and its partial reversal by narcotic antagonist. *Life Sciences*, 1977, *21*, 1645–1648.

Christensen, M. F., & Mortensen, O. Long-term prognosis in children with recurrent abdominal pain. *Archives of Disease in Childhood*, 1975, *50*, 110–114.

Chung, S. H., & Dickenson, A. Pain, enkephalin, and acupuncture. *Nature,* 1980, *283,* 243.

Clement-Jones, V., McLoughlin, L., Lowry, P. J., Besser, G. M., & Rees, L. H. Acupuncture in heroin addicts: Changes in metenkephalin and β-endorphin in blood and cerebrospinal fluid. *Lancet,* 1979, *1,* 380–382.

Clement-Jones, V., Lowry, P. J., Rees, L. H., & Besser, G. M. Metenkephalin circulates in human plasma. *Nature,* 1980, *283,* 295–297.

Cohen, M., Pickar, D., Dubois, M., Roth, Y. F., Naber, D., & Bunney, W. E. Surgical stress and endorphins. *Lancet,* 1981, *I,* 213–214.

Cox, B. M., Goldstein, A., & Li, C. H. Opioid activity of a peptide, β-lipotropin-(61-91), derived from β-lipotropin. *Proceedings of the National Academy of Sciences, U.S.A.,* 1976, *73,* 1821–1823.

Crasilneck, H. B., & Hall, J. A. Clinical hypnosis in problems of pain. *American Journal of Clinical Hypnosis,* 1973, *14,* 55–60.

Curless, R. G., & Corrigan, J. J. Headache in classical hemophilia: The risk of diagnostic procedures: *Child's Brain,* 1976, *2,* 187–194.

Dash, J. Hypnosis for symptom amelioration. In J. Kellerman (Ed.), *Psychological aspects of childhood cancer.* Springfield, Ill.: Charles C Thomas, 1980.

Degenaar, J. J. Some philosophical considerations on pain. *Pain,* 1979, *7,* 281–304.

Dennis, S. G., & Melzack, R. Pain-signalling systems in the dorsal and ventral spinal cord. *Pain,* 1977, *4,* 97–132.

Dietrich, S. L. Musculoskeletal problems. In M. W. Hilgartner (Ed.), *Hemophilia in children.* Littleton, Mass.: Publishing Sciences Group, 1976.(a)

Dietrich, S. L. Medical management of hemophilia. In D. C. Boone (Ed.), *Comprehensive management of hemophilia.* Philadelphia: Davis, 1976.(b)

Dietrich, S. L., Luck, J. V., & Martinson, A. M. Musculoskeletal problems. In M. W. Hilgartner (Ed.), *Hemophilia in the child and adult.* New York: Masson, in press.

Diggs, L. W., & Flowers, E. Sickle cell anemia in the home environment: Observations on the natural history of the disease in Tennessee children. *Clinical Pediatrics,* 1971, *10,* 677.

Dirksen, R., Otten, M. H., Wood, G. J., Verbaan, C. J., Haalebos, M. M. P., Verdouw, P. V., & Nijhuis, G. M. M. Naloxone in shock. *Lancet,* 1980, *II,* 1360–1361.

Ellenberg, L., Kellerman, J., Dash, J., Higgins, G., & Zeltzer, L. Use of hypnosis for multiple symptoms in an adolescent girl with leukemia. *Journal of Adolescent Health Care,* 1980, *1,* 132–136.

El-Sobky, A., Dostrovsky, J. O., & Wall, P. O. Lack of effect of naloxone in pain perception in humans. *Nature,* 1976, *263,* 783–784.

Epstein, M. H., & Harris, J. Children with chronic pain—Can they be helped? *Pediatric Nursing,* 1978, *4,* 42–44.

Eriksson, M. R. E., Sjölund, B. H., & Nielzen, S. Long-term results of peripheral conditioning stimulation as an analgesic measure in chronic pain. *Pain,* 1979, *6,* 335–347.

Fabrega, H., & Tyma, S. Language and cultural influences in the description of pain. *British Journal of Medical Psychology,* 1976, *49,* 349–371.

Faden, A. I., & Holaday, J. W. Opiate antagonists: A role in the treatment of hypovolemic shock. *Science,* 1979, *205,* 317–318.

Fields, H. L., Basbaum, A. I., Clanton, C. H., & Anderson, S. D. Nucleus raphe magnus inhibition of spinal cord dorsal horn neurons. *Brain Research,* 1977, *126,* 441–453.

Foley, K. M., Kourides, I. A., Inturrisi, C. E., Kaiko, R. F., Zaroulis, C. G., Posner, J. B., Huude, R. W., & Li, C. H. β-endorphin: Analgesia and hormonal effects in humans. *Proceedings of the National Academy of Sciences, U.S.A.,* 1979, *76,* 5377–5381.

Fordyce, W. E. *Behavioral methods for chronic pain and illness.* St. Louis: Mosby, 1976.

Fordyce, W. E., Fowler, R. S., Lehmann, J. F., & DeLateur, B. J. Some implications of learning in problems of chronic pain. *Journal of Chronic Diseases*, 1968, *21*, 179–190.

Fordyce, W. E., Brena, S. F., Holcomb, R. J., DeLateur, B. J., & Loeser, J. D. Relationship of patient semantic pain descriptions to physician diagnostic judgments, activity level measures and MMPI. *Pain*, 1978, *5*, 293–303.

Frid, M., & Singer, G. Hypnotic analgesia in conditions of stress is partially reversed by naloxone. *Psychopharmacology*, 1979, *63*, 211–215.

Gardner, G. G. Childhood death and human dignity: Hypnosis for David. *International Journal of Clinical and Experimental Hypnosis*, 1976, *24*, 122–139.

Goldstein, A. Opiate receptors. *Life Sciences*, 1974, *14*, 615–623.

Goldstein, A., & Hilgard, E. R. Failure of the opiate antagonist naloxone to modify hypnotic analgesia. *Proceedings of the National Academy of Sciences, U.S.A.*, 1975, *72*, 2041–2043.

Goldstein, A., Tachibana, S., Lowney, L. I., Hunkapiller, M., & Hood, I. Dynorphin-(1–13), an extraordinarily potent opiod peptide. *National Academy of Sciences, U.S.A.*, 1979, *76*, 6666–6670.

Gracely, R. H., McGrath, P., & Dubner, R. Ratio scales of sensory and affective verbal pain descriptors. *Pain*, 1978, *5*, 5–18.

Grevert, P., & Goldstein, A. Endorphins: Naloxone fails to alter experimental pain or mood in humans. *Science*, 1978, *199*, 1093–1095.

Guillemin, R., Ling, N., & Burgus, R. Endorphines, peptides, d'origine hypothalamique et neurohypophysaire à activité morphinomimetique. Isolement et structure moleculaire de l'endorphine. *C.R. Academy of Science* (Paris,), 1976, *282*, 783–785.

Guillemin, R., Vargo, T., Rossier, J., Minick, S., Ling, N., Rivier, C., Bloom, F., & Vale, W. β-endorphin and adrenocorticotropin are secreted concomitantly by the pituitary gland. *Science*, 1977, *197*, 1367–1369.

Herkenham, M., & Pert, C. B. In vitro autoradiography of opiate receptors in rat brain suggests loci of "opiotergic" pathways. *Proceedings of the National Academy of Sciences, U.S.A.*, 1980, *77*, 5532–5536.

Hilgard, E.R. The alleviation of pain by hypnosis. *Pain*, 1975 *1*, 213–231.

Hilgartner, M.W. (Ed.), *Hemophilia in children*, Littleton, Mass.: Publishing Sciences Group, 1976.

Hökfelt, T., Ljundahl, A., Terenius, L., Elde, R., & Nilsson, G. Immunohistochemical analysis of peptide pathways possible related to pain and analgesia: Enkephalin and substance P. *Proceedings of the National Academy of Sciences, U.S.A.*, 1977, *74*, 3081–3085.

Hosobuchi, Y., Adams, J. E., & Linchitz, R. Pain relief by electrical stimulation of the central gray matter in humans and its reversal by naloxone. *Science*, 1977, *197*, 183–186.

Hosobuchi, Y., Rossier, J., Bloom, F. E., & Guillemin, R. Stimulation of human periaqueductal gray for pain relief increases immunoreactive β-endorphin in ventricular fluid. *Science*, 1979, *203*, 279–281.

Hughes, J. Isolation of an endogenous compound from the brain with pharmacological properties similar to morphine. *Brain Research*, 1975, *88*, 295–308.

Hughes, J., Smith, T. W., Kosterlitz, H. W., Fothergill, L., Morgan, B. A., & Morris, H. R. Identification of two related pentapeptides from the brain with potent opiate agonist activity. *Nature*, 1975, *258*, 577–579.

Iverson, L. L. The chemistry of the brain. *Scientific American*, September 1979, *241*, 134–149.

Jaffe, B. M., & Behrman, H. R. *Methods of hormone radioimmunoassay*. New York: Academic Press, 1978.

Jeffcoate, W. J., Rees, L. H., McLoughlin, L., Ratter, S. J., Hope, J., Lowry, P. J., & Besser, G. M. β-endorphin in human cerebrospinal fluid. *Lancet*, 1978, *1*, 119–121.

Jessel, T. M., & Iverson, L. L. Opiate analgesics inhibit substance P release from rat trigeminal nucleus. *Nature*, 1977, *268*, 549–551.

Kasper, C. K. Hematologic care. In D. C. Boone (Ed.), *Comprehensive management of hemophilia*. Philadelphia: Davis, 1976.

Katz, E. R. *Distress behavior in children with leukemia undergoing medical procedures*. Paper presented at the annual convention of the American Psychological Association, New York, September 1979. ERIC Report ED 182 659.

Katz, E. R. *Beta-endorphin and acute behavioral distress in children with leukemia*. Doctoral dissertation, University of Southern California, July 1980.

Katz, E. R., Kellerman, J., & Siegel, S. E. Distress behavior in children with cancer undergoing medical procedures: Developmental considerations. *Journal of Consulting and Clinical Psychology*, 1980, *48*, 356–365.

Katz, E. R., Kellerman, J., & Siegel, S. E. Anxiety as an affective focus in the clinical study of acute behavioral distress: A reply to Shacham and Dant. *Journal of Consulting and Clinical Psychology*, 1981, *49*, 470–471.(a)

Katz, E. R., Kellerman, J., & Siegel, S. E. *Objective and subjective measurement of behavioral distress in children with leukemia*. Unpublished manuscript, 1981.(b)

Katz, E. R., Sharp, B., Kellerman, J., Marston, A. R., Hershman, J. N., & Siegel, S. E. Beta-endorphin immunoreactivity and acute behavioral distress in children with leukemia. *Journal of Nervous and Mental Disease*, 1982, *170*, 72–77.

Kellerman, J. (Ed.). *Psychological aspects of childhood cancer*. Springfield, Ill.: Charles C Thomas, 1980.

Kellerman, J., & Varni, J. W. Psychosocial aspects of pediatric hematology–oncology. In M. L. N. Willoughby & S. E. Siegel (Eds.), *Modern trends in pediatrics*. London: Butterworth, in press.

Kimura, S., Lewis, R. V., Stern, A. S., Rossier, J., Stein, S., & Udenfriend, S. Probable precursors of (Leu) enkephalin and (Met) enkephalin in adrenal medulla: Peptides of 3–5 kilodaltons. *Proceedings of the National Academy of Sciences, U.S.A.*, 1980, *77*, 1681–1685.

Kisker, C. T., Perlman, A. W., & Benton, C. Arthritis in hemophilia. *Seminars in Arthritis and Rheumatism*, 1971, *1*, 220–235.

Konishi, S., & Otsuka, M. The effects of substance P and other peptides on spinal neurons of the frog. *Brain Research*, 1974, *65*, 397–410.

Kuhar, M. H., Pert, C. B., & Snyder, S. H. Regional distribution of opiate receptor binding in monkey and human brain. *Nature*, 1973, *245*, 447–450.

La Baw, W., Holton, C., Tewell, K., & Eccles, D. The use of self-hypnosis by children with cancer. *American Journal of Clinical Hypnosis*, 1975, *17*, 233–238.

Lamotte, C., Pert, C. B., & Snyder, S. H. Opiate receptor binding in primate spine: Distribution and changes after dorsal root section. *Brain Research*, 1976, *112*, 407–412.

Lazarus, L. H., Ling, N., & Guillemin, R. β-lipotropin as a prohormone for the morphinomimetric peptides endorphins and enkephalins. *Proceedings of the National Academy of Sciences, U.S.A.*, 1976, *73*, 2156–2159.

Lehmann, H., Huntsman, R. S., Casey, R., Lang, A., Lorkin, P. A., & Comings, D. E. Sickle cell disease and related disorders. In J. W. Williams, E. Beutler, A. J. Ersler, & R. W. Rundles (Eds.), *Hematology*. New York: McGraw-Hill, 1977.

Lehmann, J. F., Warren, C. G., & Scham, S. M. Therapeutic heat and cold. *Clinical Orthopedics and Related Research*, 1974, *99*, 207–245.

Leichtman, D. A., & Brewer, G. J. Elevated plasma levels of fibrinopeptide A during sickle cell anemia pain crisis: Evidence of intravascular coagulation. *American Journal of Hematology*, 1978, *5*, 183–190.

Levine, J. D., Gordon, N. C., Jones, R. T., & Fields, H. L. The narcotic antagonist naloxone enhances clinical pain. *Nature*, 1978, *272*, 826–827.

Levine, J. D., Gordon, N. C., & Fields, H. L. Naloxone dose dependently produces analgesia and hyperalgesia in post-operative pain. *Nature*, 1979, *278*, 740–741.
Lewis, J. W., Cannon, J. T., & Liebeskind, J. C. Opioid and nonopioid mechanisms of stress analgesia. *Science*, 1980, *208*, 623–625.
Lewis, M. E., Mishkin, M., Bragin, E., Brown, R. M., Pert, C. B., & Pert, A. Opiate receptor gradients in monkey cerebral cortex: Correspondence with sensory processing hierarchies. *Science*, 1981, *211*, 1166–1169.
Lindbloom, V., & Tegner, R. Are the endorphins active in clinical pain states? Narcotic antagonism in chronic pain patients. *Pain*, 1979, *7*, 65–68.
Maccoby, E. E., & Jacklin, C. N. *The psychology of sex differences*. Stanford, Calif.: Stanford University Press, 1974.
Maddison, T. G. Recurrent abdominal pain in children. *Medical Journal of Australia*, 1977, *1*, 708–710.
Marx, J. L. Brain peptides: Is substance P a transmitter of pain signals? *Science*, 1979, *205*, 885–889.
Mason, J. W. Emotion as reflected in patterns of endocrine integration. In L. Levi (Ed.), *Emotions: Their parameters and measurement*. New York: Raven Press, 1975.
Mayer, D. J., & Liebeskind, J. C. Pain reduction by focal electrical stimulation of the brain: An anatomical and behavioral analysis. *Brain Research*, 1974, *68*, 73–93.
Mayer, D. J., & Price, D. D. Central nervous system mechanisms of analgesia. *Pain*, 1976, *2*, 379–404.
Mayer, D. J., Wolfe, T. L., & Akil, H. Analgesia from electrical stimulation in the brainstem of the rat. *Science*, 1971, *174*, 1351–1354.
Mayer, D. J., Price, D. D., Barber, J., & Rafi, A. Acupuncture analgesia: Evidence for activation of a pain inhibitory system as a mechanism of action. In J. J. Bonica & D. Albe-Fessard (Eds.), *Advances in pain research and therapy (Vol. 1)*. New York: Raven Press, 1976.
Melzack, R. *The puzzle of pain*. Harmondsworth, England: Penguin, 1973.
Melzack, R. The McGill Pain Questionnaire: Major properties and scoring methods. *Pain*, 1975, *1*, 277–299.(a)
Melzack, R. Prolonged relief of pain by brief, intense transcutaneous somatic stimulation. *Pain*, 1975, *1*, 357–373.(b)
Merskey, H. On the development of pain. *Headache*, 1970, *10*, 116–123.
Messing, R. B., & Lytle, L. D. Serotonin-containing neurons: Their possible role in pain and analgesia. *Pain*, 1978, *4*, 1–21.
Miller, A. J., & Kratochwill, T. R. Reduction of frequent stomachache complaints by time out. *Behavior Therapy*, 1979, *10*, 211–218.
Millichap, J. G. Recurrent headaches in 100 children: Electroencephalographic abnormalities and response to Phenytoin (dilantin). *Child's Brain*, 1978, *4*, 95–105.
Moe, P. G. Headaches in children: Meeting the challenge of management. *Postgraduate Medicine*, 1978, *63*, 169–174.
Nakao, K., Nakai, Y., Oki, S., Matsubara, S., Konishi, H., Nishitani, H., & Imura, H., Immunoreactive β-endorphin in human cerebrospinal fluid. *Journal of Clinical Endocrinology and Metabolism*, 1980, *50*, 230–233.
Øster, J. Recurrent abdominal pain, headache and limb pains in children and adolescents. *Pediatrics*, 1972, *50*, 429–436.
Oyama, T., Jin, T., & Yamaya, R. Profound analgesic effects of β-endorphin in man. *Lancet*, 1980, *1*, 122–124.
Parry, D. H., & Bloom, A. L. Failure of Factor VIII inhibitor bypassing activity (Feiba) to

secure hemostasis in hemophilic patients with antibodies. *Journal of Clinical Pathology*, 1978, *31*, 1102–1105.

Pert, A., & Yaksh, T. Sites of morphine-induced analgesia in the primate brain: Relation to pain pathways. *Brain Research*, 1974, *80*, 135–140.

Pert, C. B., Pasternak, G., Snyder, S. H. Opiate agonists and antagonists discriminated by receptor binding in brain. *Science*, 1973, *182*, 1359–1361.

Peterson, H. A. Leg aches. *Pediatric Clinics of North America*, 1977, *24*, 731–736.

Redwood, A. M., Williams, E. M., Desai, P., & Serjeant, G. R. Climate and painful crisis of sickle cell disease in Jamaica. *British Medical Journal*, 1976, *1*, 66.

Sank, L. I., & Biglan, A. Operant treatment of a case of recurrent abdominal pain in a 10-year-old boy. *Behavior Therapy*, 1974, *5*, 677–681.

Scranton, P. E., Hasiba, U., & Gorenc, T. J. Intramuscular hemorrhage in hemophiliacs with inhibitors. *Journal of the American Medical Association*, 1979, *241*, 2028–2030.

Sergis-Deavenport, E., & Varni, J. W. *Behavioral techniques and therapeutic adherence to proper factor replacement procedures in hemophilia*. Unpublished manuscript, 1981.

Sharp, B., Pekary, A. E., Meyer, N. V., & Hershman, J. M. Beta-endorphin in male rat reproductive organs. *Biophysical and Biochemical Research Communications*, 1980, *95*, 618–623.

Sheppard, I. M., & Sheppard, S. M. Characteristics of temporomandibular joint problems. *Journal of Prosthetic Dentistry*, 1977, *38*, 180–191.

Sherman, J. E., & Liebeskind, J. C. An endorphinergic, centrifugal substrate of pain modulation: Recent findings, current concepts, and complexities. In J. J. Bonica (Ed.), *Pain*. New York: Raven Press, 1980.

Simon, E. J., & Hiller, J. M. The opiate receptors. *Annual Review of Pharmacology and Toxicology*, 1978. *18*, 371–394.

Simon, E. J., Hiller, J. M., & Edelman, I. Stereospecific binding of the potent narcotic analgesic (^{3}H) etorphins to rat-brain homogenate. *Proceedings of the National Academy of Science, U.S.A.*, 1973, *70*, 1947–1949.

Sjölund, B., & Eriksson, M. Electro-acupuncture and endogenous morphines. *Lancet*, 1976, *2*, 1085.

Sjölund, B., Terenius, L., & Eriksson, M. Increased cerebrospinal fluids levels of endorphins after electro-acupuncture. *Acta Physiology Scandinavia*, 1977, *100*, 382–384.

Snyder, S. H., & Childers, S. R. Opiate receptors and opioid peptides. *Annual Review of Neurosciences*, 1979, *2*, 35–64.

Sokoloff, L. Biochemical and physiological aspects of degenrative joint disease with special reference to hemophilic arthropathy. *Annals of New York Academy of Science*, 1975, *240*, 285–290.

Song, J. *Pathology of sickle cell disease*. Charles C Thomas: Springfield, Ill., 1971.

Stein, L., & Belluzzi, J. D. Brain endorphins and the sense of well-being: A psychobiological hypothesis. In E. Costa & M. Trabucchi (Eds.), *Advances in biochemical psychopharmacology (Vol. 18)*. New York: Raven Press, 1978,

Stickler, G. B. & Murphy, D. B. Recurrent abdominal pain. *American Journal of Diseases of Children*, 1979, *133*, 486–489.

Stilz, R. J., Carron, H., & Sanders, D. B. Reflex sympathetic dystrophy in a 6-year old: Successful treatment by transcutaneous nerve stimulation. *Anesthesia and Analgesia*, 1977, *56*, 438–443.

Stone, R. T., & Barbero, G. J. Recurrent abdominal pain in childhood. *Pediatrics*, 1970, *45*, 732.

Suda, T., Liotta, A. S., & Krieger, D. T. β-endorphin is not detectable in plasma from normal human subjects. *Science*, 1978, *202*, 221–223.

Swanson, D. W., Marita, T., & Swenson, W. M. Results of behavior modification in the treatment of chronic pain. *Psychosomatic Medicine,* 1979, *41,* 55–61.

Swezey, R. L. *Arthritis: Rational therapy and rehabilitation.* Philadelphia: Saunders, 1978.

Terenius, L. Characteristics of the "receptor" for narcotic analgesics and synaptic plasma membrane fraction from rat brain. *Acta Pharmacology and Toxicology,* 1973, *33,* 377–384.

Terenius, L. Effect of peptides and amino acids on dihydromorphine binding to the opiate receptor. *Journal of Pharmacy and Pharmacology* (London), 1975, *27,* 450–451.

Terenius, L. Significance of endorphins in endogenous antinociception. In E. Costa & M. Trabucchi (Eds.), *Advances in biochemical psychopharmacology (Vol. 18).* New York: Raven Press, 1978.

Terenius, L., & Wahlstrom, A. Endorphins and clinical pain: An overview. *Advances in Experimental Medical Biology,* 1979, *116,* 261–277.

Tietz, N. W. (Ed.). *Fundamentals of clinical chemistry.* Philadelphia: Saunders, 1976.

Tomasi, L. G. Headaches in children. *Comprehensive Therapy,* 1979, *5,* 13–19.

Varni, J. W. Behavioral medicine in hemophilia arthritic pain management: Two case studies. *Archives of Physical Medicine and Rehabilitation,* 1981, *62,* 183–187.(a)

Varni, J. W. Self-regulation techniques in the management of chronic arthritic pain in hemophilia. *Behavior Therapy,* 1981, *12,* 185–194.(b)

Varni, J. W., & Gilbert, A. Self-regulation of chronic arthritic pain and long-term analgesic dependence in a hemophiliac. *Rheumatology and Rehabilitation,* in press.

Varni, J. W., Bessman, C. A., Russo, D. C., & Cataldo, M. F. Behavioral management of chronic pain in children: Case study. *Archives of Physical Medicine and Rehabilitation,* 1980, *61,* 375–379.

Varni, J. W., Gilbert, A., & Dietrich, S. L. Behavioral medicine in pain and analgesia management for the hemophilic child with Factor VIII Inhibitor. *Pain,* 1981, *11,* 121–126.

Varni, J. W., Kellerman, J., & Dash, J. Pediatric and adolescent pain management. In T. J. Coates (Ed.), *Behavioral medicine: A practical handbook.* Champaign, Ill.: Research Press, in press.

Vereby, K., Volavka, J., & Clouet, D. Endorphins in psychiatry. *Archives of General Psychiatry,* 1978, *35,* 877–888.

Wardlaw, S. L., & Frantz, A. G. Measurement of β-endorphin in human plasma. *Journal of Clinical Endocrinology and Metabolism,* 1979, *48,* 176–180.

Wardlaw, S. L., Stark, R. I., Baxi, L., & Frantz, A. G. Plasma β-endorphin and β-lipotropin in the human fetus at delivery: Correlation with arterial pH and pO_2. *Journal of Clinical Endocrinology and Metabolism,* 1979, *79,* 888–891.

Wasserman, R. R., Oester, Y. T., Oryshkevich, R. S., Montgomery, M. M., Poske, R. M., & Ruksha, A. Electromyographic, electrodiagnostic, and motor nerve conduction observations in patients with rheumatoid arthritis. *Archives of Physical Medicine and Rehabilitation,* 1968, *49,* 90–95.

White, J. R. Effects of a counterirritant on perceived pain and hand movement in patients with arthritis. *Physical Therapy,* 1973, *53,* 956–960.

Wilkes, M. M., Stewart, R. D., Bruni, J. F., Quigley, M. E., Yen, S. S. C., Ling, N., & Chretien, M. A specific homologous radioimmunoassay for human β-endorphin: Direct measurement in biological fluids. *Journal of Clinical Endocrinology and Metabolism,* 1980, *50,* 309–315.

Woodrow, K. M., Friedman, M. D., Siegelabu, M. S., & Collen, M. D. Pain tolerance: Differences according to age, sex, and race. *Psychosomatic Medicine,* 1972, *34,* 548–556.

Zeltzer, L. K., Dash, J., & Holland, J. P. Hypnotically induced pain control in sickle cell anemia. *Pediatrics,* 1979, *64,* 533–536.

Zimmerman, M. Neurophysiology of nociception, pain and pain therapy. In J. J. Bonica & V. Ventafridda (Eds.), *Advances in pain research and therapy.* New York: Raven Press, 1979.

7

Preparation for Surgery and Medical Procedures

Barbara G. Melamed, Rochelle L. Robbins, and Shirley Graves

There are large numbers of children for whom the hospital experience results in transient or moderately severe behavioral distrubances (Cassell, 1965). The stress of hospitalization includes fears related to separation from the parents, the distress of unfamiliar surroundings, anxiety about painful procedures, and the discomfort of the recovery from surgery or illness. Behavioral disturbances occurring in as many as 92% of hospitalized children include regressive behaviors such as increased dependency, loss of toilet training, excessive fears, and sleep and eating disturbances (Chapman, Loeb, & Gibbons, 1956; Gellert, 1958; Goffman, Buckman, & Schade, 1957). Since there is such a wide range of reported disturbances, there is a need to identify the population at risk for emotional stress related to hospitalization. Some researchers report that only about 10% to 35% of the problems precipitated by the hospital experience lead to serious long-term disturbances (Prugh, Staub, Sands, Kirschbaum, & Lenihan, 1953).

In the past decade, there has been increased research demonstrating that psychological factors influence the course of illness and bodily re-

BARBARA G. MELAMED • University of Florida, Gainesville, Florida 32610. ROCHELLE L. ROBBINS • Cornell University Medical Center, New York, New York 10021. SHIRLEY GRAVES • University of Florida School of Medicine, Gainesville, Florida 32610.

sponse to stress. It is not surprising that this recognition has led to an increase in the use of psychological preparation in hospitals. A recent survey (Peterson & Ridley-Johnson, 1980) reported that over 70% of the hospitals providing pediatric care for nonchronic conditions use psychological preparation routinely with patients undergoing diagnostic or surgical procedures. The lack of careful evaluation of the methods being used results in confusion as to which approaches are appropriate for which youngsters. In fact, the findings in our earlier work (Melamed, Meyer, Gee, & Soule, 1976) regarding the sensitization of children who are under the age of 7 and prepared too far in advance of the actual hospitalization raise the ethical question of whether some children are being made more anxious by some of these treatments.

It is often assumed that *all* children can benefit to some extent by receiving support and information about what to expect. However, children's prehospital adjustment and personality characteristics may make them more or less prone to psychological consequences, due to the hospital experience. Some children may even benefit from the mastery of their fears and anxieties following a successful experience (Sipowicz & Vernon, 1965). Studies suggest that the parents' attitudes and concerns influence the child's adjustment to the hospital (Brown, 1979; Zabin & Melamed, 1980). Often parents are neglected or lack the relevant information that would help them assist their child. In fact, the stress that a parent experiences during the hospitalization of a child has been shown to be associated with impaired physiological function and later disturbances in the adult (Wolff, Hofer, & Mason, 1964).

It is therefore the intent of this chapter to review the factors which have been thought to affect the adjustment of children undergoing elective surgeries and medical procedures. In evaluating the claims of various investigators studying therapeutic effectiveness of preparatory treatment packages, rigorous outcome criteria will be required. The studies must include objective measures in order to demonstrate that anxiety has been alleviated and that none of the children have been unduly sensitized by the information provided. The fact that some children appear to cope well during the hospital experience but suffer delayed effects suggests that longitudinal research or follow-up studies should be undertaken. There are several specific issues related to the child's adjustment that are of special interest to medical practitioners. These include the cooperation of the child with anesthetic induction or prolonged discomfort, the rate of healing, and the avoidance of dependence on pain or sleeping medications and return to normal bodily functioning. Therefore, by including postoperative indexes such as nausea, vomiting, changes in blood pressure, heart rate, and information regarding the time until first voiding, eating of solids, reports of pain, and number of analgesics required, one may command the practitioners' attention. If a child is helped to anticipate the course of the events, he or she may be more cooperative with such procedures as obtaining urine specimens, remaining still during X rays, insertion of catheters or intravenous tubes, and administration of fluids.

METHODOLOGICAL AND THEORETICAL CONSIDERATIONS

Many of the procedures being employed such as puppet therapy, bibliotherapy, and film modeling have developed out of the practical experience of nursing and pediatric patient education specialists. Their aim is to inform the child about what will happen during the hospitalization. Decisions about how much information to give and at what time prior to surgery appear to be based upon normative experience of the staff with children of different ages. Hospital regimens also determine the availability of personnel and time for the implementation of such programs. There are few studies in the research literature which were derived from specific theoretical positions. Instead, different treatment packages are compared with one another. This does not encourage sequential research and replications which would allow us to converge upon a network of findings that can guide selection. The current review will attempt to assess the studies with regard to three considerations: (1) *patient variables,* including the age of the child, prehospital adjustment, previous experience and coping styles; (2) *treatment variables,* including anxiety reduction techniques focusing on information, support, behavioral rehearsal, film modeling, coping strategies, or systematic desensitization; and (3) *context variables,* including nursing/medical staff, inpatient setting, and parents' involvement.

In evaluating the efficiency of a psychological treatment package, the psychologist can provide special skills. The questions that must be raised include:

1. How do we know if the child can understand and make use of the information being presented?
2. How do we measure the effectiveness of treatment? What measures of anxiety should be obtained and when?
3. What determines the effective component of treatment? What factors related to the children's age or anxiety level may influence their ability to benefit from one approach over another? Should parents be involved in the preparation and care of their children in the hospital?

Although research has not been generally theoretically oriented, psychological theories exist within the framework of learning and personality that can be useful in leading to predictions that will answer the necessary questions. Although a complete review would not be possible within this chapter, those theories relevant to understanding hospital preparation is to reduce anxiety. The thought of denying a child some (persons, procedures, events) will be mentioned.

There appears to be a general assumption that the goal of surgery preparation is to reduce anxiety. The thought of denying a child some form of preparation is often rejected as unethical. Yet, the theoretical and research literature make predictions regarding the function of anxiety in helping the patient to cope with the stressful event. Lazarus (1966) postulated that in the face of threat an individual experiences arousal and that his or her appraisal of the situation can motivate adaptive coping behavior. Janis (1958) predicted that a moderate level of anxiety prior to surgery facilitated a more satisfactory postsurgical adjustment. The process by which this occurs is described as the "work of worrying," in which thoughts and fantasies about the outcome of surgery are elicited and dealt with. Johnson and Leventhal (1974) examined the influence of accurate expectations regarding sensory and procedural information on patients' cooperation with and recovery from surgery. Melamed and Siegel (1975) applied Bandura's (1969) social learning theory, which predicted that if individuals are exposed to a model of someone dealing with the fearful event in the absence of adverse consequences, their fear would be vicariously extinguished. Research questions have been slow in being devised to test these theories because many of the principal statements, including "moderate arousal," "vicarious extinction," "appraisal and coping mechanisms," have been so poorly operationalized.

The attempt to define anxiety in such a way that there can be consensus regarding adaptive functioning has been fraught with problems. Despite Lacey's (1976) plea that anxiety be defined by more than one measure, the research literature abounds with studies that measure heart rate and label it anxiety. A broad, systematic, multidimensional approach prompted by Lang (1968, 1977), Rachman (1976), and others has rarely been undertaken despite the finding that there is generally poor intercorrelation between different measures of anxiety (Kozak & Miller, 1981). Some investigators use paper and pencil tests such as Spielberger's (Spielberger, Gorsuch, & Lushene, 1970) State–Trait Anxiety questionnaire or Taylor's (1953) Manifest Anxiety Scale, and then fail to observe the patients' behavior. Others rely largely on retrospective patient/parent reports or global ratings and then fail to measure the physiology (Janis, 1958).

Even when broad measurement is undertaken, it is difficult to understand which index is relevant. Rather than evaluate the patterning of responses over time within a particular setting, investigators attempt to interpret single measures. Thus, Cataldo and his colleagues (Cataldo, Bessman, Parker, Pearson, & Rogers, 1979) reported that reduced mobility of children in intensive care resulted in a lack of interpersonal responding or activity. Perhaps, for these youngsters restrained by lifesaving and monitoring devices, immobilization and lack of initiating contact help them to minimize pain and discomfort. Patient compliance may be preferred by the hospital staff. Sedation often produces cooperation more rapidly than psychological explanation. If we are to provide health care practitioners with a reason to consider psychological preparation, we must provide better measures of the effectiveness of therapy.

Measures of Therapeutic Effectiveness

In order to compare one procedure with another, different investigators must use reliable and valid measures of criteria rather than developing new tools. The current potpourri of measures to define *anxiety* and *pain* includes a wide range of self-report inventories, behavioral ratings, physiological indexes, and physical symptoms. Rather than replicating and providing cross-validation of a given procedure, new techniques are presented, often combining several ingredients and making it wholly impossible to interpret the specific palliative agent. In fact, several studies exist that implicate beneficial effects of placebo (Barber, 1977; Kleinknecht & Bernstein, 1979; Rachman & Philips, 1980) as a nonspecific cure for discomfort.

Physiological

The concept of adaptive functioning is difficult to define. One implication is that the organism must maintain a certain degree of homeostasis within its biological system. Selye's (1956) hypothetical notion of a general adaption syndrome postulated that it is the relationship between continued stress and eventual chronic overuse of a system which leads to breakdown. Others (Holmes & Masuda, 1974) implicate physiological stress in making individuals more prone to succumbing to later physical illness. Yet, physiological measures are often omitted in psychosocial evaluations of surgery preparation since they are thought to be too difficult to obtain. However, there are simple observable correlates of autonomic activity that could easily be included, such as flushing, sweaty palms, irregular breathing patterns, and pulse rate. A measure of palmar sweating can be obtained and reliably scored by a plastic impression technique (Johnson & Dabbs, 1976) or by simply collecting a sample of the sweat and determining electrical conductivity (Strahan, Todd, & Ingolis, 1974).

Behavioral

Measurement of anxiety by observation of an individual's behavior during stressful procedures should be done by someone who is uninformed regarding the nature of preparation. Several rating scales exist (Johnson & Melamed, 1979) which could be used to rate the degree of cooperation of the patient during medical procedures. Children as young as preschoolers show motoric behaviors that reflect anxiety (Glennon & Weisz, 1978). The Observer Rating Scale of Anxiety, developed by Melamed & Siegel (1975) specifically to reflect anxiety-related behaviors of children during hospitalization for elective surgery, has been cross-validated in several studies (Ferguson, 1979; Melamed & Siegel, 1980). Peterson, Schultheis, Ridley-Johnson, Miller, and Tracey (1981) used global rating scales by parents and nurses, in addition to a behavior checklist, to evaluate treatment.

Subjective and Cognitive Factors

Self-report measures of anxiety in children are somewhat problematic. Younger children (less than 7 years) have difficult in reporting gradations in fear and tend to score higher than older children in any given situation. This may indeed reflect less conceptual ability to grasp the meaning of fear as well as fewer mastery experiences in these situations. Johnson and Melamed (1979) reviewed the literature on children's fear measures and concluded that self-report measures tend to be less reliable than observational methods and not as valid in predicting behavior problems, particularly with very young children. New measurement instruments, such as the Self-Assessment Mannequin (Dearborn & Melamed, 1981; Lang, 1980; Silverman & Melamed, 1981), allow children to use a multidimensional scale on which to rate pleasure, arousal, and dominance of a given situation. Low pleasure, high arousal, and low dominance ratings accompany subjective feelings of fear.

Physical Recovery Indexes

The physical recovery factors which nurses and physicians must monitor throughout recovery should be recorded in addition to measuring psychological aspects of surgery. They are as important to the child's well being as psychological health. Vomiting, swelling, length of recovery, and use of pain medication, as well as data regarding physical status (i.e., blood pressure, heart rate, wound healing), can be measured and have been found to be influenced by psychological factors (George, Scott, Turner, & Gregg, 1980; Melamed & Siegel, 1975; Knight, Atkins, Eagle, Evans, Finkelstein, Fukushima, Katz, & Weiner, 1979). Evidence for change in these biological factors may provide motivation to even the most skeptical physician to pay attention to the value of preparation techniques.

A recent study (George, *et al.*, 1980) found that psychological factors such as patients' expectations and anxiety about recovery, trait anxiety, coping style, and health locus of control influence postoperative indexes such as swelling rate and duration and difficulty in eating or resuming normal activities. The effects of psychological preparation are shown to be independent of the actual physical trauma of surgery.

In summary, the type and range of measures which have been used may have contributed to the lack of convergence of findings across studies. The use of broad-based therapeutic strategies may have the effect of differentially altering behaviors that are reflected to varying extents by subjective, behavioral, or physiological indexes. Thus, in order to determine which type of approach would be of particular benefit to a given individual, it is necessary to assess each of the response systems and then apply that treatment procedure which has proved most effective in reducing stress of that particular nature. Thus, a profile of the patient's disturbance must be determined prior to applying a specific procedure.

Time of Measurement

In addition to careful selection of appropriate measures, the time at which responses are evaluated is important in the consideration of treatment effectiveness. Prehospital and posthospital experiences are often as important in evaluating change as the actual measures taken during the hospitalization. It is important to have a baseline of anxiety against which to evaluate the amount of change. It has often been reported that prehospital levels of anxiety of both the child and their parents are predictive of coping. The rate at which different systems may reflect the psychological preparation may be desynchronous (Hodgson & Rachman, 1974). For instance, the initial effect of preparation may be to heighten subjective anxiety. However, this may then lead to coping behaviors which may ultimately yield less distress, as measured by the behaviors required during medical procedures or the autonomic indexes of stress. Subjective feelings of relief from stress may lag behind behavioral adjustment. Although the patient may be sleeping better and healing well, requests for medication for pain may continue. If recovery indexes are taken too soon following surgery, their interpretation may be confounded by residual effects of anesthesia or pain medication. It is therefore a good idea to reexamine the effects of hospital preparation during a follow-up procedure. If patients must return for suture removal or postsurgery examination, their medical concerns may once again be heightened. Their ability to handle this new stressor provides an opportunity to assess the generalization of treatment effects.

Additional Hospital Practices in Effect

It is difficult to measure the effect of a given preparatory procedure in the absence of other practices used by the hospital staff in preparing patients. Therefore, it is necessary to evaluate the behavior of children in a control group which does not receive the psychological preparation but is matched on all other variables, such as age, sex, and type of surgery, that may influence anxiety measures. The effectiveness of the proposed psychological intervention must then be demonstrated to produce change above and beyond that already brought about by the current practice.

It is important to define in which aspects the proposed treatment clearly differs from existing practices. For example, there is likely to be overlap in the amount of staff investment, supportive relationships, and general information provided. These potentially therapeutic factors must be carefully matched so that only the "effective ingredient" is operating in the preparation group.

PATIENT CHARACTERISTICS

It is a widely held belief that all children can benefit from some form of hospital preparation. However, data exist to indicate that patients can be sensitized by advanced preparation (Melamed, *et al.*, 1976; Shipley, Butt, Horwitz, & Farbry, 1978; Shipley, Butt, & Horwitz, 1979). It is also true that some children get through the hospital experience with no adverse effects. It is necessary to optimize our therapeutic efficiency by predicting which patients are most likely to benefit. Thus, the relationships between individual characteristics such as age, intelligence, trait anxiety, and flexibility of coping styles must be examined to determine their interaction with specific interventions.

Cognitive Development

There have been recent efforts to understand the development of children's concepts of illness (Campbell, 1975; Peters, 1978; Simeonsson, Buckley, & Monson, 1979). There is extensive evidence that children of different ages have different conceptions of illness causality. Nagy (1951) asked 350 healthy children, ranging from 3 to 12 years of age, "What makes us ill?" The responses she obtained could be separated into four age-related stages of causal belief. Children under six based their cause and effect connections in the time contiguity between events. Thus, if a child becomes sick after drinking milk, the child concludes that milk makes him sick. Six- and seven-year-olds made reference to an unspecified infection. Eight- to ten-year-olds believed all illnesses were caused by microorganisms but did not differentiate between types. A full understanding of the fact that different organisms lead to different illnesses characterizes the responses of eleven- and twelve-year-olds. It has been clearly demonstrated in a number of studies that the conceptual ability of children regarding their understanding of what a doctor does increases with age (Steward & Regalbutto, 1975).

The application of the Piaget theory of stage development has clarified the relationship between age and conceptions of illness (Neuhauser, Amsterdam, Hines, & Steward, 1978; Simeonsson *et al.*, 1979). The hypothesis that the concrete operational children would be able to give more accurate causal statements about healing inside the body was demonstrated by dividing the groups of children into their stage of cognitive development by a conservation task and interviewing them about their concepts of illness. Concrete operational children were able to give more accurate causal statements then preoperational children.

Simeonsson *et al.* (1979) used hospitalized children of three age groups (5, 7, 9). A significant age progression was found to exist across several cognitive tasks with responses becoming more abstract. Children's conceptions of illness were also significantly related to performance in conservation, egocentrism, and physical causality measures. The children in this study were all hospitalized, and the generalizability of these findings to nonchronically ill or healthy children must be explored, since other research has suggested that children with chronic illness have a lower level of cognitive development on Piaget conservation tasks than age-matched healthy controls but have illness causality notions equal to their peers (Myers-Vando, Steward, Folkins, & Hines, 1979).

Individual Coping Predispositions

When faced with threat, individuals may exhibit vigilance, denial, or a combination of these and other emotions. The interaction between coping style and type of preparation is important. Research on hospital preparation with adults demonstrates different therapeutic effects depending upon coping style (Kendall, in press). Patients, for example, who were found to exhibit repression showed a heightened heart rate during endoscopic insertion if they had viewed a single presentation of a coping model videotape, whereas sensitizers had reduced heart rate (Shipley *et al.*, 1978).

The literature on children is less clear because of the lack of stability in their coping styles and the general unavailability of inventories by which to classify their predisposing coping tendencies. Burstein and Meichenbaum (1979) did find less hospital toy and doll play in children rated high in defensiveness. However, the relationship between this characteristic and postsurgery recovery was not significant.

Knight *et al.* (1979) classified children as to their degree of defense effectiveness based on interview data and the Rorschach test. Types of defenses included intellectualization, denial, intellectualization and isolation, denial and isolation, mixed, displacement, and projection. The degree of defense reserve was measured by the patients' capacity to mobilize greater defenses in the face of increasingly threatening situations. All children were prepared by being told what procedures they would encounter during their hospital stay. During the discussion of needles, surgery, and other frightening procedures, the clinician would challenge their defenses to judge their defense reserve. It was found that a significant relationship existed between Rorschach anxiety rating and cortisol production rates. There was a relationship between defenses used and rated effectiveness of defenses during the outpatient clinic appointment. Two weeks prior to hospitalization for surgery, there was no relationship between cortisol production and effectiveness of defense. However, during the hospitalization, the cortisol production rates obtained the day before surgery were inversely related to defense effectiveness. Children who used certain types of defenses, such as intellectualization with or without isolation, mixed, and flexible defenses, appeared to cope more successfully than children who used denial, with or without isolation, displacement, or projection in a rigid defense structure. These findings have implications for the ways in which doctors and parents prepare children for their hospital experience. The authors suggest:

> Careful attention must be paid to the way a child copes with stimuli in his environment before he is prepared and hospitalized for surgery. While the children who intellectualized wanted to hear every detail of the upcoming experience, the children who denied often covered their ears, trying to block out all the information. These latter children would probably do best with

> little information and a great deal of supportive, nurturing care in the out-patient department and in the hospital.

The results indicated that children who use denial had higher cortisol production rates than children who use other defenses. This differs from results found in which adults who used denial had lower 17-hydroxy-corticosteroid excretion rates than adults using other defenses (Wolff, Friedman, Hofer, & Mason, 1964). It may be that children ages 7 to 11 are striving for mastery and cope better when they feel they can intellectually and physically master new situations, and that denial does not allow them consciously to process information and subsequently master new situations.

Thus, the importance of carefully assessing each child's psychological and physiological reactions to the stress of hospitalization is indicated in determining appropriate preparation.

TREATMENT COMPONENTS

Most of the preparation procedures that have been employed with children provide information about what will happen in the hospital. The context within which this information is provided varies depending upon the behavioral strategy employed. It is the goal of this chapter to elaborate on the effective mechanisms of change that may contribute to the child's successful experience with the hospitalization or medical procedure. Thus, one might postulate that traditional puppet therapy leads to a catharsis of emotional feelings that alleviates a child's anticipatory anxiety. Behavioral programs, which are increasingly being used, specify the mechanism by which anxiety is brought under control. Modeling procedures often provide the child with a prototype for handling the stressful situation by portraying other children undergoing similar experiences. The modeling format may include relaxation instructions, graded exposure, or specific examples of self-control or cognitive coping strategies. Systematic desensitization usually involves the training of a response to compete with anxiety such as relaxation or emotive imagery. The patient is then exposed in imagination or in the real-life setting to a hierarchy of situations ranked from least to most threatening and instructed to substitute relaxation should fear be evoked. Exposure therapy involves having the patient stay in the situation that evokes anxiety. It is based on an extinction notion that the presence of fear-related stimuli leads to arousal without adverse consequences, thus facilitating habituation. Within each of these treatments the degree of support from the therapist (or parents' presence) may influence treatment outcome.

Unfortunately, most of the research reviewed presented a horse race, in which one compound treatment package was compared to another. Even if one package is demonstrated to be more effective, it is difficult to parcel out the separate treatment components. This chapter includes research studies in the existing literature that have been selected because they have made at least an attempt to control for patient variables (age, sex, race) and hospital procedures (type of surgeries, short-term hospital stay, general anesthesia vs. local anesthetic). The studies selected must have included at least two different response systems by which to evaluate the results. The majority of these studies have not grown out of theoretical viewpoints but have been developed through practical use in hospitals. A final section will present a guide to the systematic evaluation of hospital preparation by showing the manner in which a sequence of studies was carried out to evaluate the theoretically viable components of modeling procedures. The influence of age, previous experience, and autonomic patterning of responses in determining the amount of information acquired through these modeling procedures was demonstrated.

Preoperative Medication

Anesthesiology is one of the most widely used techniques to assure cooperation of children about to undergo surgery. The very young child is thought to be overwhelmed by the stimulation involved in surgery preparation, and arguments regarding the usefulness of quieting children for their own good are often raised. A sedated child is thought to cooperate more with the procedure and suffer less psychic trauma. Older children, as a rule, are affected less by events surrounding anesthesia and surgery. However, many children resist injections, suppositories, or unpleasant smells associated with gaseous inductions. Points of view vary from surgeons who believe that children should be put under even before they are in the operating room to others who feel that there is no beneficial effect. There are those who feel that some children use the hospital experience as a way to master fears and adapt to brief separations from the family. Rather than an adverse event, the hospital experience can be a learning experience. The effects of premedication may dampen an individual's capacity to take in information provided to them at this time; it may even inhibit the retrieval of information that was acquired during preparation and could be useful for handling stress.

The preoperative visit by the anesthesiologist is an important part of the psychological preparation the child receives. According to Bothe and Galdston (1972), the children they observed used the relationship with the anesthesiologist in a surrogate manner. The visit allows the anesthesiologist to establish rapport with the child and the opportunity to explain in simple, honest, and realistic terms the events that will occur with anesthesia. It is important to remember that the younger child thinks in concrete terms and that explanations should be geared to the child's level of understanding. Explanations that involve physical props such as an anesthesia mask may be useful. One should endeavor to gain the child's confidence and assure him that the induced sleep will be temporary.

In addition to careful prehospitalization and preoperative education of the child, sedation may be of value in reducing anxiety. Preoperative sedation has been used widely in pediatric patients to reduce the emotional impact and psychological consequences of anesthesia and surgery. Waters (1938), in his original and classic paper on preanesthetic medication in children, recommended morphine and atropine. Eckenhoff (1953a) reported a decreased incidence of crying in children given preanesthetic sedation. Leigh and Belton (1949) reported that the advantages of preanesthetic sedation outweighed the disadvantages. Although much has been written since these early efforts, little has been accomplished toward finding a drug that produces tranquility without side effects. Many combinations of drugs have been used. The most commonly used combination is a barbiturate, an opiate, and belladonna. This combination was suggested by Freeman and Bachman (1959). They found that a barbiturate used in conjunction with belladonna provided significantly greater sedation than the other drugs studied. The addition of morphine significantly decreased the incidence of postoperative or emergence delirium. However, only 75% of children (Smith, 1980) will be sedated satisfactorily with any combination currently used. This fact merely emphasizes our inability to measure emotional differences in our patients. If an objective equation to quantitate anxiety could be derived, then a more effective drug dose and combination could be chosen. Smith suggested five factors that predisposed children to increased anxiety:

1. Age 1 to 4 years
2. Inability to communicate
3. Emotional or neurotic patients

4. Previous traumatic hospital experience
5. Fear of expected operation—amputation, cardiac surgery, or other extensive procedures

It is helpful to consider these factors during the preoperative evaluation.

Some have preferred to use no preoperative sedation, relying entirely on a positive psychological approach to reduce anxiety. Beeby and Hughes (1980) recently reported on the behavior of 344 unsedated children. They confirmed Doughty's (1959) observation that children 7 years and older showed less anxiety. Also, the frequency of satisfactory behavior in their unsedated population compared favorably with published trials of sedative premedication. When no sedation is used, parents are frequently allowed to be present during the induction of anesthesia. This technique has limitations when a major procedure such as a cardiac operation is to be undertaken.

Given the present state of research on the question, a combination of psychological preparation and pharmocological premedication is generally recommended. Administration of the premedication by the oral or rectal route usually causes less fear in the child than do injections. If premedication is given, timing is important. Adequate time for the onset of action and maximal effect of the drug prior to transporting the patient to the operating room must be allowed. No premedication is far better than an injection on the way out of the door to the operating room. For the very anxious patient, a sedative the night prior to surgery will promote a restful night and provide some residual sedation in the morning. If the child is going to surgery late in the day, a sedative in the morning may prove useful in decreasing the anxiety that naturally accompanies the wait.

The induction of anesthesia has been thought to be a particularly vulnerable time for the child. The theory that a quiet, peaceful induction or "steal" induction (induction without the child's being awakened) leads to less emotional trauma is most popular (Eckenhoff, 1953b). However, there are others who believe that crying and screaming satisfy the child's need to express himself. Agitated behavior may be a way of coping with stress and preventing sequelae for some children. Meyers and Muravchick (1977) looked at the behavior consequences of different induction techniques. Their data showed that hospitalized children not undergoing anesthesia and surgery had as many problems as children carefully prepared and gently anesthetized. They did show that "the asleep induction technique is associated with a significantly greater proportion of children free from apparent behavioral upset than are awake induction techniques" (p. 541).

In summary, the care and preparation of each child must be individualized. Reliance on personal communication and interaction between the patient and hospital personnel is most certainly important in reducing behavioral complications of hospitalization, anesthesia, and surgery. Premedication and sedation may not be necessary in all situations but may be extremely helpful in others. The use of behavioral approaches may be beneficial.

Behavioral Approaches

Systematic Desensitization

Children and adults have been treated successfully by exposing them either in imagination or in real life to medical events and concerns. Fear of injections (Taylor, Ferguson, & Wermuth, 1977), dental treatment (Gale & Ayer, 1969), and intravenous procedures and hemodialysis phobias (Katz, 1974; Nimmer & Kapp, 1974) have been treated by desensitization. The basic procedure is to train the individual in a response antagonistic to anxiety, such as relaxation, and then gradually to expose the patient to the fear hierarchy. Lang (1969) reviewed group studies conducted to evaluate the effective components. Neither the ordering of the hierarchy nor relaxation alone turned out to promote as much fear reduction as the pairing of exposure and relaxation. However, shorter techniques of general coping strategies (Goldfried, 1971) as well as exposure in the absence of relaxation (modeling and flooding procedures) have produced similar effects.

The long-term maintenance of reduction of fears of dental treatment does show an advantage for systematic desensitization over modeling (Shaw & Thoreson, 1974) and over general coping and stress management training (Krop, Jackson, Mealiea, 1976). The treatment of dental-avoidant patients was judged by the criterion of whether the patient returned for dental work. Although modeling produced a greater number of patients completing dental work, this procedure, like desensitization, included both relaxation and a graded exposure. Desensitization patients showed a greater reduction in subjective fear and a significant decrease in their negative attitudes before and after treatment.

Children have also been able to benefit from being introduced to the dental setting via systematic desensitization. Again, the specific component that promotes change is not clear. When Machen and Johnson (1974) compared film modeling and desensitization, both were found to be equally effective and better than a control group. Sawtell, Simon, and Simeonsson (1974) demonstrated that the dental assistant's friendly chat with the child outside of the dental operatory also reduced fear behavior. Again, the results suggest that supportive treatment within the context of the relevant situation may be sufficient to reduce anxiety. Kleinknecht and Bernstein (1979) reported unexpected positive results with a placebo condition of merely providing the patient with a supportive dentist with no pretreatment.

Self-control group applications of desensitization have been successful with adult patients (Gatchel, 1980). Data on group desensitization of children do not exist.

Modeling

The presence of a cooperative peer or older sibling has led to less disruption in younger children and led to the adoption of this technique within a modeling context. Modeling provides the observer with an exposure to the aspects of the feared situation through the eyes of another child. The question of what modeling conveys beyond observation has been subjected to careful studies using children receiving injections during dental restorative work or preoperative procedures. When 4- to 8-year-old children (White & Davis, 1974) were exposed to the dental situation without a model, they were more cooperative than those receiving no exposure at all. When modeling subjects were compared with those provided with familiarization (no model), they were equally cooperative during treatment. However, patients were less likely to hide or avoid treatment after exposure to a model.

In a more explicit attempt to evaluate the effective component of modeling, Melamed, Yurcheson, Fleece, Hutcherson, and Hawes (1978) found that peer modeling was superior to an information only (no model) preparatory videotape in reducing the disruptive behavior of children undergoing dental restorative procedures including novocaine injection. Follow-up work has related this phenomenon to the autonomic reactivity of the children to the fearful segment during the videotape observation (Melamed, 1982). The children viewing the peer model showed increased sympathetic activation, accompanied by cardiac deceleration which facilitated the intake of information from the videotape. Children watching the demonstration tape, on the other hand, showed sympathetic activation (palmar sweating) accompanied by cardiac acceleration. These children rejected information being presented. In fact, they scored much lower on the test of information recalled than those children who had viewed a peer model. Thus, it does appear that an active ingredient in the modeling situation is learning how to behave by watching another child cope with the stress.

Peterson *et al.* (1981) compared the effectiveness of a puppet model, a local modeling film accurate for procedures in a particular setting, *Ethan Has An Operation* (Melamed & Siegel, 1975), and a control group receiving minimal preparation from the nursing staff. The surgery procedure involved, for example, excisions of nonmalignant tumors in the gums and impacted teeth and required general anesthetic and a hospital stay of less than 24 hours. The children were 2 to 11 years of age, with a mean age of 4.77 years. Ratings consisted of behavior checklists compiled by parents, nurses, and observers both before and after surgery. Results demonstrated a lack of difference among the three experimental groups (each facilitated reduction of anxiety more than the informal preparation). Thus, there is less support for the hypothesis that the preparation must provide information specific to the hospital experience; rather, there is a suggestion that exposure to medical situations may suffice.

Early work by Vernon (1973) suggested that film modeling did facilitate the cooperation of children during the anesthetic induction procedures. Unfortunately, the measure of anxiety was strictly a global rating of fear and an adequate control condition was lacking.

In our laboratory we replicated the beneficial effect of film modeling for hospitalized children using a peer who narrates his feelings and fears from the time he enters the hospital until he goes home. Using a wide range of measures of anxiety, we demonstrated that for children with no prior hospital experience, film modeling was more effective than routine in-hospital preparation (Melamed & Siegel, 1975). Although we replicated this finding for older children, when preparation was presented a week in advance of hospitalization, the effect of the film on younger children (under 7 years) appeared to be sensitization. They had higher levels of palmar sweating upon admission and at preoperative and postoperative assessments, than younger children who were shown videotape only at the time of admission. Ferguson (1979), with a different videotape, demonstrated a similar age effect. Children prepared at home by a visiting nurse were compared with those receiving film preparation during hospital admission. With a wide range of measures, it was found that the younger child (3 to 5) benefited from the videotape modeling exposure at the time of admission, whereas older children (6 to 7 years) did as well with simple verbal preparation at home. The need to consider the age of the child, then, relates to both the amount of arousal such exposure might generate and the length of time over which the child can deal with knowing what will happen.

The need to evaluate the effect of a child's prior hospital experience on the preparation via modeling was suggested by the fact that the children who had experience with hospitals already knew what to expect. Siegel (Melamed and Siegel, 1980) demonstrated that the hospital videotape during an admission for a second operation produced no greater effect than showing an unrelated film. The level of autonomic arousal, as measured by palmar sweat activity, was significantly higher before the film for experienced children than it was for inexperienced children. This may reflect their having been sensitized by the exposure to staff and equipment as well as conditioned to painful stimulation during the prior hospitalization. The autonomic component of orienting was not optimal for information-processing.

In a recent study (Melamed, 1982) which focused more on examining the children's readiness to process the information presented, we recorded the children's pulse rate and sweat activity (palmar sweat index) prior to exposing them to a slide-tape preparation made specifically for children at Shands Teaching Hospital. Table I summarizes the patient characteristics. Each child was individually prepared the night before scheduled surgery. By this time, many of the preoperative procedures, including blood tests and x-rays, had been completed.

The child was read the hospital fears rating scale so that the prefilm level of medical concerns could be obtained. An independent rater observed the frequency of verbal and nonverbal anxiety behavior during the assessment following the slide-tape when pulse rate measurement and palmar sweating were reassessed.

Table I. Major Population Characteristics

Age	Previous experience	Type of surgery	Sex	Race
6	No	Hydroureter	M	W
7	No	Cabitus varus of left elbow	M	W
7	No	Aortic stenosis	F	NW
7	Yes	Umbilical hernia	F	NW
7	Yes	Hernia	M	NW
9	No	Femeral anteversion	F	W
10	Yes	Tonsillectomy	M	W
10	Yes	Catheterization of aorta	M	W
10	No	Rectal mass	M	NW
10	No	Urinary tract	M	W
10	Yes	Hernia	M	W
10	Yes	Heart valve closure	M	NW
11	Yes	Renal staghorn calculous	M	W
15	Yes	Congenital scoliosis	F	W
15	Yes	Urinary retention	M	W

The results are encouraging, for they support the theoretical position of Lacey (1967) regarding the pattern of autonomic responding and the individual's readiness to acquire information. The correlations (Table II) revealed the fractionation of the heart rate and sweat gland response that was previously found in children who observed the peer modeling version of the dental injection videotape. The higher the prefilm level of sweat gland activity, and the lower the heart rate, the greater the amount of information retained on a postpreparation questionnaire concerning the content of the slide-tape. It was revealing to examine the age and prefilm level of subjective anxiety as they influenced information processing. In general, the younger the child, the less the information retained. In addition, those children who reported a low degree of medical concern had higher levels of sweat gland activity and remembered less information. This discordance between self-report and level of sympathetic activation has been found to underlie the repressor disposition in adult patients watching distressing film (Opton & Lazarus, 1967). Perhaps, the issue of coping style can be readdressed in terms of the individual's autonomic response characteristics that set him or her to receive information from the environment. Support for this notion comes from the results of Knight *et al.* (1979), in which children who used denial to block out information about their surgery (held their ears) showed higher cholesterol rates than children who used intellectualization to master this situation. The stress reaction, as measured by its psychophysiological counterparts, may mediate the child's information-processing capabilities.

The above suggests that Lacey's (1967) theory of fractionation as it is related to information-processing can lead to testable predictions about

Table II. Intercorrelations between Measures of Autonomic Arousal, Self-Report of Hospital Fears, Age, and Percent of Information

	Percentage information	Hospital fears
Prefilm heart rate	−.23	.002
Postfilm heart rate	−.31	.09
Prefilm palmar sweat index	.52[a]	−.52[a]
Postfilm palmar sweat index	.10	−.26
Age	.758[b]	−.32
Hospital fears rating scale	−.51[a]	−.32

[a] $p < .07$.
[b] $p < .001$.

who is likely to learn from preparatory information. Perhaps a calibration can be made of the degree of receptiveness which the patient is likely to display when faced with noxious information by measuring the pattern of autonomic responsivity when confronted by a loud noxious tone or anticipation of aversive stimulation. Then the information can be presented in a manner and at a time when it is most likely to be needed. The need to assess what information is actually obtained is clear. Unfortunately, there are very few studies that have obtained such a measure. Interpretations of treatment effectiveness, therefore, rely on the inference that if the children exposed to preparatory information improved in behavior and if anxiety was reduced, it was because of something they had learned from the treatment package.

Behavioral Rehearsal

Behavioral rehearsal is a procedure which includes many aspects of modeling but encourages the participant actually to try out new behaviors either in imagination, through role playing, or in real life. Ayer (1973) used emotive imagery to improve children's cooperation in "mouth-opening" prior to the injection of local anesthetic. Chertok and Bornstein (1979) used covert rehearsal to reduce dental fears in children from 5 to 13 years. It was found that imagining another person coping or mastering fear was no more effective than having the children imagine the relevant stimuli associated with the dental operatory. Anecdotal reports indicated that younger children were not able to follow the instructions as well as one might have expected. Therefore, it is likely that there was no real difference in what the children actually imagined. Further, it is not surprising that the different cognitive instructions were equally effective.

Although cognitive instructions have been reported to help children reduce fears of the dark (Kanfer, Karoly, & Newman, 1975) and improve task performance (Kendall, 1979), there is as yet little evidence that is has helped children cope with hospitalization and surgery.

Operant Reinforcement Procedures

Although frequently applied to increase complaint behaviors in medically related situations (i.e., drug-taking and oral hygiene), few efforts have been directed at reducing children's fears in medical or dental situations by varying reinforcement contingencies. The most recent data generated in our laboratory suggest that the health professional's style of interacting with children in the dental operatory does influence the children's degree of anxiety and disruptiveness. In general, aversive procedures in which punishment or criticism is made contingent on inappropriate behavior lead to more disruptiveness, although there is a reduction of disruption in subsequent sessions. Thus, information was obtained regarding required behavior. The degree of experience the child has had in the situation determines which form of reinforcement is most useful. For children with no prior experience, directive guidance without affective feedback or positive reinforcement of appropriate behavior and punishment of inappropriate responses appear to be the most pleasant and effective approaches. On the other hand, children who have a broader repertoire of responses from prior encounters must learn to discriminate the desirable from the undesirable behavior. Since they must unlearn inappropriate behavior, it appears that a combination of positive reinforcement for appropriate compliance and punishment for inappropriate behavior produces the least fear and the most cooperative behavior in the child patients who have already had experience (Melamed *et al.*, 1982).

CONTEXT VARIABLES

Parent Involvement

Few hospitals or clinics have developed programs that might be directed to the reduction of stress in the child and his family from the time the appointment for the medical procedure is made through post-procedure recuperation. The following review focuses on the available literature concerning the preparation of parents of children undergoing medical procedures and hospitalization.

Focus on the Parent

Hospital admission not only is stressful to the child but may also be quite anxiety-producing and disruptive to the parents, who must often cope with their uncertainty about outcome; try to mitigate the child's fears, pains, and discomforts; juggle their own expectations, past experiences, and needs; and maintain as well their continuing familial, occupational, marital, and personal roles.

A number of factors are likely to influence the nature of the parental involvement. These include the parents' psychological state concerning illness and hospitalization (Mahaffy, 1965; Roskies, Bedard, Gauvreau-Guilbault, & Laforture, 1975; Roskies, Mongeon, & Gragnon-Lefebvre, 1978; Skipper, Leonard, & Rhymes, 1968); their relationship with the child (Robertson, 1958, 1976); support available to the parent both from within the family and from the larger community (Skipper & Leonard, 1968; Visintainer & Wolfer, 1975); competing demands on the parents' time; and amount of supportive information available to the parents concerning the role that they may play in their child's hospitalization (Eckhardt & Prugh, 1978; Prugh *et al.*, 1953; Roskies *et al.*, 1978).

It is a question of great concern to doctors and medical staff whether participation of parents should be encouraged or whether parental presence is more of a burden on the staff and not likely to contribute to the welfare of the child. This question is answered in several ways, depending on the parental response to stress and on the hospitals' beliefs about the deleterious effects of separating children from their families, particularly their mothers, during times of great pain, discomfort, and stress. In either case, parents must be "prepared" for their role in their child's hospitalization.

Theoretically, one could posit a range of possible points on a continuum that express the degree of parental involvement in the child's hospitalization. The parent may be the agent who delivers and retrieves the child from the hospital. The parent may be the noninvolved visitor who arrives periodically and remains uninvolved with the ongoing treatment or care of the child. The parent may dutifully visit during the prescribed hours and entertain the child, but still remain separate from the care of the child. Or the parent may become an intricate and involved member of the care team, becoming involved in and responsible for the nonmedical care of the child.

The degree of parental involvement may depend on a variety of factors, not the least of which is the policy, both implicit and explicit, of the hospital or clinic toward that involvement.

Importance of Parent–Child Relationship

Much clinical literature has focused on the child's relationship with the mother during the time of hospitalization. It appears that if a parent is uncomfortable and anxious about the child's welfare, the child will experience the parent's anxiety in interaction with the parent. This will in turn increase the child's level of stress and anxiety. Escalona (1953) discussed how such feelings may be communicated between mother and child on a nonverbal as well as verbal level. This may occur with children of all ages and may not be under the voluntary control of the mother. Escalona called this nonverbal communication "contagion."

Data exist (Brown, 1979) which have also demonstrated that a child's response to hospitalization varies with the child's relationship to the family at home, the mother–child interactions, and the mother's anxiety. Through the use of personality tests, structured interviews, and direct observation of children 3 to 6 years old and their mothers during a short stay in the hospital, Brown observed that those children who were close to their families were most likely to withdraw and show stress during their hospital experience. Further, mothers' attitudes and behavior in the hospital were found to relate to the degree of visual contact and immobility which the child displayed on the wards. Finally, mothers who were themselves anxious and highly accepting of hospital authorities tended to have children who were distressed and withdrawn.

Vardaro (1978) examined the anxiety of the parent prior to the hospitalization of the preschool child. She found that anxiety in the parent was correlated with anxiety in the child. She concluded that one method of decreasing the preschool child's anxiety prior to hospitalization is by assisting the mother in coping with anxiety. One method of preadmission teaching would be through the parent because of the ability of the mother to affect the child's emotional response. Vardara's study was methodologically quite weak but did point out an important focus for intervention, namely the mother and her relationship to her child.

Preparation—Parent's Role

Vernon, Foley, Sipowicz, and Schulman (1965), in their discussions of psychological preparation of children for hospitalization, have focused on three major themes. These are: (a) the information imparted to the child, (b) the encouragement of emotional expression, and (c) the establishment of relationships of trust and confidence with the hospital staff by means of psychological preparation.

There are many good reasons for examining the usefulness of the parent as an important agent who can contribute to the child's recovery from the necessary medical procedures. Parental preparation may enhance the individual parent's ability to be supportive of the child and may be important to the parent's own emotional adjustment during times of stress. The parent as well as the child may be in need of ventilation of feelings and the opportunity for developing a trusting relationship with the staff.

Oremland and Oremland (1973) discussed the role of the parent in this process of preparation. The parent is viewed by some as so possessed by anxiety as to be immobilized into ineffectiveness. Alternatively, the parent may be the key figure in the preparation process and the central behavioral and identification model for the sick child. The majority of preparation techniques in use in hospitals focus on the child.

At times, the mother is left to learn about the medical procedures from the books and discussions provided for her child. Her own questions and concerns are left unmet. Such haphazard approaches to the parents are likely to increase their anxiety and interfere with their coping strategies. The preparation of parents is too important to be left to the discretion of the physician alone. The physician may be without the necessary time, skills, or initiative to make sufficient contact with the parents to allay their anxiety or to provide necessary information about the nature of the medical procedures.

Gofman, Buckman, and Schade (1957) suggest that emotional trauma from the hospital experience may be prevented by an adequate parent–child relationship, the proper preparation of the parent and child, and modification of the hospital experience as much as possible to meet the endurance of the child and the parents.

Research Evaluation of Parental Involvement

Recent literature concerned with the examination of parental involvement in preparation of children for surgery and medical procedures is sparse indeed. As early as the mid-1950s professionals had acknowledged that the parent–child relationship was a critical one to be maintained during the child's hospitalization. Subsequent literature has focused on the parental involvement in terms of the parent's ability to affect the child's response to hospitalization.

Vernon *et al.* (1965) presented a comprehensive review of articles and books concerning the psychological responses of children to hospitalization and illness. Of the over 200 references cited, Visintainer and Wolfer (1975) point out that "only six of the studies reviewed were some form of clinical experiment where preparation was given to children or their parents along with an attempt to determine if the preparatory communication had a positive outcome" (p. 188). A review of these studies before 1965 provides six noncomparable investigations which yield tentative neutral to positive findings for the varied attempts to provide psychological preparation for children and their families.

The present reviewers have chosen seven generally well-controlled studies to examine the best and most recent research concerning parental involvement in preparation for children's hospitalization.

Prugh *et al.* (1953) have focused on the use of an experimental program of ward management with 100 2- to 12-year-old children. They compared the comparative effects of the traditional ward practices (weekly 2-hour parent visits, and limited encouragement of parental ward care of the child) with an experimental program that included daily visiting periods, a play program, early ambulation of patients, psychological preparation for and support during potentially emotionally traumatic diagnostic or treatment procedures, an attempt at clearer definition and integration of the parent's role in the care of the child, and weekly ward conferences for staff that focused on the adjustment of each child. They reported that children who

> showed the most successful adjustment on the ward were those who seemed to have the most satisfying relationship with their parents, especially the mother, and whose parents accomplished the most balanced adaptation themselves to the experience of illness and hospitalization on the part of their child. (pp. 81–82)

They tentatively concluded that the most diffusely supportive aspects of the experimental program were most prophylactic in effect (p. 102). The Prugh *et al.* study contrasted the most limited form of parental involvement and staff support with a varied, nonspecific array of techniques designed to decrease child stress and parental distance from the child. This study set the stage for studies to identify more specifically techniques that might be employed to decrease the potentially traumatic effects from pediatric hospitalization.

In a related series of well-controlled studies by Mahaffy (1965), Peterson and Shigetomi (1981), Skipper, Leonard, and Rhymes (1968), Wolfer and Visintainer (1975), and Visintainer and Wolfer (1975), the effect of preparing the mothers of children having minor surgery was evaluated. This work bears on the emotional contagion hypothesis (Campbell, 1957; VanderVeer, 1949), which holds that a parent's emotional state may be transmitted to a young child, and the clinical observation that emotionally upset or uninformed parents often are unable to assist their children in coping with stress. These studies will be presented briefly for historical and theoretical interest.

Mahaffy (1965) investigated "the possibility of improving the hospital care for children by involving the parents." The author supported parental involvement by referring to the mother's unique understanding of her own child. He emphasized the importance of the mother's feeling comfortable herself so that she could more adequately meet her child's needs. He felt that most hospitals made the parents feel uncomfortable because the parent's role in the child's care during hospitalization was not well specified.

Peterson and Shigetomi (1981) compared a variety of preparation techniques for parents and children, to test their hypothesis "that the use of a cognitive coping procedure would further improve previously used techniques" (p. 4). Parents were taught how to help their children.

Sixty-six children, aged $2\frac{1}{2}$ to $10\frac{1}{2}$ (mean age $5\frac{1}{2}$) and their parents were treated in small groups using minimal information preparation and one of the following procedures: control (minimal preparation only), self-coping, filmed modeling, or self-coping plus filmed modeling.

The results indicated that children receiving coping plus modeling techniques were more calm and cooperative during the blood test and presurgical injection than children receiving coping or modeling alone. Other measures indicated that the coping procedures were more effective than modeling only procedures for both parents and children.

The authors failed to control for previous hospitalization, although none of the children had been hospitalized within the last year. The client population was white and middle-class. It is not clear that the youngest children could perform the self-coping procedures. It was not specified how involved the parents became in this procedure. It was not specified what effect group viewing (by parents and children) of the Melamed film would have on its previous effectiveness. Also, the fact that all groups received a hospital tour and party made it difficult to interpret which ingredient was responsible for change.

Peterson and Shigetomi felt that the consistency of their results suggests that "coping skills" may be a useful addition to a presurgical preparation program. With younger children especially, they felt that the parents could be actively involved in helping the children to employ the coping techniques.

Unfortunately, the follow-up of parents of children in this study at the end of one year (Peterson & Shigetomi, in press) revealed that few parents reported using any of this coping training once their child left the hospital. It was reported, however, by 60% of the original participants, that their children recalled more of the positive aspects of their hospital experience than negative aspects, such as pain or preoperative injections. The results do not clearly support the contention that coping skills were a useful presurgical preparation in that only 30% of the mothers reported that they had made use of this training.

Nursing Involvement

One aspect of the hospital experience that may especially have to be reviewed is the relationship between the mother and the nurse. Vernon *et al.* (1965) discuss the "confusion and anxiety" which parents "often feel regarding their child's illness" and the problems associated with the process of the mother's turning the child over to the care of the staff and the nurse's being threatened by the presence of the mother.

Jacobs (1979) found that the existence of hospital practices such as visiting hours, giving nurses task assignments or shifts rather than assigning them to patients, and off-ward living arrangements work against the establishment of rapport among child, parents, and staff. She postulated that the denial of personal responsibility may be a mechanism that enables the staff to maintain their emotional detachment and maintain a closed system. The staff seemed to be reinforcing compliance, obedience, and unobtrusiveness in children and parents.

The Mahaffy (1965) study employed an experimental nurse to work with the parents. The nurse was to deal with the parents' emotional response to their child's hospitalization and make them feel more capable and competent and also was to help teach the parents to care for their hospitalized child. It was predicted that "experimental nursing" would reduce distress in the child. The specific techniques which were only generally specified, focused on increasing free communication between parent and nurse.

Mahaffy assigned 43 children, 2 to 10 years of age, who had been admitted for tonsillectomy and adenoidectomy into two groups. Twenty-one of the patients and their mothers received the "experimental nursing" program described above; 22 received routine ward care. For this second group, a staff nurse answered the parent's questions but seldom volunteered information or initiated conversation.

The author compared the children's vital signs, ability to take fluids, time of first voiding, vomiting, incidence of crying, and parental responses to a post-hospitalization questionnaire. It was reported that children whose mothers had received the experimental nursing program had lower means on their vital signs, higher mean oral fluid intake, earlier time of first voiding, less vomiting, and less crying during the hospitalization. Posthospitalization questionnaire reports indicated that in the experimental nurse group children incurred fever less frequently after discharge and "recovered" from the operation more quickly. The author concluded that the introduction of the experimental nursing services to the parents might be a valuabe method of reducing the physical and psychological distress of the hospitalized child.

This is a preliminary study that was completed with a small sample of children. Many factors, such as amount of parent contact during the hospitalization, were not controlled. The author sees this study as suggestive for further research in this area.

Skipper, Leonard, and Rhymes (1968) also examined the differences of specific nursing interactions with mothers. The authors felt that the addition of communication of information and emotional support to mothers through social interaction with an authoritative person was an effective way to reduce the parents' level of stress and allow them to make a more successful adaptation to the medical procedures and hospitalization of their child. In a related study by Skipper and Leonard (1968), 3- to 9-year-old children hospitalized for tonsillectomies with mothers exposed to such stress-reducing nursing care showed less physiological stress than children whose mothers did not receive this specific nursing care.

The Wolfer and Visintainer (1975) studies were the best methodologically and will be elaborated here. They hypothesized that children and parents who received special psychological preparation and continued supportive care, in contrast to a control group of children and mothers who did not, would show less upset behavior and better coping and adjustment as indicated on the following 10 dependent variables: (1) blind observer ratings of the children's upset behavior and cooperation with procedures at five potential stress points; (2) pulse rates at admission and before and after the blood test and preoperative injections; (3) resistance to induction; (4) recovery room medications; (5) ease of fluid intake; (6) time of first voiding; (7) posthospital adjustment on Vernon *et al.* 1966 Post-hospital Behavior Questionnaire; (8) mothers' self-ratings of anxiety at potential stress points throughout the hospitalization; (9) mothers' rated satisfaction with various aspects of the nursing and medical care they had received; and (10) mothers' ratings of adequacy of the information they had received.

The experimental condition was a combination of psychological preparation and supportive care for mother–child dyads which was provided at six stress points: (1) admission, (2) shortly before the blood test, (3) late in the afternoon the day before the operation, (4) shortly before the preoperative medications, (5) before transport to the operating room, and (6) upon return from the recovery room. The preparation with the mothers was intended to explore and clarify their feelings and thoughts, to provide accurate information and appropriate reassurance, and to explain how they could help care for their children. The child component of "stress point nursing care" included information, sensory expectation, role identification, rehearsals, and support.

The results of the study supported the hypothesis that children and parents who received systematic psychological preparation and continued supportive care, in contrast to those who did not, would show less upset behavior and more cooperation in the hospital and fewer posthospital adjustment problems. The experimental group had significantly lower mean upset ratings and higher mean cooperation ratings at each of the stress points than did the control group. Children in the prepared condition demonstrated greater ease of fluid intake, manifested significantly lower heart rates after the blood test and before and after the preoperative medication, had a significantly lower incidence of resistance to anesthesia induction, and obtained significantly lower post-hospital adjustment scores. Younger children (3–6) showed greater upset and less cooperation than older children (7–14). Parents in the experimental group had significantly lower self-ratings of anxiety, higher ratings of the adequacy of the information received, and greater satisfaction with care than parents in the control group. Thus, there is evidence of the beneficial effects of systematic preparation and support for hospitalized children and their parents. Three methodological problems were apparent: (1) observer bias was introduced in that the nurse rater occasionally observed the nurse researcher in the act of preparation; (2) parents who were given specialized attention by the nurse may have felt obliged to rate their satisfaction higher; and (3) the use of one nurse researcher allows us to postulate that it was a warm, trusting relationship rather than the specific preparation that was responsible for the results. Although no children had been hospitalized within the past year, it is not clear that the samples were balanced as to prior hospital experience.

In a second study, Vistintainer and Wolfer (1975) attempted to parcel out the specific contribution of the supportive relationship, the role of information alone, or the combination of information and a supportive relationship. Eighty-four children, ages 3 to 12, admitted for tonsillectomies, were randomly assigned to one of three treatment conditions or to a control group. The groups included:

1. A combination of systematic preparation, rehearsal, and supportive care conducted prior to each stressful procedure
2. A single-session preparation conducted after admission (new in this study)
3. Consistent supportive care given by one nurse at the same points as in the first condition, but including no systematic preparation or rehearsal (new in this study)

It was hypothesized "that the stress-point preparation would be the most effective stress-reducing condition followed by the single-session preparation, then the consistent supportive care condition, and finally, the control condition" (p. 190).

Children in the stress-point preparation and single-session preparation conditions tended to have lower upset and higher cooperation and ease of fluid intake ratings than those in the supportive care and control groups. Children in stress-point preparation showed this improvement only on the preoperative medication upset and ease of fluid intake ratings. It would appear that the information given in both the stress-point preparation and single-session preparation was an important condition to the child's behavioral adjustment. In terms of posthospital adjustment, however, the single-session preparation was not significantly different from the supportive care condition, which in turn was not significantly different from the control condition. Parents in the stress-point preparation were less anxious, more adequately informed and more satisfied with the care they received, than were parents in the other three conditions. Thus, the benefit of combining information with the opportunity of a supporting relationship with a nurse at particular stressful points during the hospital experience was demonstrated.

A more recent study by Roskies *et al.* (1978) focused on the reduction of the trauma of hospitalization for children by increasing their mothers' presence and participation in their care and by modifying the attitudes and behavior of the nursing staff. The mothers of 48 children, aged 1 to 5, to be admitted for elective surgery to a large, metropolitan pediatric hospital, were assigned either to a control group or to an experimental group which had an extra half-hour session in a preadmission interview focusing on visiting, during which specific suggestions were made about the frequency and timing of visits as well as about the role of the mother during her visits. During the experimental period, weekly meetings were held with the nursing staff to enlist their support for this change in visiting patterns. Results indicated that duration and timing of visits and behavior during visits were significantly modified for the experimental group of mothers. In contrast, the nurses did not significantly change their relationship with the mothers or the children.

The authors introduced the importance of educating the parents and nurses to provide for the parents' place in the hospital. The nurses were asked to focus on the mother–child unit and to aid the mother in providing psychological care for her child.

> The intervention program may have helped the nurses become more willing to admit the mother to the hospital, but the role designated to her was clearly that of a subordinate performing the least desirable tasks, rather than of a collaborator sharing responsibility. (p. 773)

The authors point to the many difficulties and suggest that future programs to seek change in nursing roles should focus more on supervisory level personnel.

The previous clinical research has pointed to the potentially beneficial effects of including the parent as an active and involved caregiver during the child's hospitalization. The parent, in turn, must be prepared by the staff to be a knowledgeable and calm liaison between the hospital staff and the patient. The hospital routines, the role of the parent *vis à vis* the nursing and medical staff, and the parent's own concerns and coping strategies must be examined to promote this shift toward greater parent involvement.

The research to date has not identified any specific or reliable techniques that might be used to accomplish attitudinal or behavioral changes on the part of the hospital or staff. Several broad approaches to seeking such changes have been attempted and tested. The role of providing information to parents in the preparation process appears to be critical. The emotional contagion theory appears to be holding up under continued examination in terms of parental attitudes and behaviors influencing their children. The research also appears to be confirming the importance of parental involvement with the preschool child.

Home Preparation

Giving information to both the parent and the child in advance of hospitalization has been the focus of attention in three studies (Ferguson, 1979; Newswanger, 1974; Wolfer & Visintainer, 1979). The optimal time to prepare children for approaching hospitalization for surgery has been a theoretical debate for many years. Mellish (1969) suggested that younger children, particularly very anxious ones, need only a few days of preparation since longer intervals may only increase anticipatory anxiety. Heller (1967) speculated that older children need a more detailed and lengthy preparation lasting several days or weeks. Preparing younger children too far in advance may lead to an increase in fear. Most of the investigators used nurse–teachers to deliver preadmission information in the child's own home on the assumption that the supportive relationship with the parent would help in the preparatory process.

Newswanger (1974) used a two-day preparatory period to prepare mothers and their preschool children for hospitalization for tonsillectomy and adenoidectomy. Home visits were made to the children, and either various hospital procedures were demonstrated (experimental) or an unrelated story was read (control sample). Observational data was obtained in the hospital during critical medical procedures. Children who had received relevant preparation showed significantly more positive mastery behavior and less behavior classified as negative or passive than did those children in the control group. However, there were no longer any group differences found at the two-week posthospital interview, when a questionnaire was filled out by the parents to indicate whether any behavioral problems had occurred. This study failed to assess the children's degree of understanding of the materials presented at the home visit, and it is difficult to interpret their in-hospital findings. Perhaps parental anxiety was reduced and mothers therefore were able to furnish more appropriate preparation.

The effect of home preparation on parents' attitudes and anxiety was assessed in addition to the children's response to hospitalization in a study by Wolfer and Visintainer (1979). Three to four days prior to admission, children were presented with materials developed in the authors' earlier hospital studies. In this study, 163 children between 2 and 12 years of age were assigned to one of five groups: advanced preparatory materials, stress-point preparation in the hospital (no home preparation), home preparatory materials plus stress-point preparation, home preparation plus supportive care from a single nurse, and a routine care control condition.

The result of using only the preparatory materials at home was as effective as preparation plus supportive care of stress point preparation. These children were rated as less upset and more cooperative during the blood test, venipuncture, and transport to the operating room. However, parental ratings of satisfactory care and anxiety reduction were improved only for those mothers for whom home preparation was combined with stress-point preparation or consistent supportive nursing care in the hospital. In fact, data indicated that 20% of the parents did not use the home preparatory materials, either because they felt that they already knew what to expect or because their children did not seem to need preparation. It was surprising that there were no differences due to age interacting with type of preparation, although younger children were rated as more upset than older children during the blood test, preoperative medication, and the time of first voiding. The authors report a lack of differences between the groups as to the effects of previous experience in the hospital influencing the treatment results.

Ferguson (1979) compared a procedure involving home contact for preadmission as opposed to exposure to film model upon admission, and a combination of the two. All children were experiencing their first hospital admission for an elective tonsillectomy and ranged in age from 3 to 7 years. The children's adjustment in the hospital was measured by behavioral ratings by the parents and nurses. Parents' anxiety was measured on a self-report rating scale and by observations of their degree of anxiety and satisfaction. Preadmission visits lessened maternal anxiety during and after the child's operation and led to greater satisfaction. The incidence of reported negative posthospital behavior was also reduced, particularly in the 6- to 7-year-old children. However, the increase of undesirable behavior posthospital was more effectively reduced in the younger group by a peer-modeling film shown at admission, whereas 6- to 7-year olds responded as positively to the preadmission visit alone.

PREVENTION

Primary prevention of anxiety induced by medical procedures may ultimately be accomplished efficiently with population-wide interventions with normal children. Although many public schools have introduced health units as part of the educational curriculum, few studies have been generated from these programs. Earlier, we reviewed studies which investigated children's conceptions of illness behavior in normal and chronically ill children. Yet, despite the recognition that attitudes toward doctors and medical procedures may influence health care behavior, there has been a lack of research evaluation of interventions designed to introduce children to general medical care. The pediatrician is the first health care practitioner with whom the child interacts. His or her ability to manage the child's anxiety during routine visits may predispose the child's receptivity to later medical care. Behavioral pediatrics is a recognized area in which the collaboration between the pediatrician and the psychologist has yielded successful treatment of childhood problems including enuresis, school phobia, hyperactivity, and aggressive behavior problems (Melamed, 1980). The ability of psychologists to intervene in the preparation of children for medical treatment has not been tapped to any great extent. The few studies which exist do provide evidence that the systematic presentation of information about medical procedures, personnel, and equipment does alter the attitude of children toward medical practitioners in a desirable manner.

Penticuff (1976) presented third-grade children with *Ethan Has an Operation* and measured their attitudes and physiological responses as they changed after viewing this hospital preparation film. The 57 children rated their attitudes toward doctors, nurses, and medical procedures on a semantic differential rating scale. The immediate effect of seeing the film was to increase their autonomic sweat gland activation. There was a correlated significant improvement in their attitudes toward medical procedures.

In a more recent study, Roberts and his colleagues (1981) also presented information about medical procedures to children in the second through fourth grades. This study provided a control group in which a slide-tape unrelated to the hospital was presented. The state of anxiety, degree of medical concerns, and hospital fear of the children were assessed one week prior to the film presentations. The children viewed the film in small groups. Immediately following the slide presentation, each child again completed the children's manifest anxiety scale, the Scherer–Nakamura modified Children's Fear Survey, the hospital fears questionnaire, and an 11-item multiple-choice questionnaire for assessing the recall of various medical terms and procedures which had been included in the hospital package. There was a repeat assessment battery two weeks later. Unfortunately, the measure of acquisition of information was not repeated. No medical procedures were actually involved in assessing the effects of preparation.

Results revealed that the children who received the hospital package significantly reduced their fears on the self-report measures of hospital concerns at the postpresentation and follow-up assessments compared to their own prepackage level and to the levels in children receiving the travelogue package. The results also indicated that the prepared subjects showed positive changes of an even greater magnitude at the follow-up. The children who had viewed the hospital package had an increased knowledge base regarding medical procedures relative to the control sample. Although the study is a good contribution, it does not truly demonstrate prevention because no measure of the children's behavior in an actual medical situation (doctor's examination) or hospitalization was obtained.

The research in which children not preselected for fearful behavior were shown either dental preparatory videotapes (Melamed *et al.*, 1978) or hospital videotapes (Melamed & Siegel, 1975; Melamed *et al.*, 1976) more clearly demonstrated preventive applications during medical procedures. Children in these studies who had not had previous experience in the situation (medical or dental) showed a significant reduction of fear and improved behavioral cooperation following exposure, as compared with children viewing unrelated materials. Siegel and Peterson (1980) demonstrated that by providing young dental patients with coping skills or sensory information, behavioral control and anxiety reduction were enhanced.

On the other hand, modeling studies of children who have already experienced hospitalization, surgery, or dental treatment do not yield such consistent results (Melamed *et al.* 1978; Klorman, Hilpert, Michael, LaGana, & Sveen, 1980). Often children with previous experience have conditioned emotional responses and may be further sensitized by exposure to preparatory materials. Thus, in the area of primary prevention, a vicarious prototype of what these experiences involve and how to cope with them may reduce the anxiety prior to the initial exposure and result in more cooperative patients.

SUMMARY AND CONCLUSIONS

The general impression from the critical review of the literature is that psychological factors have been shown to influence cooperation with medical procedures and recovery from surgery. The difficulty in making treatment recommendations is that because the research to date does not have a strong theoretical base, it is difficult to show that specific factors can be predicted to have a functional relationship with behavior. For example, many of the studies have been exploratory and served more to generate hypotheses than to answer any practical questions about which children to prepare and in what manner. Samples have tended to range over a wide age distribution and are so heterogeneous with regard to the specific medical interventions as to limit the conclusions drawn. However, there does exist enough substantial evidence to suggest that depending on certain individual characteristics, surgical and medical preparation must take into account the patient's age and prehospital level of anxiety and to consider seriously the role of parent.

It has been found in a number of studies that younger children (under age 7) must be prepared just at the time of the impending procedure. Similarly, preparation must be geared to their level of understanding. Children's conceptions of illness and medical procedures change with their developmental maturity. There are obvious developmental influences on their capability to cope with hospitalization. Unfortunately, few longitudinal studies exist which have attempted to assess this interaction. The data available have often been based on patient interviews collected in a nonstandardized manner. Most follow-up studies to assess the long-term benefit of hospitalization preparation have been of less than two months duration and have been done by telephone inquiry, which adds a selective bias in that only those families who have had a positive experience may be willing to share their reactions.

At present there exist reliable and acceptable measurement instruments so that the field is fertile for undertaking studies to assess psychological treatments that propose to prepare families for hospitalization. Observational recording devices and behavior checklists are being employed by several investigators. Physiological measures are being obtained through nursing records or by actual assessment of such indices as pulse rate or palmar sweating.

Self-report measures are least reliable in differentiating the responses of young children, although several new measures are being tried. The older child does appear to be able to report consistently on the degree of medical concerns. The realization that the behavioral change programs may effect various individual systems (subjective, behavioral, physiological) to different extents had led to the general adoption of a multidimensional approach. There is also a realization that surgery and hospitalization are a process and that measurement must include follow-up data in addition to information about the patients' behavior in the hospital.

As to prediction in the matter of which children are at risk, studies have related the prehospital factors of the children's age, the parents' anxiety, and the coping styles of both children and parents. This has led to the suggested importance of measuring these predisposing factors prior to undertaking preparation.

The actual implications of psychological treatments which have proved effective for preparing individuals facing stress depend upon their integration with what is already established routine. Because the hospital is designed primarily to accomplish a medical goal, one must demonstrate that the psychological preparation will be compatible with the ongoing practice and produce greater efficiency in that individual management problems will be reduced if patient understanding and compliance are improved. Investigators must turn to the staff to determine what the patient is required to do and must include indexes such as reduced complaints of vomiting and nausea, decreased sleep problems, and shorter hospital stays. The data collection that must be undertaken should involve the staff in a minimal fashion so that they do not view this as another task demand. Wherever possible, psychologists or pediatric education specialists should develop instruments that can be applied by the parent or volunteer personnel under their guidance.

Perhaps the greatest shortcoming in the current literature is the tendency for compound treatments to be offered to unselected children. This makes it difficult to determine individual treatment components. If we merely match statistically for such predisposing factors as age or prehospital anxiety level, then we can not assess their interactions with the various treatments. Instead, hypothesis-testing must be generated to answer questions regarding the optimal preparation time for children of different ages. The level or pattern of autonomic functioning which facilitates the taking in of information from the package needs further study. Does information alone necessarily lead to appropriate behavior, or must certain individuals (i.e., perhaps those without previous experience) be taught specific coping strategies.

Instead of comparing one treatment package with another to demonstrate that something works better than nothing, we must identify what we feel are the active therapeutic ingredients. There is much overlap between systematic desensitization, modeling, and behavioral rehearsal, in that they all provide information to the patient about what will occur. It is not surprising, then, that many outcome studies report that they are equally effective in reducing anxiety. One must parcel out the common ingredients, such as supportive relationships and information, and examine their separate and then their combined results. Thus, in modeling, are we teaching a coping strategy or merely presenting information? If relaxation is included, how would this differ from systematic desensitization? Is exposure alone sufficient for some children to evoke the use of coping mechanisms? The selection of appropriate control groups is very important. The ethical concerns of depriving children of preparation must be answered with data that show that many children need little or no preparation and do quite well.

The need for psychological outcome studies in answering such questions as whether conjoint preparation with the family improves adjustment and how mothers' anxiety effects children's behavior is paramount before hospital policies such as rooming-in or open visiting can be promoted.

There is a growing trend for same-day ambulatory surgery. This would allow the child to experience minimal hospital stress and return immediately to the care of the parent. Few studies exist which would permit any recommendations as to what type of information parents need regarding postoperative care for their children and when this should be imparted so that they will understand what is required. Again, prior to recommending that ambulatory care be the future goal, one must ascertain whether in fact there is a reduction in psychological problems with this innovation.

We have made much progress in the past decade. Statistics reveal that over 70% of nonchronic care pediatric hospitals provide some type of preparation. Thus, the research arena is set. The data we may derive from studying our behavior methods of reducing fear in this setting may provide information that is relevant to many other situations in which an individual must cope with life stress. It is an excellent place for developmental psychologists to apply their theories about the ability of children to cope as they change over time and develop greater cognitive control. It is an ideal problem area in which to bridge the gap between medicine and psychology. The reciprocal contributions of professionals from the different specialties will promote a holistic and probably most beneficial approach. The next decade will provide us with specific answers to the questions about how to individualize treatment in an efficient manner so that the appropriate people will get the most advantageous preparation.

REFERENCES

Ayer, W. A. Use of visual imagery in needle phobic children. *Journal of Dentistry for Children,* March–April 1973, 1–3.

Bandura, A. *Principles of behavior modification.* New York: Holt, Rinehart, & Winston, 1969.

Barber, J. Rapid induction analgesia: A clinical report: *American Journal of Clinical Hypnosis,* 1977, *19,* 138–147.

Beeby, D. A., & Hughes, J. O. M. Behavior of unsedated children in the anesthetic room. *British Journal of Anaesthesiology,* 1980, *52,* 279–281.

Bothe, A., & Galdston, R. The child's loss of consciousness: A psychiatric view of pediatric anesthesia. *Pediatrics,* 1972, *50,* 252–263.

Brown, B. Beyond separation. In D. Hall & M. Stacey (Eds.), *Beyond separation.* London: Routledge & Kegan Paul, 1979, pp. 18–53.

Burstein, S., & Meichenbaum, D. The work of worrying in children undergoing surgery. *Journal of Abnormal Child Psychology,* 1979, *7,* 121–132.

Campbell, E. *Effects of mothers' anxiety on infants' behavior.* Unpublished doctoral dissertation, Yale University, 1957.

Campbell, J. Illness as a point of view: The development of children's concepts of illness. *Child Development,* 1975, *46,* 92–100.

Cassell, S. Effects of brief puppet therapy upon the emotional responses of children undergoing cardiac catheterization. *Journal of Consulting Psychology,* 1965, *29,* 1–8.

Cataldo, M. F., Bessman, C. A., Parker, L. H., Pearson, J. E., & Rogers, M. C. Behavioral assessment for pediatric intensive care units. *Journal of Applied Behavior Analysis,* 1979, *12,* 83–98.

Chapman, A. H., Loeb, D. G., & Gibbons, M. J. Psychiatric aspects of hospitalization of children. *Archives of Pediatrics,* 1956, *73,* 77–88.

Chertok, S. L., & Bornstein, P. Covert modeling treatment of children's dental fears. *Child Behavior Therapy,* 1979, *1,* 249–255.

Dearborn, M. J., & Melamed, B. G. *Lang's self-assessment mannequin as a measure of emotion.* Unpublished manuscript, 1981.

Doughty, A. G. The evaluation of premedication in children. *Proceedings of the Royal Society of Medicine,* 1959, *52,* 823–834.

Eckenhoff, J. E. Preanesthetic sedation of children. American Medical Association, *Archives of Otalaryngology,* 1953, *57,* 411–415.(a)

Eckenhoff, J. E. The relationship of anesthesia to postoperative personality changes in children. *American Journal of Diseases of Children,* 1953, *86,* 587–592.(b)

Eckhardt, L. O., & Prugh, D. G. Preparing children psychologically for painful medical and surgical procedures. IN E. Gelbert (Ed.), *Psychosocial aspects of pediatric care.* New York: Grune & Stratton, 1978.

Escalona, S. Emotional development in the first year of life. *Transactions of the Sixth Annual Conference on Problems of Infancy and Childhood.* New York: Josiah Macy, Jr. Foundation, 1953.

Ferguson, B. G. Preparing young children for hospitalization: A comparison of two methods. *Pediatrics,* 1979, *64,* 656–665.

Freeman, A., & Bachman, L. An evaluation of preoperative medication. *Anesthesia & Analgesia—Current Researches,* 1959, *38,* 429–437.

Gale, E., & Ayer, N. M. Treatment of dental phobias. *Journal of the American Dental Association,* 1969, *73,* 1304–1307.

Gatchel, R. J. Effectiveness of two procedures for reducing dental fear: Group-administered desensitization and group education and discussion. *Journal of American Dental Association,* 1980, *101,* 635–637.

Gellert, E. Reducing the emotional stress of hospitalization for children. *American Journal of Occupational Therapy*, 1958, *12*, 125–129.

George, J. M., Scott, D. S., Turner, S. P., & Gregg, J. M. The effects of psychological factors and physical trauma on recovery from oral surgery. *Journal of Behavioral Medicine*, 1980, *3*, 291–310.

Glennon, B., & Weisz, J. R. An observational approach to the assessment of anxiety in young children. *Journal of Consulting and Clinical Psychology*, 1978, *46*, 1246–1257.

Gofman, H., Buckman, N., & Schade, G. The child's emotional response to hospitalization. *American Journal of Diseases of Children*, 1957, *93*, 157–164.

Goldfried, M. R. Systematic desensitization as training in self-control. *Journal of Consulting and Clinical Psychology*, 1971, *37*, 228–234.

Heller, J. A. *The hospitalized child and his family* Baltimore: Johns Hopkins, 1967.

Hodgson, R., & Rachman, S. Desynchrony in measures of fear. *Behavior Research and Therapy*, 1974, *12*, 319–326.

Holmes, T. H., & Masuda, M. Life change and illness susceptibility. In B. S. Dohpenwend & B. P. Dohpenwend (Eds.), *Stressful life events: Their nature and effects*. New York: Wiley, 1974.

Jacobs, R. Tbe meaning of hospital: Denial of emotions. In D. Hall & M. Stacey (Eds.), *Beyond separation*. London: Routledge & Kegan Paul, 1979, pp. 82–108.

Janis, I. L. *Psychological stress*. New York: Wiley, 1958.

Johnson, J. E., & Leventhal, H. Effects of accurate expectations and behavioral instructions on reactions during a noxious medical examination. *Journal of Personality and Social Psychology*, 1974, *29*, 710–718.

Johnson, R., & Dabbs, J. M. Enumeration of active sweat glands: A simple physiological indicator of psychological changes. *Nursing Research*, 1967, *16*, 273–276.

Johnson, S. B., & Melamed, B. G. The assessment and treatment of children's fears. In B. Labey & A. Kazdin (Eds.), *Advances in clinical child psychology* (Vol. 2). New York: Plenum Press, 1979, pp. 107–139.

Kanfer, F., Karoly, P., & Newman, A. Reduction of children's fear of the dark by competence-related and situational threat-related cues. *Journal of Consulting and Clinical Psychology*, 1975, *43*, 251–258.

Katz, R. C. Single session recovery from a hemodialysis phobia: A case study. *Journal of Behavior Therapy and Experimental Psychiatry*, 1974, *5*, 205–206.

Kendall, P. C. On the efficacious use of verbal self-instructional procedures of children. *Cognitive Therapy and Research*, 1977, *1*, 331–341.

Kendall, P. C. Stressful medical procedures: Cognitive-behavioral strategies for stress management and prevention. In D. Meichenbaum & M. Jaremko (Eds.), *Stress reduction and prevention: A cognitive-behavioral perspective*. New York: Plenum Press, in press.

Kleinknecht, R., & Bernstein, D. Comparative evaluation of three social learning approaches to the reduction of dental fear. In B. Ingersoll & R. McCutcheon (Eds.), *Proceedings of second national research conference on behavioral dentistry*. Morgantown: West Virginia University Press, 1979.

Klorman, R., Hilpert, P., Michael, R., LaGana, C., & Sveen, D. Effects of coping and mastery modeling on experienced and inexperienced pedodontic patients' disruptiveness. *Behavior Therapy*, 1980, *11*, 156–168.

Knight, R., Atkins, A., Eagle, C., Evans, N., Finkelstein, J. W., Fukushima, D., Katz, J., & Weiner, H. Psychological stress, ego defenses, and cortisol production in children hospitalized for elective surgery. *Psychosomatic Medicine*, 1979, *41*, 40–49.

Kozak, M. J., & Miller, G. A. Hypothetical constructs vs. intervening variables: A reappraisal of the three-systems model of anxiety assessment. *Behavioral Assessment*, in press.

Krop, H., Jackson, E., & Mealiea, W. *Effects of systematic desensitization and stress management training in reducing dental phobia.* Paper presented at the meeting of the Association of the Advancement of Behavior Therapy, Washington, D.C., December 1976.

Lacey, J. I. Somatic response patterning and stress: Some revisions of activation theory. In M. H. Appley & R. Trumbull (Eds.), *Psychological stress: Issues in research.* New York: Appleton–Century–Crofts, 1967.

Lang, P. J. Fear reduction and fear behavior: Problems in treating a construct. *Research in Psychotherapy,* 1968, *3,* 90–102.

Lang, P. J. The mechanics of desensitization and the laboratory study of human fear. In C. M. Franks (Ed.), *Behavior therapy: Appraisal and status.* New York: McGraw-Hill, 1969, pp. 160–191.

Lang, P. J. The psychophysiology of anxiety. In H. Akiskal (Ed.), *Psychiatric diagnosis: Exploration of biological criteria.* New York: Spectrum, 1977.

Lang, P. J. Behavioral treatment and bio-behavioral assessment: Computer applications. In J. B. Sidowski, J. H. Johnson, & T. A. Williams (Eds.), *Technology in mental health care delivery systems.* Norwood, N.J.: Ablex, 1980.

Lazarus, R. S. Some principles of psychological stress and their relation to dentistry. *Journal of Dental Research,* 1966, *45,* 1620–1626.

Leigh, M. D., & Belton, K. Premedication in infants and children. *Anesthesiology,* 1949, *7,* 611–615.

Machen, J., & Johnson, R. Desensitization, model learning, and the dental behavior of children. *Journal of Dental Research,* 1974, *53,* 83–89.

Mahaffy, P., Jr. The effects of hospitalization on children admitted for tonsillectomy and adenoidectomy. *Nursing Research,* 1965, *14,* 12–19.

Melamed, B. G. Psychology and behavioral pediatrics. In S. J. Rachman (Ed.), *Contributions of medical psychology (Vol. 2).* Oxford: Pergamon, 1980.

Melamed, B. G. Reduction of medical fears: An information processing analysis. In J. Boulougouris (Ed.), *Learning theories in psychiatry.* New York: Wiley, in press.

Melamed, B. G., & Siegel, L. J. Reduction of anxiety in children facing hospitalization and surgery by use of filmed modeling. *Journal of Consulting and Clinical Psychology,* 1975, *43,* 511–521.

Melamed, B. G., & Siegel, L. J. *Behavioral medicine: Practical applications in health care.* New York: Springer, 1980.

Melamed, B. G., Meyer, R., Gee, C., & Soule, L. The influence of time and type of preparation on children's adjustment to hospitalization. *Journal of Pediatric Psychology,* 1976, *1,* 31–37.

Melamed, B. G., Yurcheson, R., Fleece, E. L., Hutcherson, S., & Hawes, R. Effects of film modeling on the reduction of anxiety-related behaviors in individuals varying in level of previous experience in the stress situation. *Journal of Consulting and Clinical Psychology,* 1978, *46,* 1357–1367.

Melamed, B. G., Bennett, C. G., Ross, S., Bush, J., Hill, C., Ronk, S., Courts, F., & Jerrell, G. *The effects of dentist reinforcement strategies on pediatric patient compliance and fear.* Unpublished manuscript, 1982.

Mellish, R. W. P. Preparation of a child for hospitalization and surgery. *Pediatric Clinics of North America,* 1969, *16,* 543–553.

Meyers, E. F., & Muravchick, S. Anesthesia induction techniques in pediatric patients: A controlled study of behavioral consequences. *Anesthesia and Analgesia Current Researchers,* 1977, *56,* 538–542.

Myers-Vando, R., Steward, M., Folkins, C., & Hines, P. The effects of congential heart

disease on cognitive development, illness causality concepts, and vulnerability. *American Journal of Orthopsychiatry*, 1979, *49*, 617–625.
Nagy, M. Children's ideas of the origin of illness. *Health Education Journal*, 1951, *9*, 6–12.
Nimmer, W. H., & Kapp, R. A. A multiple impact program for the treatment of an injection phobia. *Journal of Behavior Therapy and Experimental Psychology*, 1974, *5*, 257–258.
Neuhauser, C., Amsterdam, B., Hines, P., & Steward, M. Children's concepts of healing: Cognitive development and locus of control factors. *American Journal of Orthopsychiatry*, 1978, *48*, 335–341.
Newswanger, A. *Hospital preparation of pre-school children as related to their behavior during and immediately following hospitalization for tonsillectomy and adenoidectomy* Unpublished master's thesis, University of Delaware, May 1974.
Opton, E. M., Jr., & Lazarus, R. S. Personality determinants of psychophysiological response to stress: A theoretical analysis and an experiment. *Journal of Personality and Social Psychology*, 1967, *6*, 291–303.
Oremland, E. K., & Oremland, J. D. *The effects of hospitalization on children: Models for their care*. Springfield, ILL.: Charles C Thomas, 1973.
Penticuff, J. H. *The effect of filmed peer modeling, cognitive appraisal, and autonomic reactivity in changing children's attitudes about health care procedures and personnel*. Unpublished Ph.D. Dissertation, Case Western Reserve University, 1976.
Peters, B. M. School-aged children's beliefs about causality of illness: A review of the literature. *Maternal Child Nursing Journal*, 1978, *7*, 143–154.
Peterson, L., & Ridley-Johnson, R. Pediatric hospital response to survey on prehospital preparation for children. *Journal of Pediatric Psychology*, 1980, *5*, 1–7.
Peterson, L., & Shigetomi, C. The use of self-control techniques to minimize anxiety in hospitalized children. *Behavior Therapy*, 1981, *12*, 1–14.
Peterson, L., & Shigetomi, C. One year follow-up of elective surgery child patients receiving pre-operative preparation. *Journal of Pediatric Psychology*, in press.
Peterson, L., Schultheis, K., Ridley-Johnson, R., Miller, D. J., & Tracy K. *Comparison on three modeling procedures in the presurgical preparation of children*. Unpublished manuscript, 1981.
Prugh, D. G., Staub, E. M., Sands, H. H., Kirschbaum. R. M., & Leniban, E. A. A study of the emotional reactions of children and families to hospitalization and illness. *American Journal of Orthopsychiatry*, 1953, *23*, 70–106.
Rachman, S. J. *The meaning of fear*. London: Penguin, 1976.
Rachman, S. J., & Phillips, C. *Psychology and behavioral medicine*. New York: Cambridge University Press, 1980.
Roberts, M. C., Wurtele, S. R., Boone, R. R., Ginther, L. J., & Elkins, P. D. Reduction of medical fears by use of modeling: A preventive application in a general population of children. *Journal of Pediatric Psychology*, 1981, *6*(3), 293–300.
Robertson, J. *Young children in hospitals*. New York: Basic Books, 1958.
Robertson, J. The child in hospital. *South African Medical Journal*, November 1976, 749–752.
Roskies, E., Bedard, P., Gauvreau-Guilbault, H., & Laforture, D. Emergency hospitalization of young children. Some neglected psychological considerations. *Medical Care*. 1975, *13*(7), 570–581.
Roskies, E., Mongeon, M., & Gagnon-Lefebvre, B. Increasing maternal participation in the hospitalization of young children. *Medical Care*, 1978, *16*(9), 767–777.
Sawtell, R., Simon, J., & Simeonsson, R. The effects of five preparatory methods upon child behavior during the first dental visit. *Journal of Dentistry for Children*, 1974, *41*, 37–45.

Selye, H. *The stress of life.* New York: McGraw–Hill, 1956.
Shaw, C., & Thoreson, C. Effects of modeling and desensitization in reducing dental phobia. *Journal of Counseling Psychology,* 1974, *21,* 415–420.
Shipley, R. H., Butt, J., Horwitz, B., & Farbry, J. Preparation for a stressful medical procedure: Effect of amount of prestimulus exposure on coping style. *Journal of Consulting and Clinical Psychology,* 1978, *46,* 499–507.
Shipley, R. H., Butt, J. H., & Horwitz, B. A. Preparation to reexperience a stressful medical examination: Effect of repetitious videotape exposure and coping style. *Journal of Consulting and Clinical Psychology,* 1979, *47,* 485–492.
Siegel, L. J., & Peterson, L. Stress reduction in young dental patients through coping skills and sensory information. *Journal of Consulting and Clinical Psychology,* 1980, *48,* 785–787.
Silverman, W., & Melamed, B. G. Measurement of emotion in children. Unpublished manuscript, 1981.
Simeonsson, R., Buckley, L., & Monson, L. Conceptions of illness causality in hospitalized children. *Journal of Pediatric Psychology,* 1979, *4,* 77–84.
Sipowicz, R. R. & Vernon, D. T. Children, stress and hospitalization. *Journal of Health and Social Behavior,* 1965, *9,* 275–287.
Skipper, J., Leonard, R. C., & Rhymes, J. Child hospitalization and social interaction: An experimental study of mothers' feelings of stress, adaptation, and satisfaction. *Medical Care,* 1968, *6,* 496–506.
Smith, R. *Anesthesia for infants and children* (4th ed.). St. Louis: Mosby, 1980.
Spielberger, C. D., Gorsuch, R. L., & Lushene, R. E. *Manual for the state–trait anxiety inventory.* Palo Alto, Calif.: Consulting Psychologists Press, 1970.
Steward, M., & Regalbutto, G. Do doctors know what children know? *American Journal of Orthopsychiatry,* 1975, *45,* 146–149.
Strahan, R. F., Todd, J. B., & Ingolis, G. B. A palmar sweat measure particularly suited for naturalistic research. *Psychophysiology,* 1974, *11,* 715–720.
Taylor, C. B., Ferguson, J. M., & Wermuth, B. M. Simple techniques to treat medical phobias. *Postgraduate Medical Journal,* 1977, *53,* 28–32.
Taylor, J. A. A personality scale of manifest anxiety. *Journal of Abnormal and Social Psychology,* 1953, *48,* 285–290.
Vanderveer, A. A psychopathology of physical illness and hospital residence. *Quarterly Journal of Child Behavior,* 1949, *1,* 55–71.
Vardaro, J. A. R. Preadmission anxiety and mother–child relationships. *Journal of the Association of Care of Children in Hospitals,* 1978, *7,* 8–15.
Vernon, D. T. Use of modeling to modify children's responses to a natural potentially stressful situation. *Journal of Applied Psychology,* 1973, *58,* 351–356.
Vernon, D. T., Foley, J. M., Sipowicz, R. R., & Schulman, J. L. *The psychological responses of children to hospitalization and illness.* Springfield, Ill.: Charles C Thomas, 1965,
Visintainer, M., & Wolfer, J. Psychological preparation for surgical pediatric patients: The effects on children's and parents' stress responses and adjustment. *Pediatrics,* 1975, *56,* 187–202.
Waters, R. M. Pain relief for children. *American Journal of Surgery,* 1938, *39,* 470.
White, W., & Davis, M. Vicarious extinction of phobic behavior in early childhood. *Journal of Abnormal Child Psychology,* 1974, *2,* 25–32.
Wolfer, J. A., & Visintainer, M. A. Pediatric surgical patterns and parents' stress responses and adjustment as a function of psychological preparation and stress-point nursing care. *Nursing Research,* 1975, *24,* 244–255.
Wolfer, J. A. & Visintainer, M. A. Tonsillectomy patients: Effects on children's and parents' adjustment. *Pediatrics,* 1979, *64,* 646–655.

Wolff, C. T., Friedman, S. B., Hofer, M. A., & Mason, J. W. Relationship between psychological defenses and mean urinary 17–hydroxycorticosteriod excretion rates. *Psychosomatic Medicine,* 1964, *26,* 576–591.

Wolff, C. T., Hofer, M. A., & Mason, J. W. Relationship between psychological defenses and mean urinary 17-OHCS excretion rates: Part 1. A predictive study of parents of children with leukemia. *Psychosomatic Medicine,* 1964, *26,* 592–609.

Zabin, M., & Melamed, B. G. The relationship between parental discipline and children's ability to cope with stress. *Journal of Behavioral Assessment,* 1980, *2,* 17–27.

IV
SETTINGS AND ENVIRONMENTS

8

Behavioral/Environmental Considerations in Pediatric Inpatient Care

Michael F. Cataldo, Harvey E. Jacobs, and Mark C. Rogers

Behavioral pediatrics has typically emphasized two areas of interface between the behavioral sciences and pediatric medicine. One is the identification, treatment, and prevention of childhood behavior problems through routine clinic and well-child visits (Christophersen & Rapoff, 1979). The second is the use of behavioral procedures to correct primary medical problems and those secondary to particular medical conditions. For example, behavioral procedures have been demonstrated to be effective in reducing seizures (Cataldo, Russo, & Freeman, 1979; Zlutnick, Mayville, & Moffat, 1975), in the treatment of neuromuscular disorders (Cataldo & Bird, in press; Cataldo, Bird, & Cunningham, 1978), in treating fecal incontinence secondary to myelomeningocele (Whitehead, Parker, Masek, Cataldo, & Freeman, 1981) and in dealing with pediatric chronic pain (Varni, Bessman, Russo, & Cataldo, 1980).

MICHAEL F. CATALDO • John F. Kennedy Institute; Johns Hopkins University School of Medicine, Baltimore, Maryland 21205. HARVEY E. JACOBS • Rehabilitation Medical Center, Los Angeles, California 90073. MARK C. ROGERS • Johns Hopkins University School of Medicine, Baltimore, Maryland 21205. Preparation of this paper and conduct of the work on which it is based were supported in part by Grant 917 from Maternal and Child Health and Grant 9 MPH 1077 JHH from the Morris Goldseker Foundation. This work has been a joint effort of the Kennedy Institute and the Behavioral Medicine Program of the Johns Hopkins University School of Medicine.

This chapter will explore a third problem area to be addressed by a behavioral pediatrics approach: the iatrogenic effect of appropriate medical and surgical care. Because the relationship between environmental events and their biobehavioral consequences has been the subject of considerable scientific inquiry, we will explore the hypothesis that events necessary for the medical management of hospitalized children can result in unwanted behavioral and biological effects. Primary consideration will be given to that body of research concerned with the principles of learning, because such principles have identified extremely powerful methods for the control of human behavior and have documented the interface between environmental/biological interaction among many species in numerous settings. The present chapter proposes to extend previous findings to the ill child in the hospital environment.

HOSPITAL ENVIRONMENTS

There are at least two categories of events which may affect the behavioral and biological status of patients: (1) the daily hospital routine developed by the need to provide institutional-based services to many dependent people and (2) dramatic and often painful medical and surgical procedures which occur to the patient or are observed by the patient as they occur to others. Pertinent to the first category, Goffman (1961) provided a landmark sociological analysis of the effects that institutional management procedures may have on human behavior.

In his analysis, Goffman defined the institutional environment as "a place of residence and work where a large number of like-situated individuals, cut off from the wider society for an appreciable period of time, together lead an enclosed, formally administered round of life" (p. XIII). He further pointed out that the key fact which led to this circumstance of a "total institution" was the handling of the many needs of large groups of people by a bureaucratic organization—whether or not the methods chosen are necessary or effective means of social organization. While Goffman's focus of analysis was primarily mental patients and other inmates, the first of his five groupings of institutions applies to hospital environments as well—that is, "an institution established to care for persons who are both incapable and harmless" (p. 4). Further, while not originally based upon an analysis of hospitals, Goffman's characteristics of total institutions are surprisingly descriptive of hospitals, as evidenced from the points abstracted below in which the only alteration has been the substitution of the word patient for inmate.

- All aspects of life are conducted in the same place and under the same single authority.
- Daily activity is conducted in the presence of others, all treated alike and required to do the same thing.
- All phases of the day's activities are tightly scheduled, with the sequence imposed from above by a system of explicit formal rulings and body of officials.
- There is a basic split between a large managed group and a small supervisory staff: (a) patients have restricted contact with the outside world; (b) staff operate on an 8-hour shift; (c) staff often feel superior and righteous while patients tend, in some ways at least, to feel inferior, weak, blameworthy and guilty; (d) social distance between the two groups is typically great and often formally prescribed, including information about the staff's plans for patients.
- For the patient who is work-oriented on the outside and often untrained in leisurely pursuits, the result is often boredom and demoralization.
- Visitors are a privilege.
- Physical indignities abound.
- The patient is obliged to take oral or intravenous medications, whether desired or not, and to eat his or her food, however unpalatable.
- Spheres of life are desegregated, so that a patient's conduct in one scene of activity is thrown up to him or her by the staff as a comment or check on conduct in another context.
- Patients must request permission or supplies for minor activities that one can execute on one's own on the outside.
- Statements made by the patient may be discounted, with staff giving attention to nonverbal aspects of the reply.

The most striking aspect of Goffman's analysis is not only how accurate a description it is of hospital routine, but also that each point can be well justified in terms of efficient management practices. Although medical and surgical procedures are specifically designed with careful attention to the patient, aspects of hospital activities not directly related to the physical status of the patient are designed for efficacy of hospital operation. Without a clear indication that such time-honored hospital routines are detrimental to the patients' physical or psychological status and in the absence of an effective alternative set of routines without risks to patients, there is no reason for current practices to change.

Children may not be as affected as adolescents and adults by the characteristics Goffman has identified, primarily because children are already in similarly structured situations at the home and at school. In their comprehensive review, Vernon, Foley, Sipowicz, and Schulman (1965) identified four variables most often considered as causes of psychological upset in hospitalized children: (1) unfamiliarity of the hospital setting, (2) separation from parents, (3) age, and (4) prehospitalization personality. Of these factors, unfamiliarity and separation are most relevant to posthospital psychological problems, whereas separation and age are the two most important variables to upsets during hospitalization. Vernon *et al.* conclude that the age relationship is curvilinear, with children six months to four years of age being most susceptible. Since age and prehospitalization personality cannot be manipulated, the primary area of intervention for most studies has been with unfamiliarity and separation. The conceptual and practical aspects of successful intervention strategies in these areas have been well discussed by Melamed and Siegel (1980). From a behavioral point of view, it can be said that unfamiliarity and separation interventions attempt to change stimulus properties of hospital events so as to make them less strange and, therefore, less aversive. Intervention strategies of preparing the child for hospitalization have also been used to help mitigate the effects of aversive procedures as well as unfamiliar hospital routine.

Many aspects of pediatric inpatient care result in unfamiliar and painful but necessary experiences for children. However, the most dramatic examples occur on an intensive care unit (ICU). For this reason, this setting provides an excellent opportunity to study effects of hospitalization on children and to explore opportunities to mitigate undesirable biobehavioral effects.

ICU ENVIRONMENTS AND AVERSIVE STIMULATION

Although some controversy does exist (Davenport & Werry, 1970; Holland, Sgroi, Marwit, & Solkoff, 1973), the consensus from the literature is that intensive care units provide potential for psychiatric and psychological trauma. The nonmedical problems associated with such hospitalization may be the result of at least three factors: (a) organic variables related to medical and surgical interventions, (b) the patient's perceived physical status and fear of sequelae, and (c) the ecological characteristics of the hospital environment, including the relationship of aversive medical procedures to the behavior of the patient. While medical interventions could be considered for each of these factors, each would also suggest a unique intervention approach. For example, one would expect that behavioral and psychological problems caused by organic factors would be remediated when the patient's physical status was again stable and normal. The patient's perceived physical status and fears as the result of hospitalization might best be resolved through psychiatric counseling. Behavioral and psychological problems which result from characteristics of the hospital's physical and social environment might best be resolved by altering the characteristics of the environment.

In this environment, children may be isolated from parents and other family members, be exposed to a number of children with life-threatening illnesses or injuries, witness critical medical interventions, or experience a number of necessary but painful medical procedures. Children are typically connected to a number of vital-sign monitoring devices, given fluids intravenously, and restrained from moving freely in their beds. Exposure to this traumatic and unfamiliar environment may result in a decrease in age-appropriate behavior and an increase in problem behavior, making good medical care harder to provide and transition to home more difficult (Haller, Talbert, & Dombro, 1967; Lindheim, Glaser, & Coffin, 1972). Despite the importance of the nonmedical problems associated with intensive care, very few studies have addressed pediatric populations. Therefore, in referring to the literature on intensive care units for adult patients, the nonmedical problems have been categorized in a variety of ways. One method is on the basis of psychiatric problems, which include referrals for mood, denial, anxiety, depression, delirium, organic brain syndrome, psychosis, and suicide attempts (Hackett, Cassem, & Wishnie, 1968; Hale, Koss, Kerstein, Camp, & Barush, 1977). Alternatively, Kornfeld (1977) has suggested that nonmedical patient-related problems can be grouped as: (1) reactions to serious medical-surgical interventions (Blachly & Starr, 1964; Bowden, 1975; Danilowicz & Gabriel, 1971; Davenport & Werry, 1970; Egerton and Kay, 1964; Kornfeld, Zimberg, & Malm, 1965; Sveinsson, 1975); (2) problems associated with the hospital environment (Langford, 1961; McKegney, 1966; Tomlin, 1977); and (3) delayed patient actions, manifested after discharge (Druss & Kornfeld, 1967; Klein, Kliner, Zipes, Troyer, & Wallace, 1968).

Nonmedical problems which occur as the result of intensive care have been jointly termed the *ICU syndrome*. The syndrome has been related to various parameters of the social and physical environment: observation of other patients, exposure to equipment, length of stay in the ICU, decreased sensory stimulation, separation from families, lack of staff attention other than to medical status, psychological predisposition (Egerton & Kay, 1964; Kornfeld, 1971; Maron, Bryan-Brown, & Shoemaker, 1973; Vernon *et al.*, 1965), and sleep deprivation (Kornfeld *et al.*, 1965; Lazarus & Hagens, 1968). Despite efforts to document the ICU syndrome, few empirically based observational studies have been reported. Two exceptions are those by Bruhn, Thurman, Chandler, and Bruce (1970), and Lazarus and Hagens (1968). Bruhn *et al.* (1970) obtained data from the vital-sign monitors of patients observing the death of another patient and found increases in heart rate and blood pressure. Lazarus and Hagens (1968) found that modifications in sensory monotony, sleep deprivation, and physical design decreased the incidence of delirium in postcardiotomy patients. Other studies have relied on subjective data from ward psychiatrists, staff, and families to describe patients' behavior and its etiology. Although interesting, these data should be considered anecdotal and the resulting conclusions equivocal.

To date, the major intervention consideration for ICUs has been the result of psychiatric referrals. The common psychiatric interventions have included providing the patient with privacy without isolation; placing monitoring equipment outside the patient's room; administering psychiatric counseling, medication (including phenothiazines and minor tranquilizers), and ECT for depression; and providing a supportive external environment to the patient.

Investigators have reported implementation of preventive and treatment strategies such as environmental redesigning (Abbott, Hansen, & Lewis, 1970; Lindheim *et al.*, 1972), preparation of children for surgery and hospitalization (Petrillo, 1968; Rie, Boverman, Grossman, & Ozoa, 1968), and normalization of the environment through social interaction and provision of play materials (Blumgart & Korsch, 1964; Hott, 1970; Jolly, 1968) to mitigate the traumatic effects of the hospital environment. Treatment strategies have involved both behavioral and counseling approaches to such problems as attention-seeking (Efthymiou, 1976; Zide & Pardoe, 1976) and verbalizations of stress. Unfortunately, only one of the 12 studies reviewed provided objective data (Rie *et al.*, 1968).

The two criticisms which arise when considering the literature on hospitalization of children and the effects of pediatric intensive care units (PICUs) are the lack of experimental rigor (Davenport & Werry, 1970) and the failure to relate findings to basic processes or mechanisms of behavior. As indicated in the introduction, our bias will be to consider the effects of hospitalization in terms of learning principles. From such principles would follow a behavioral analysis attempting to identify: (1) the environmental events which affect behavior and (2) the functional relationship between these events and behavior. Our opinion is that events described by others, such as separation, unfamiliarity, and painful and frightening procedures, are likely to be especially aversive to children. Accordingly, we have turned to the basic animal and human literature on aversive consequences and the behavioral sequelae which proceed from contingent relationships to such consequences. In selectively reviewing this literature, we began with two important assumptions: first, that medically necessary yet aversive procedures probably occur because of variables other than the child's behavior but functionally (because of serendipitous contingent relationships) modify behavior and second, that *both* overt observable behavior and internal states of the patient (physiology and biochemistry) must be considered because of the importance of medical status in the hospital setting. The following review of the literature is significant in terms of the dramatic behavioral and biological changes which result under contingencies employing aversive stimulation.

CONDITIONED SUPPRESSION AND LEARNED HELPLESSNESS

Previous research has shown that aversive events produce distinctive behavioral and biological reactions in humans and animals (Brady, 1979; Miller, 1980). Behavioral reactions include general suppression of behavior, unconditioned skeletal responses, escape and avoidance behavior, and aggression (Blackman, 1977; Mackintosh, 1974). Whereas a low-intensity aversive stimulus must occur in close temporal proximity in order to suppress a particular behavior, a very intense, aversive stimulus may suppress behavior regardless of the delay. The profound effect of this simple relationship between the environment and behavior is attributable to the related empirical finding that stimuli associated with aversive events can come to produce similar reactions. The classic experimental conditioned suppression paradigm, developed by Estes and Skinner (1941), explicates this stimulus control phenomenon. In this paradigm, an unrestrained animal (rat) is initially trained to press a bar in order to obtain food. After stable responding on the bar is established, a conditioned stimulus, CS, (light, tone, etc.) is presented, and termination of the stimulus is followed by an aversive unconditioned stimulus, UCS (shock). Subsequent to this pairing, presentation of the CS alone has been shown to produce the conditioned response, CR, namely, reductions in the rate of bar-pressing. A clear relationship has been demonstrated between the reliability with which the CS results in the CR and the degree to which the CS is correlated with, and therefore predictive of, the UCS (Dweck & Wagner, 1970; Gamzu & Williams, 1971; Rescorla, 1968). However, in circumstances during which the intensity of the aversive stimulus is varied and/or the probability of the UCS is as great in the absence of the CS as in its presence (i.e., the occurrence of the UCS is unpredictable), behavioral suppression is increased. In addition, during conditions wherein aversive events are uncontrollable, many responses come to be suppressed relative to conditions when aversive events can be partially or totally controlled (Seligman, Maier, & Solomon, 1971). During conditions of uncontrollable aversive events, animals are speculated to *learn* that their behavior is independent of aversive stimulation (Seligman & Maier, 1967; Maier, Seligman & Solomon, 1969) and this situation has been ascribed the label *learned helplessness* (Maier, Seligman, & Solomon, 1969). Particularly relevant to developmental aspects is the finding that performance in learning situations is adversely affected subsequent to such a learned helplessness experience. For example, dogs who received inescapable shock in a Pavlov harness were severely retarded in the ability to learn escape responses to shock in an entirely different situation relative to a control group which had no prior shock history (Overmier & Seligman, 1967; Seligman & Maier, 1967). These findings have been replicated across shock frequencies, intensities, durations and temporal distributions (Overmier & Seligman, 1967; Seligman & Maier, 1967), across laboratories (Carlson & Black, 1960; Leaf, 1964), for different types of learning situations (Mullin & Mogenson, 1963), and for various types of animal and human subjects (Behrend & Bitterman, 1963; Pinckney, 1967). Further, subsequent research has shown that learning decrements in a learned helplessness paradigm are clearly not the result of some learned motor response such as freezing (Maier, 1970).

Biological reactions to aversive stimulation have been extensively noted in regard to endocrine, cardiovascular, and gastrointestinal changes. For example, systematic increases in plasma 17-hydroxycorticosteroid (17–OH–CS) levels have been noted as a function of acquisition of conditioned suppression with monkeys (Mason, Brady, & Tolson, 1966). A twofold to fourfold increase has also been noted in 17–OH–CS levels during exposure to a conditioned avoidance paradigm (Brady, 1965, 1966). Anticipating a shock avoidance schedule has been shown to result in elevated blood pressure with a decrease or no change in heart rate (Anderson & Brady, 1971; Anderson & Tosheff, 1973). Both rhesus (Forsyth, 1969) and squirrel monkeys (Herd, Morse, Kelleher, & Jones, 1969) have developed hypertensive blood pressure levels under conditions of responding to avoid aversive stimulation. In other studies, gastric stomach lesions have occurred as a result of animals' receiving unavoidable shocks (Moot, Cebulla, & Crabtree, 1970; Weiss, 1971).

Unavoidable aversive stimulation has also been shown to result in decreased levels of norepinephrine in the brain (Weiss, Stone, & Harrell, 1970; Weiss, Glazer, & Pohorecky, 1976) and subsequent decrements in ability to learn (Glazer, Weiss, Pohorecky, & Miller, 1975; Weiss, Glazer, Pohorecky, Brick, & Miller, 1975).

CONDITIONS WHICH MITIGATE THE EFFECTS OF AVERSIVE STIMULATION

In a classic experiment from his laboratory, Pavlov (1927) reported that stimulus–response relationships are ordered hierarchically depending on their physiological strength and biological importance and that this observation could be used to design procedures to mitigate the effects of aversive stimulation. Presenting strong noxious stimuli such as a strong electric current, or wounding or cauterization of the skin usually results in a strong defense reaction. However, by pairing such a stimulus with the presentation of food, the usual defense reaction is replaced with an alimentary response (salivation). Further, during conditions when food is paired with the aversive stimulation, no appreciable change in pulse or respiration occurs, whereas such changes are always present when the noxious stimulus has not been converted into an alimentary conditioned stimulus (Erofeeva, 1916; 1921). This early example of a counterconditioning paradigm has been extensively replicated and elaborated (Dickinson & Pearce, 1976, 1977; Marukhanyan, 1954; Pearce & Dickinson, 1975). Employing an operant paradigm, Brady (1955) also demonstrated the hierarchical and mitigating effects of positive events. The rate of extinction of a conditioned "fear" reaction (response suppression) was shown to be a function of the reinforcement schedule under which the operant had been established. In another series of studies, investigators have shown that effects of aversive stimulation can be decreased by the provision of a stimulus which makes more predictable the occurrence of an aversive event. Pigeon key-pecking for food was suppressed to a greater degree under a condition of variable-interval (VI) as opposed to fixed-interval (FI) shock (Azrin, 1956). Similarly, rats' bar-pressing for food is greater during conditions of signaled as opposed to unsignaled shock (Brimer & Kamin, 1963).

Increasing a subject's ability to predict the occurrence of an aversive event also mitigates the biological effects of the aversive stimulus. For example, animals receiving signaled shocks demonstrated up to a sixfold decrease in the length of stomach lesions compared to rats receiving unsignaled shock (Weiss, 1970; Miller, 1972). Corticosteroid responses in conditioned avoidance studies are markedly reduced when warning stimuli are provided (Mason *et al.*, 1966).

In summary, in situations in which strong, unpredictable, unavoidable, and inescapable aversive stimulation is programmed, the following reactions have been noted: (a) appetitive and goal-oriented behavior is suppressed; (b) under prolonged conditions of exposure to such aversive stimulation, defense behavior (avoidance and escape responses) may also cease; (c) biological reactions related to increases in 17-hydroxycorticosteroid and catecholamine levels, increases in gastric secretions resulting in increased stomach lesions, and changes in brain amines, specifically decreases in norepinephrine levels, to name a few, can occur. Conditions which mitigate these effects, when aversive events cannot be avoided or escaped, are primarily two: (a) providing a physiologically stronger and biologically more important positive (nonaversive) event and (b) making the occurrence of aversive events more predictable, usually by a preceding warning stimulus.

BEHAVIOR ANALYSIS RESEARCH ON PEDIATRIC INTENSIVE CARE

Our preliminary empirical investigations on the PICU at Johns Hopkins Hospital (Cataldo, Bessman, Parker, Pearson, & Rogers, 1979) indicate that: (1) the behavioral characteristics of children on a PICU support a hypothesis characterizing this environment in terms of aversive events, (2) the children's behavior is not necessarily the result of their medical status, and (3) should a characterization of PICUs in terms in aversive events be valid, then unintended behavioral and biological effects could be better understood and intervention strategies could be designed, both based on the extensive basic animal and human research on aversive stimulation. Two previous pilot studies on the Hopkins PICU (Cataldo *et al.*, 1979; Pearson, Cataldo, Tureman, Bessman & Rogers, 1980) suggest that at least one basic research-based intervention strategy (providing positive events) could be effective. The details of this research are worth noting, for they represent a method of analyzing learning principles which may be interacting with events on PICU to affect children's biobehavioral status.

The problem first considered in assessing pediatric intensive care was that behavior to be recorded and analyzed could not be readily predicted. For situations in which little information is available on what is likely to occur during an observation, behavioral assessment can be based on subcomponent categories of behavior, as demonstrated by the Resident Activity MANIFEST (Cataldo & Risley, 1974). The MANIFEST was designed for use in residential institutions where few planned activities occurred and children's behavior was often restricted or bizarre. Use of the MANIFEST in the assessment of institutions for the retarded allowed observations to be made on subcomponent categories consisting of residents' location, body position, focus of visual attention, and what residents were touching. Unlike behavioral observation procedures in which observers note specific precoded behavior (e.g., hitting, out-of-seat behavior, attending to teacher), observers using the MANIFEST note in a descriptive phrase what the resident is doing. The only restriction is that a phrase or word be provided for each category. In this manner, the broadest possible latitude is permitted for behavioral assessment, with reliability based on the degree of agreement between observers when making simultaneous but independent observations. By observing each of the residents in succession at numerous times during the day (time sampling), the investigator can obtain a quantitative profile of activities and interactions. Further, the measure is economical in terms of staff time, taking usually three to four minutes to observe and record the behavior of each individual.

Employing an assessment methodology similar to the Resident Activity MANIFEST, a behavioral observation study (Cataldo *et al.*, 1979) was made of the Pediatric Intensive Care Unit of the Johns Hopkins Hospital. Ninety-nine patients, ranging in age from two days to 22 years (mean 6 years, median 4 years), were observed over a two-month period. The length of stay for the observed patients ranged from one day to 11 months (mean 4.9 days, median 2 days). The behavior observed is presented in Table I. As indicated in the table, observers had considerable latitude in making behavioral descriptions. However, interobserver agreement with this data collection method was found to be high (mean, 93.5%; range, 89% to 97%).

Data on descriptive characteristics are presented in Figures 1 and 2. While data on the percentage of children sitting or lying (Figure 1B) are quite predictable, it is significant to note that over one third (35%) of the patients were judged to be alert when observed (Figure 1A). This means that they were able to be influenced by the sights and sounds (stimuli) in the PICU, as well as by the medical/surgical procedures administered to them. On the basis of prior basic research, continued exposure to such stimuli, if aversive to the children, would likely lead to suppression of behavior. An aversive environmental characterization of the PICU tends to be supported by the results shown in Figure 1C for affective state. Whereas negative affect would be expected in a sit-

Table I. Definitions of Behavior for Demographic Analysis

Category	Symbol	Definition
Waking state	Al	Alert: the child's eyes are open and the child is not in a coma; or the child's eyes are closed or not visible, as with bandages, but the child is engaged in some verbal or physical activity.
	A	Awake: the child's eyes are open but fixed on one thing with no accompanying verbal or physical activity, and the child is not in a coma; or the child is waking up and falling asleep, opening and closing eyes.
	S	Asleep: the child's eyes are closed and the child is not engaged in any verbal or physical activity, and the child is not in a coma.
	C	Coma: the child is in a coma, as indicated by a verbal report from a nurse or other medical person.
	Pav	Pavulon: the child has been administered the drug Pavulon, which relaxes all the muscles of the body, as indicated by a verbal report from the medical people.
Position	S	Sitting: the child is in an upright or semi-upright position, at least 45%, even if being supported by the bed. This includes infants in an infant seat, or being held in an upright position.
	L	Lying: the child is in a prone or supine position, or is flat on the bed on side, or is being held in a prone or supine position.
Verbalizations: from the child	√	Verbalization: the child utters any words during the interval.
	—	No verbalization: the child does not utter any words during the interval.
Verbalizations: to the child	√(T)	To: if the statement is directed to the child.
	√(A)	About: if the statement is to someone else about the child, and the interaction occurs in close proximity to the child.
	—	No verbalization: no verbal interaction occurs from others during the interval.

Attention on/eye contact	Scan	Scanning: the child's eyes are moving about and do not fixate on any object or person.
	Space	Space: the child does not appear to be looking at any object or person, but eyes are open.
	Other words	Record whatever object(s) or person(s) the child makes eye contact with during the interval.
Affective state	P	Positive: the child smiles or laughs during the interval.
	N	Negative: the child cries or whines during the interval, or verbalizes fear or unhappiness (VF), or pain (VP)
	Nu	Neutral: the child exhibits no overt behavioral response to indicate an affective state during the interval.
	Unknown	Unknown: the child has tape or bandages on face, or is wearing a face mask or other respiratory apparatus that interferes with determination of an affective state.
Number and type of people within 1m of bedside	M	Medical: doctors, respiratory and physical therapists, laboratory and X-ray technicians.
	N	Nurses and nurse aides (nursing staff were scored as a separate category of medical staff because of the significant degree of their interaction with PICU children).
	P	Parents and grandparents.
	C	Child Life workers.
	O	Others: any person not otherwise categorized.
Activity	√	Occurs: the child engages in a play activity during the interval—*e.g.*, painting, puzzles, blocks, reading books or being read to, watching TV, looking at or touching a mobile. This also includes the presence of a radio or other musical toy that is turned on and in the bed of an alert child.
	—	Does not occur: the child does not engage in a play activity during the interval.

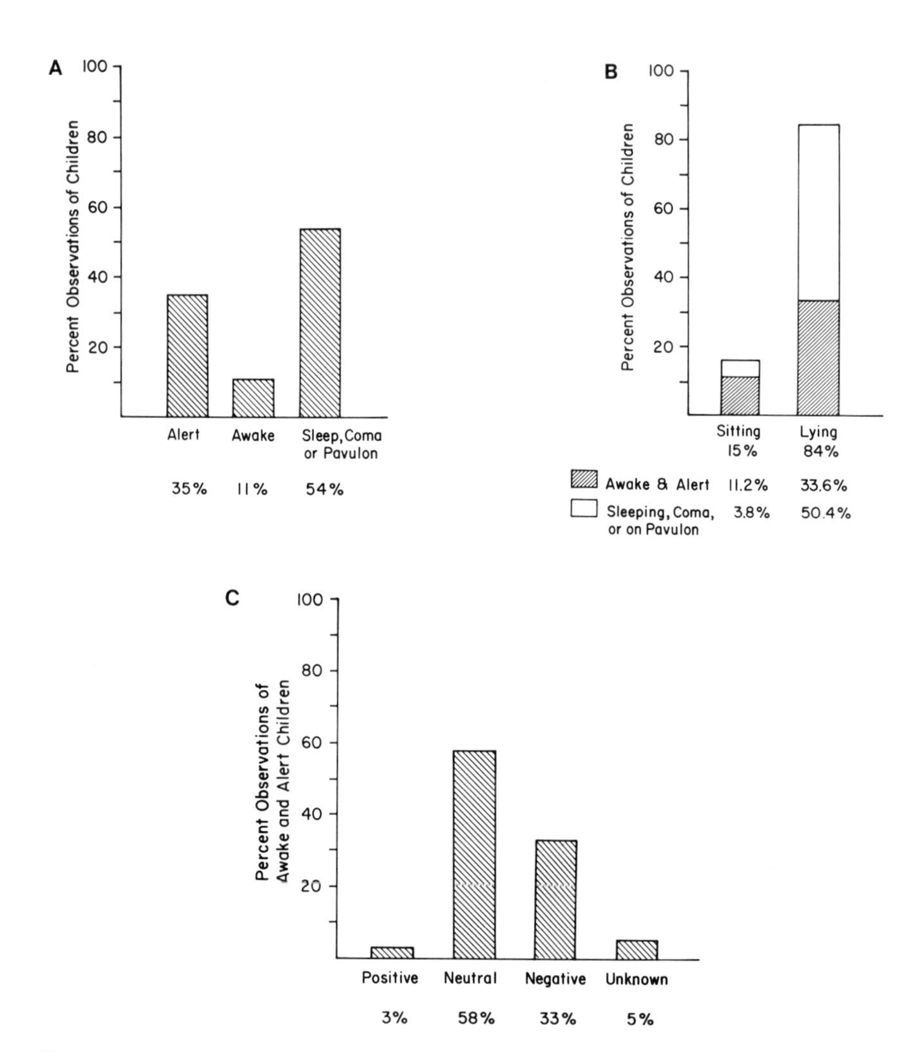

Figure 1. Percent observations of children in each of the (A) waking states, (B) positions, and (C) affective states. In Figure 1B, data for 1% of the children are not presented because they could not be categorized as sitting or lying. Percentages in Figure 1C do not sum to 100% due to downward rounding. From "Behavioral Assessment for Pediatric Intensive Care Units" by M. F. Cataldo, C. A. Bessman, L. H. Parker, J. E. R. Pearson, and M. C. Rogers, *Journal of Applied Behavior Analysis*, 1979, *12*, 83–97. Copyright 1979 by the *Journal of Applied Behavior Analysis*. Reprinted by permission.

uation in which children are frightened or hurt, the predominant affect was judged to be neutral (58%). Further, even though verbalization to children was as high as 55% (Figure 2A), the children's verbal responsiveness was considerably lower, 16% (Figure 2B). Similarly, interaction judged by attention (eye contact) was only 63% (Figure 2C) and 46% (Figure 2D) as judged by engagement in any type of activity. Stated inversely the data are more revealing, in that children did not verbalize at all 82% of the time observed, did not attend to anything in the room 37% of the time, and were not engaged with any activity 54% of the time. Such neutral affect and noninteraction is characteristic of aversive stimulation research findings, particularly as described in the conditioned suppression and learned helplessness literatures.

Of course, the observed behavior could have been the result of the patients' medical status rather than caused by environmental circumstances. In general, patients' length of stay is related to their medical status such that patients are discharged from the ward as their status improves. Thus, if expression of affect and engagement with the environment were solely influenced by medical status, this child behavior would be positively correlated with length of stay. Whereas, if these behaviors were significantly affected by aversive events on the unit, then they would be negatively correlated with length of stay. Analysis of alert children's affect and engagement across days on the unit are presented in Figures 3 and 4. Since none of the children in the study who were alert during our observation remained on the unit longer than 6 days, the data in these two figures represent their entire length of stay. The results shown in Figure 3 indicate that neutral affect increased and expression of affect (i.e., positive and negative) decreased over days on the unit, both at all times observed (a) and at times other than during a medical procedure (b). Similarly, as shown in Figure 4, engagement decreased across days on the unit. Because these data represent the averaged behavior noted by time samples for all alert children on each day since their admission, the findings could be biased by many factors. However, these data do tend to support the aversive stimulation characterization of the PICU setting.

Another method for determining whether children's behavior is solely influenced by medical status, as opposed to being affected by environmental variables also, would be experimentally to manipulate either medical status or the environment. Since, by definition, PICU staff do all that can be done to improve medical status and to do anything less would be unethical, the most reasonable strategy is to manipulate environmental variables.

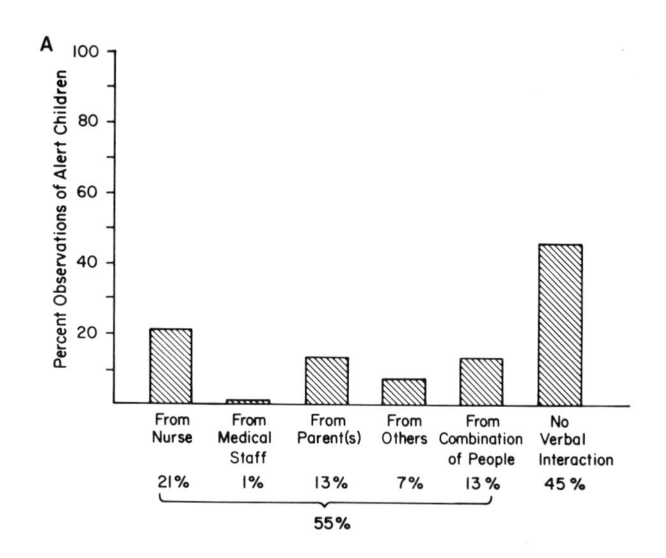

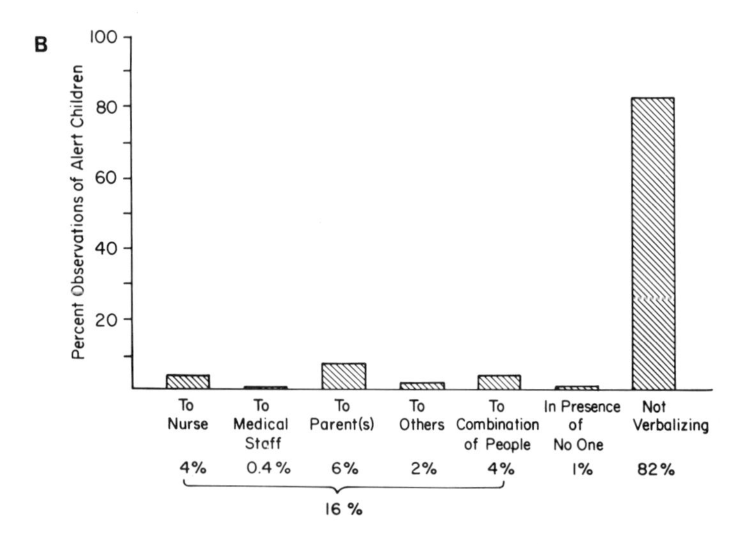

Figure 2. Percent observations of alert children in each of the subcategories of the following categories: (A) verbalizations to the child, (B) verbalizations from the child, (C) eye contact, and (D) engagement in activities. Percentages in Figure 2B do not sum to 100% due to downward rounding. From "Behavioral Assessment for Pediatric Intensive Care Units" by M. F. Cataldo, C. A. Bessman, L. H. Parker, J. E. R. Pearson, and M. C. Rogers, *Journal of Applied Behavior Analysis*, 1979, *12*, 83–97. Copyright 1979 by the *Journal of Applied Behavior Analysis*. Reprinted by permission.

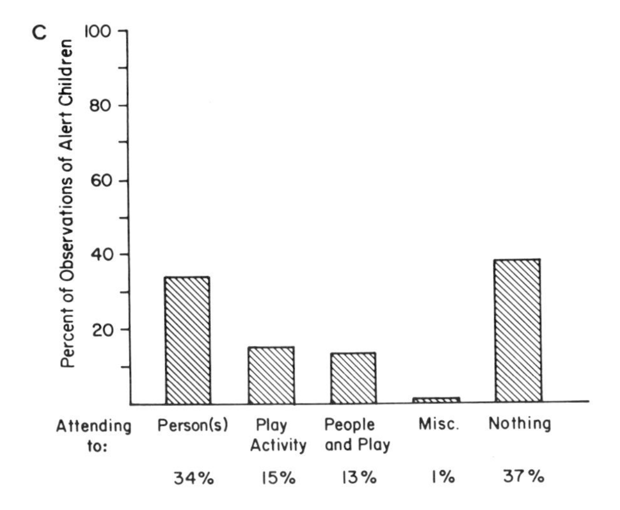

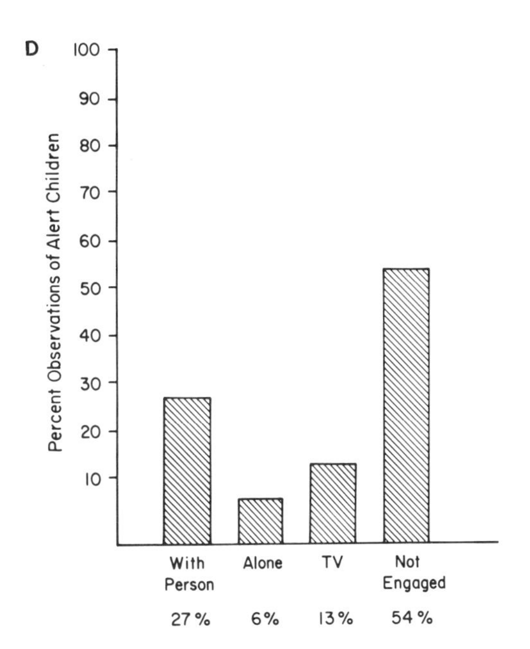

In a second study on the Hopkins PICU (Cataldo *et al.*, 1979) such an environmental manipulation was conducted with 11 children ranging in age from 1 to 21 years of age. Using each child as his own control, a reversal design was employed to evaluate the effects of planned play-related activities on seven patient behaviors. A condition of "no toys"

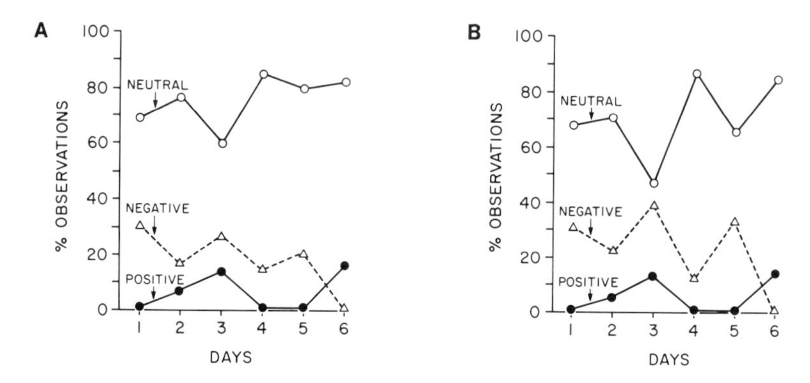

Figure 3. Children's affect on the PICU as a function of the number of days of stay on the unit (A) at all times and (B) at times other than medical procedures.

was followed by a condition of "person and toys" and then the "no toys" condition was repeated. During the "no toys" condition, the child was observed as found in his bed. During the "person and toys" condition, one of the hospital staff interacted with the child using a specially selected play material. Each condition lasted 5 minutes and coded behaviors were recorded by an observer every 20 seconds after 10 seconds of observation. Three condition trials of "no toys—person and toys—toys" and "toys—person and toys—toys" were also conducted. Obviously, if engagement and affect were solely the result of medical status, no change between conditions should have been noted.

Data on the results of this study are presented in Table II for all

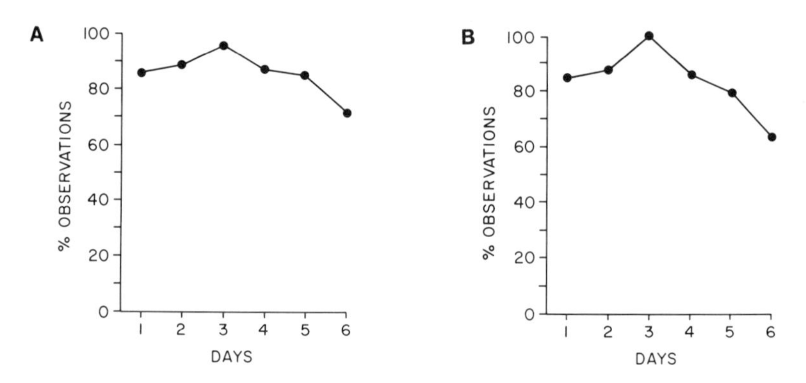

Figure 4. Children's engagement with the PICU environment as a function of days of stay on the unit (A) at all times and (B) at times other than in the presence of a medical procedure.

Table II. Group Means of Percent Observations for Attention on Person or Activity, and Interaction with Materials, across Conditions for Each of the Interventions

Intervention	*n*	Condition 1		Condition 2		Condition 3	
		Att.[d]	Int.[e]	Att.	Int.	Att.	Int.
A NT/P+T/NT[a]	3	26.60	0	97.80	93.40	24.46	8.90
B NT/P+T/T[b]	6	24.43	0	95.10	79.16	52.63	48.93
C T/P+T/T[c]	5	67.96	53.64	96.04	68.08	75.24	68.00

[a]no toys/person + toys/no toys.
[b]no toys/person + toys/toys.
[c]toys/person + toys/toys.
[d]attention on person or activity.
[e]interaction with materials.

subjects and in Figure 5 for six different individuals within subject comparisons. Data from Table II indicate a great increase between conditions comparing "no toys" with "person and toys." Both types of comparison clearly demonstrate that children's behavior on a PICU is affected by these environmental events. Further, the degree of specificity of this behavior change is clearly shown in Figure 5, in that affect (smiling), interaction, and other behavior (finger-sucking, squeezing the blood pressure bulb and playing with EKG electrodes) were clearly altered by this simple environmental manipulation.

These findings were systematically replicated in a third study (Pearson *et al.*, 1980), in which the duration of each condition was increased. During the intervention a play activity was provided by a staff member. Before and after the intervention the child was only observed. The intervention period (I) was always 20 minutes, while preintervention and postintervention periods (P_1 and P_2 respectively) varied from one half to two hours depending upon unit routines. In this study toys remained with the child after the staff member departed. The results are shown in Figure 6. For all eleven patients in the study, interaction was greater during the intervention than either before or after (A). Further, of the five children who demonstrated any positive affect, four of them showed increases when the intervention occurred (B).

SUMMARY AND CONCLUSIONS

In our original paper (Cataldo *et al.*, 1979), we speculated that most medical procedures are aversive to children, and in environments such as pediatric intensive care units the procedures are performed dependent

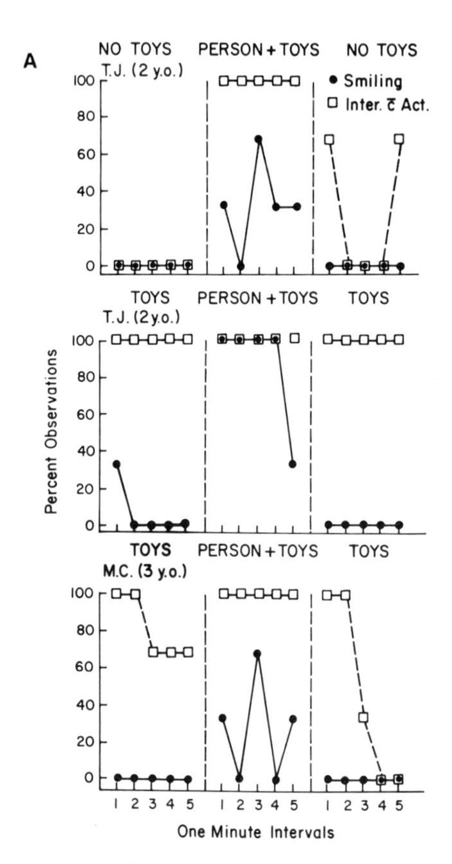

Figure 5. (A) Percent observations of smiling and interaction with activities for T.J. in the no toys, person + toys, no toys intervention, and the toys, person + toys, toys intervention; and for M.C. in the toys, person + toys, toys intervention. (B) Percent observations of attention on people, interaction with activities, and inappropriate or self-stimulatory behaviors for J.F. and H.G. Finger-sucking is presented in the top panel for J.F. in the no toys, person + toys, no toys intervention. Squeezing of the bulb on the sphyg-

on staff time and children's medical status rather than children's overt behavior. We further hypothesized that this would likely lead to the behavioral results observed in our study because children's overt behavior would be ineffective in controlling their environment. We concluded that the environmental solution is not to reduce necessary, albeit aversive, medical procedures, but rather to provide positive events as a contrast. In such circumstances, staff become discriminative stimuli

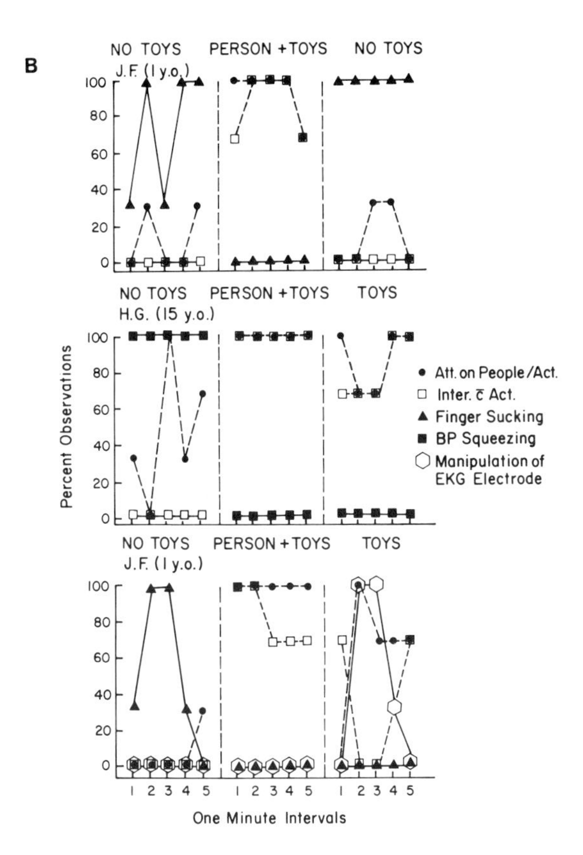

momanometer for H.G. in the no toys, person + toys, toys condition is presented in the middle panel. In the bottom panel finger-sucking and manipulation of EKG electrodes is presented for J.F. for the no toys, person + toys, toys condition. From "Behavioral Assessment for Pediatric Intensive Care Units" by M. F. Cataldo, C. A. Bessman, L. H. Parker, J. E. R. Pearson, and M. C. Rogers, *Journal of Applied Behavior Analysis*, 1979, *12*, 83–97. Copyright 1979 by the *Journal of Applied Behavior Analysis*. Reprinted by permission.

for both negative and positive consequences, thereby engendering behavioral repertoires typical of environments with differential consequences.

We mean to suggest here that the effects on children of an inpatient admission are profound, resulting in important biological as well as behavioral changes. Should the aversive characterization of hospital environments in general, and PICUs in particular, prove true, then con-

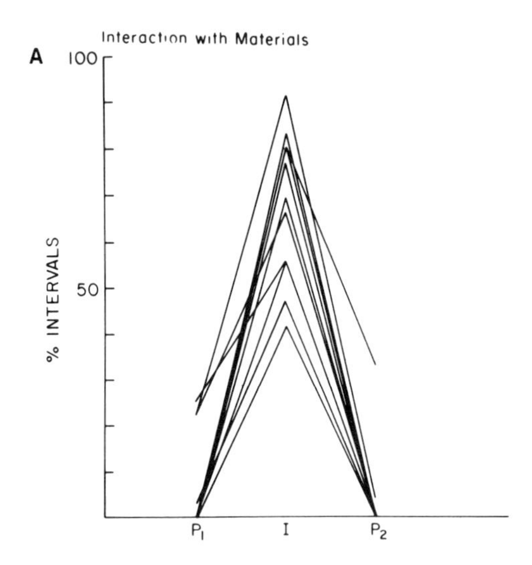

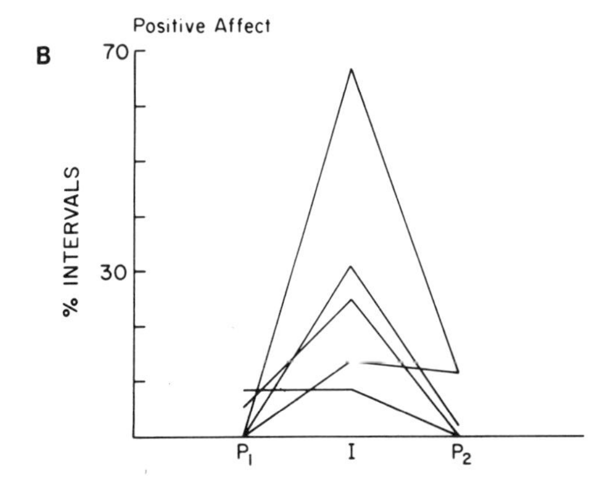

Figure 6. Percentage of intervals of interaction for (A) all 11 subjects across conditions and (B) for the 5 subjects who showed positive affect during any condition. From "Pediatric Intensive Care Unit Patients: Effects of Play Intervention on Behavior" by J. E. R. Pearson, M. F. Cataldo, A. Tureman, C. Bessman, and M. C. Rogers, *Critical Care Medicine*, 1980, *8*, 64–67. Copyright 1980 by the Williams & Wilkins Company. Reprinted by permission.

siderable evidence can be found for biobehavioral changes in the basic animal and human literature on the effects of aversive events.

Study of the biological effects of hospitalization is facilitated by the fact that in many instances children are catheterized, attached to vital sign monitors, and have arterial lines in place, allowing for nonintrusive collection of 24-hour urine, heart rate, blood pressure, and respiration measures, and collection of a 24-hour plasma pool for hydrocorticosteroid and catecholamine analysis. Presentation of aversive events and concomitant changes in 17–OH–CS and catecholamine levels may be particularly important to investigate because both are thought to be associated with increased risk factors in critical care medicine. Briefly, as an example with regard to this point, response to trauma includes increased metabolic activity (Cuthbertson, 1930), which is believed to occur primarily for the restoration of homeostasis, particularly blood volume (Gann, 1976; Jarhult, 1973; Jarhult, Lundvall, Mellander, & Tibblin, 1972; Pirkle & Gann, 1976). However, this is not without cost to the patient, including increased oxygen consumption, energy need, and protein breakdown (Baue, 1974). This hypermetabolic state, while necessary, does call upon finite oxygen and nutritional resources, resources which, for the critically ill patient, should not have to be used in response to environmental trauma as well. Although the mechanisms are not completely understood, catecholamines have a direct relationship in mediating the hypermetabolic state (Griffith, 1951; Harrison, Seton, & Feller, 1967; Wilmore, Long, Mason, Skreen, & Pruitt, 1974). Therefore, elevations in catecholamine levels in response to environmental events may indicate an unwanted metabolic demand better reserved for response to physical trauma.

While still quite speculative, this approach to studying iatrogenic effects of hospitalization suggests that solutions can be found in the basic literature on operant and classical conditioning, and from studies on learned helplessness. Specifically, positive events should be provided and the occurrence of aversive events should be made more discriminable, such that periods when aversive events will *not* occur can be clearly identified by children. An environment in which positive events occur would permit behavioral repertoires associated with such consequences to be maintained. Further, identification of periods during which the child is "safe" from aversive medical procedures (of course with the exception of emergencies) would allow reversal or relaxation of behavioral and biological responses associated with conditions signaling aversive events.

Such speculation is not yet the basis for program implementation or for changes in unit routine, but does offer a fertile area for further investigation in behavioral pediatrics.

REFERENCES

Abbott, N. C., Hansen, P., & Lewis, K. Dress rehearsal for the hospital. *American Journal of Nursing*, 1970, *70*, 2360–2362.

Anderson, D. E., & Brady, J. V. Pre-avoidance blood pressure elevations accompanied by heart rate decreases in the dog. *Science*, 1971, *172*, 595–597.

Anderson, D. E., & Tosheff, J. Cardiac output and total peripheral resistance changes during pre-avoidance periods in the dog. *Journal of Applied Psychology*, 1973, *34*, 650–654.

Azrin, N. H. Some effects of two intermittent schedules of immediate and non-immediate punishment. *Journal of Psychology*, 1956, *42*, 3–21.

Baue, A. E. The energy crisis in surgical patients. *Archives of Surgery*, 1974, *109*, 349–357.

Behrend, E. R., & Bitterman, M. E. Sidman avoidance in the fish. *Journal of the Experimental Analysis of Behavior*, 1963, *6*, 47–52.

Blachly, P. H., & Starr, A. Post-cardiotomy delirium. *American Journal of Psychiatry*, 1964, *121*, 371–375.

Blackman, D. Conditioned suppression and the effects of classical conditioning on operant behavior. In W. K. Honig & J. E. R. Staddon (Eds.), *Handbook of operant behavior*. Englewood Cliffs, N.J.: Prentice–Hall, 1977.

Blumgart, E., & Korsch, B. M. Pediatric recreation: An approach to meeting the emotional needs of hospitalized children. *Pediatrics*, 1964, *34*, 133–136.

Bowden, P. The psychiatric aspects of cardiac intensive care therapy: A review. *European Journal of Intensive Care Medicine*, 1975, *1*(2), 89–91.

Brady, J. V. Extinction of a conditioned "fear" response as a function of reinforcement schedules for competing behavior. *Journal of Psychology*, 1955, *40*, 25–34.

Brady, J. V. Experimental studies of psychophysiological responses to stressful situations. In *Symposium on Medical Aspects in the Military Climate*, Walter Reed Army Institute of Research. Washington, D.C.: United States Government Printing Office, 1965.

Brady, J. V. Operant methodology and the production of altered physiological states. In W. Honig (Ed.), *Operant behavior: Areas of research and application*. New York: Appleton-Century Crofts, 1966.

Brady, J. V. Learning and conditioning. In O. F. Pomerleau & J. P. Brady (Eds.), *Behavioral medicine: Theory and practice*. Baltimore: Williams & Wilkins, 1979.

Brimer, C. J., & Kamin, L. J. Disinhibition, habituation, sensitization and the conditioned emotional response. *Journal of Comparative and Psychological Psychology*, 1963, *56*, 508–516.

Bruhn, J. G., Thurman, E., Jr., Chandler, B. C., & Bruce, T.A. Patients' reactions to death in a coronary care unit. *Journal of Psychosomatic Research*, 1970, *14*, 65–70.

Carlson, N. J., & Black, A. H. Traumatic avoidance learning: The effects of preventing escape responses. *Canadian Journal of Psychology*, 1960, *14*, 21–28.

Cataldo, M. F., & Bird, B. L. Behavioral medicine treatment of neuromuscular disorders. In T. J. Coates (Ed.), *Behavioral medicine: A practical handbook*. New York: Research Press, in press.

Cataldo, M. F., & Risley, T. R. Infant day care. In R. Ulrich, T. Stachnik, & J. Mabry (Eds.), *Control of human behavior* (Vol. III). Glenview, Ill.: Scott, Foresman, 1974, pp. 44–50.

Cataldo, M. F., Bird, B. L., & Cunningham, C. E. Experimental analysis of EMG feedback in treating cerebral palsy. *Journal of Behavioral Medicine*, 1978, *1*, 311–322.

Cataldo, M. F., Bessman, A., Parker, L. H., Pearson, J. E., & Rogers, M. C. Behavioral assessment for pediatric intensive care units. *Journal of Applied Behavior Analysis*, 1979, *12*, 83–97.

Cataldo, M. F., Russo, D. C., & Freeman, J. M. A behavioral analysis approach to high-rate myoclonic seizures. *Journal of Autism and Developmental Disorders*, 1979, *9*, 413–427.

Christophersen, E. R., & Rapoff, M. A. Behavioral pediatrics. In O. F. Pomerleau & J. Brady, *Behavioral medicine: Theory and Practice*. Baltimore: Williams & Wilkins, 1979.

Cuthbertson, D. P. The disturbance of metabolism produced by bony and non-bony injury, with notes of certain abnormal conditions of bone. *Biochemical Journal*, 1930, *24*, 1244–1263.

Danilowicz, D. A., & Gabriel, H. P. Post-operative reactions in children: "Normal" and abnormal responses after cardiac surgery. *American Journal of Psychiatry*, 1971, *128*, 185–188.

Davenport, H. T., & Werry, J. S. The effect of general anesthesia surgery and hospitalization upon the behavior of children. *American Journal of Orthopsychiatry*, 1970, *40*, 806–824.

Dickinson, A., & Pearce, J. M. Preference and response suppression under different correlations between shock and a positive reinforcer in rats. *Learning and Motivation*, 1976, *7*, 66–85.

Dickinson, A., & Pearce, J. M. Inhibitory interactions between appetitive and aversive stimuli. *Psychological Bulletin*, 1977, *84*, 690–711.

Druss, R. G., & Kornfeld, D. S. The survivors of cardiac arrest: A psychiatric study. *Journal of the American Medical Association*, 1967, *201*, 291–296.

Dweck, C. S., & Wagner, A. R. Situational cues and correlation between CS and US as determinants of the conditioned emotional response. *Psychonomic Science*, 1970, *18*, 145–147.

Efthymiou, G. Management of attention-seeking patients—2. *Nursing Times*, 1976, *72*, 1490–1491.

Egerton, N., & Kay, J. H. Psychological disturbances associated with open heart surgery. *British Journal of Psychiatry*, 1964, *110*, 433–439.

Erofeeva, M. N. Contributions a l'étude des réflexes conditionnels destructifs. *Compte rendu de la société de biologie, Paris*, 1916, *79*, 239–240.

Erofeeva, M. N. Additional data on nocuous conditioned reflexes. *Izvestiya Petrogradskovo Nauchnavo Instituta im P. F. Lesgafta*, 1921, *3*, 69–73. (Russian)

Estes, W. K., & Skinner, B. F. Some quantitative properties of anxiety. *Journal of Experimental Psychology*, 1941, *29*, 390–400.

Forsyth, R. P. Blood pressure responses to long-term avoidance schedules in the restrained rhesus monkey. *Psychosomatic Medicine*, 1969, *31*, 300–309.

Gamzu, E., & Williams, D. R. Classical conditioning of a complex skeletal response. *Science*, 1971, *171*, 923–925.

Gann, D. S. Endocrine control of plasma protein and volume. *Surgical Clinics of North America*, 1976, *56*, 1135–1145.

Glazer, H. I., Weiss, J. M., Pohorecky, L. A., & Miller, N. E. Monoamines as mediators of avoidance-escape behavior. *Psychosomatic Medicine*, 1975, *37*, 535–543.

Goffman, E. *Asylums*. New York: Doubleday, 1961.

Griffith, F. R. Fact and theory regarding the calorigenic action of adrenaline. *American Journal of Physiology*, 1951, *31*, 151.

Hackett, T. P., Cassem, N. H., & Wishnie, H. A. The coronary-care unit: An appraisal of its psychologic hazards. *The New England Journal of Medicine*, 1968, *279*(25), 1365–1370.

Hale, M., Koss, N., Kerstein, M., Camp, K., & Barush, P. Psychiatric complications in a surgical ICU. *Critical Care Medicine*, 1977, *5*(4), 199–203.

Haller, J. A., Talbert, J. L., & Dombro, R. H. The hospitalized child and his family. Baltimore: Johns Hopkins, 1967.

Harrison, T. S., Seton, J. F., & Feller, I. Relationship of increased oxygen consumption to catecholamine excretion in thermal burns. *Annals of Surgery*, 1967, *165*, 169.

Herd, J. A., Morse, W. H., Kelleher, R. T., & Jones, L. G. Arterial hypertension in the squirrel monkey during behavioral experiments. *American Journal of Physiology*, 1969, *217*, 24–29.

Holland, J., Sgroi, S. M., Marwit, S. J., & Solkoff, N. The ICU syndrome: Fact or fancy. *Psychiatry in Medicine*, 1973, *4*, 241–249.

Hott, J. Rx: Play PRN in pediatric nursing. *Nursing Forum*, 1970, *9*, 288–309.

Jarhult, J. Osmotic fluid transfer from tissue to blood during hemorrhagic hypotension. *Acta Physiologica Scandinavica*, 1973, *89*, 213.

Jarhult, J., Lundvall, J., Mellander, S., & Tibblin, S. Osmolar control of plasma volume during hemorrhagic hypotension. *Acta Physiologica Scandinavica*, 1972, *85*, 142.

Jolly, R. H. Play and the sick child. *Lancet*, 1968, *2*, 1286–1287.

Klein, R. F., Kliner, V. S., Zipes, D. P., Troyer, W. G., Jr., & Wallace, A. G. Transfer from a coronary care unit. *Archives of Internal Medicine*, 1968, *122*, 104–108.

Kornfeld, D. S. Psychological problems of intensive care units. *Medical Clinics of North America*, 1971, *55*, 1353–1363.

Kornfeld, D. S., Zimberg, S., & Malm, J. R. Psychiatric complications of open-heart surgery. *New England Journal of Medicine*, 1965, *273*, 282–287.

Langford, W. S. The child in the pediatric hospital: Adaptation to illness and hospitalization. *American Journal of Orthopsychiatry*, 1961, *31*, 667.

Lazarus, H. R., & Hagens, J. H. Prevention of psychosis following open-heart surgery. *American Journal of Psychiatry*, 1968, *124*, 1190–1195.

Leaf, R. C. Avoidance response evocation as a function of prior discriminative fear conditioning under curare. *Journal of Comparative and Physiological Psychology*, 1964, *58*, 446–449.

Lindheim, R., Glazer, H., & Coffin, C. *Changing hospital environments for children*. Cambridge: Harvard, 1972.

Mackintosh, N. J. *The psychology of animal learning*. New York: Academic Press, 1974.

Maier, S. F. Failure to escape traumatic shock: Incompatible skeletal motor responses or learned helplessness? *Learning and Motivation*, 1970, *1*, 157–170.

Maier, S. F., Seligman, M. E. P., & Solomon, R. L. Pavlovian fear conditioning and learned helplessness. In B. A. Campbell & R. M. Church (Eds.), *Punishment and aversive behavior*. New York: Appleton, 1969.

Maron, L., Bryan-Brown, C., & Shoemaker, W. C. Toward a unified approach to psychological factors in the ICU. *Critical Care Medicine*, 1973, *1*, 81–84.

Marukhanyan, E. V. The effect of the duration and intensity of a conditioned electroshock stimulus upon the magnitude of the conditioned food and acid reflexes. *Zhurnal Vysshei Nervio Deyatel'nosti*, 1954, *4*, 684–691. (Russian)

Mason, J. W., Brady, J. V., & Tolson, W. W. Behavioral adaptations and endocrine activity. In R. Levine (Ed.), *Proceedings of the Association for Research in Nervous and Mental Diseases*. Baltimore: Williams & Wilkins, 1966.

McKegney, F. P. The intensive care syndrome: The definition, treatment, and prevention of new "diseases of medical progress." *Connecticut Medicine*, 1966, *30*, 633–636.

Melamed, B. G., & Siegel, L. J. *Behavioral medicine: Practical applications in health care*. New York: Springer, 1980.

Miller, N. E. Interactions between learned and physical factors in mental illness. *Seminars in Psychiatry*, 1972, *4*, 239.

Miller, N. E. Applications of learning and biofeedback to psychiatry and medicine. In H. I. Kaplan, A. M. Freeman, & B. J. Sadock (Eds.), *Comprehensive textbook of psychiatry III*. Baltimore: Williams & Wilkins, 1980.

Moot, S. A., Cebulla, R. P., & Crabtree, J. M. Instrumental control and ulceration in rats. *Journal of Comparative Physiological Psychology*, 1970, *71*, 405–410.

Mullin, A. D., & Mogenson, G. J. Effects of fear conditioning on avoidance learning. *Psychological Reports*, 1963, *13*, 707–710.

Overmier, J. B., & Seligman, M. E. P. Effects of inescapable shock upon subsequent escape and avoidance responding. *Journal of Comparative and Physiological Psychology*, 1967, *63*, 28–33.

Pavlov, I. P. *Conditioned reflexes*. New York: Dover, 1927.

Pearce, J. M., & Dickinson, A. Pavlovian counterconditioning: Changing the suppressive properties of shock by association with food. *Journal of Experimental Psychology: Animal Behavior Processes*, 1975, *1*, 170–177.

Pearson, J. E. R., Cataldo, M. F., Tureman, A., Bessman, C., & Rogers, M. C. Pediatric intensive care unit patients: Effects of play intervention on behavior. *Critical Care Medicine*, 1980, *8*, 64–67.

Petrillo, M. Preventing hospital trauma in pediatric patients. *American Journal of Nursing*, 1968, *68*, 1469–1473.

Pinckney, G. Avoidance learning in fish as a function of prior fear conditioning. *Psychological Reports*, 1967, *20*, 71–74.

Pirkle, J. C., & Gann, D. S. Restitution of blood volume after hemmorrhage: role of the adrenal cortex. *American Journal of Physiology*, 1976, *230*, 1683–1687.

Rescorla, R. A. Probability of shock in the presence and absence of CS in fear conditioning. *Journal of Comparative Physiological Psychology*, 1968, *66*, 1–5.

Rie, H. E., Boverman, H., Grossman, B., & Ozoa, N. Immediate and long-term effects of intervention early in prolonged hospitalization. *Pediatrics*, 1968, *41*, 755–764.

Seligman, M. E. P., & Maier, S. F. Failure to escape traumatic shock. *Journal of Experimental Psychology*, 1967, *74*, 1–9.

Seligman, M. E. P., Maier, S. F., & Solomon, R. L. Unpredictable and uncontrollable aversive events. In F. R. Bush (Ed.), *Aversive conditioning and learning*. New York: Academic Press, 1971.

Sveinsson, I. S. Postoperative psychosis after heart surgery. *Journal of Thoracic and Cardiovascular Surgery*, 1975, *70*, 717–726.

Tomlin, P. J. Psychological problems in intensive care. *British Medical Journal*, 1977, *2*, 441–443.

Varni, J. W., Bessman, C. A., Russo, D. C., & Cataldo, M. F. Behavioral management of chronic pain in children: Case study. *Archives of Physical Medicine and Rehabilitation*, 1980, *61*, 375–379.

Vernon, D. T. A., Foley, J. M., Sipowicz, R. R., & Schulman, J. L. *The psychological responses of children to hospitalization and illness*. Springfield, Ill.: Charles C Thomas, 1965.

Weiss, J. M. Somatic effects of predictable and unpredictable shock. *Psychosomatic Medicine*, 1970, *32*, 397–409.

Weiss, J. M. Effects of coping behavior in different warning signal conditions on stress pathology in rats. *Journal of Comparative Physiological Psychology*, 1971, *77*, 1–13.

Weiss, J. M., Stone, E. A., & Harrell, N. Coping behavior and brain norepinephrine level in rats. *Journal of Comparative Physiological Psychology*, 1970, *72*, 153–160.

Weiss, J. M., Glazer, H. I., Pohorecky, L. A., Brick, J., & Miller, N. E. Effects of chronic exposure to stressors on avoidance–escape behavior and on brain norepinephrine. *Psychosomatic Medicine*, 1975, *37*, 522–534.

Weiss, J. M., Glazer, H. I., & Pohorecky, L. A. Coping behavior and neurochemical changes: An alternative explanation for the original "learned helplessness" experiments. In G. Serban & A. Kling (Eds.), *Animal models in human psycho-biology*. New York: Plenum Press, 1976.

Whitehead, W. E., Parker, L. H., Masek, B. J., Cataldo, M. F., & Freeman, J. M. Biofeedback treatment of fecal incontinence in myelomeningocele. *Developmental Medicine and Child Neurology*, 1981, *23*, 313–322.

Wilmore, D. W., Long, J. M., Mason, A. D., Skreen, R. W., & Pruitt, B. A. Catecholamines: Mediator of the hypermetabolic response to thermal injury. *Annals of Surgery*, 1974, *180*, 653–669.

Zide, B., & Pardoe, R. The use of behavior modification therapy in a recalcitrant burned child: Case report. *Plastic and Reconstructive Surgery*, 1976, *57*, 378–382.

Zlutnick, S., Mayville, W. J., & Moffat, S. Modification of seizure disorders: The interruption of behavioral chains. *Journal of Applied Behavior Analysis*, 1975, *8*, 1–12.

9

Research in Ambulatory Pediatrics

Edward R. Christophersen and Janice E. Abernathy

INTRODUCTION

Traditionally, the practice of pediatrics not only has concentrated upon the infectious, physical, metabolic, nutritional, and congenital problems of children, but also has emphasized the prevention of these problems. Until the last decade, however, little effort has been directed toward the recognition and treatment of the emotional and social aspects of normal growth and development (Friedman, 1970). Within the last 10 years, consumer demand and the increasing body of knowledge of growth and development have prompted the evolution of an important pediatric primary care subspecialty, that is, ambulatory pediatrics. The practitioners in this subspecialty provide care for the ever-increasing well-child population and consequently have discovered that the health status of children cannot be separated into that which is physical and that which is emotional; rather, a relationship exists between these two components of child health.

The purpose of this chapter is to evaluate the literature on the psychosocial, developmental, and behavioral (biosocial) components of ambulatory pediatrics, in terms of (1) the historical development of the

---

EDWARD R. CHRISTOPHERSEN • Department of Pediatrics, University of Kansas Medical Center, Kansas City, Kansas 66103; Bureau of Child Research, University of Kansas, Lawrence, Kansas 66045. JANICE E. ABERNATHY • Platte Medical Clinic, Platte City, Missouri 64079. Preparation of this manuscript was partially supported by a grant from NICHD (HD 03144) to the Bureau of Child Research, University of Kansas and by the Platte Medical Clinic.

recognition of these components as within the scope of practice of ambulatory pediatrics; (2) the preparation of physician-practitioners to manage these components; and (3) the current status of management of these components of child health by pediatric practitioners. In addition, a methodological critique of the literature and suggestions for future research will be discussed. The literature reviewed for this chapter is concerned primarily with physician primary care providers in ambulatory pediatric settings. There are some references cited which also discuss allied health professionals; however, these references are cited only to illustrate intervention strategies.

DEFINITION OF AMBULATORY PEDIATRICS AND ITS HISTORICAL DEVELOPMENT

Definition of Ambulatory Pediatrics

Ambulatory pediatrics is the practice of child advocacy. The role of child advocate takes many forms. Practitioners of ambulatory pediatrics* not only care for the sick child but also encourage healthy and maximal growth and development in all children. These pediatricians are called upon to provide counseling, anticipatory guidance, developmental appraisals, referrals, consultations, and screening for medical and psychosocial problems, and to mobilize community resources when necessary to attain those things which are in children's best interests. Pediatricians are challenged to manage behavioral and developmental disorders such as school phobias and learning disabilities as well as guide children through crises such as physical abuse, divorce, and serious illness or death of other family members (Task Force on Pediatric Education, 1978). Combining the practice of managing the disease processes of physically ill children with the identification of and intervention in the biosocial problems of all children and their families is what distinguishes the primary care provider in ambulatory pediatrics as a child advocate in this modern society.

Historical Development of Ambulatory Pediatrics

The relative containment of infectious disease and the increased knowledge of psychosocial aspects of health have influenced the move-

*Ambulatory pediatric practitioners include all physicians, nurse practitioners, and physician assistants who provide primary health care in ambulatory pediatric settings. The term *pediatrician* will be used throughout this paper in the generic sense to refer to all these practitioners.

ment toward the development of ambulatory pediatrics. Hoekelman (1975) estimates that the pediatrician spends 25–60% of his professional time engaging in well-child supervision. The well-child visit provides practitioners with opportunities to give anticipatory guidance, supervise children's developmental progress, assist family members with child-rearing problems, and deal with minor relational problems within the family (Korsch, Negrete, Mercer, & Freemon, 1971). As early as 1930, Anderson acknowledged the need for pediatricians to incorporate into their practices an emphasis upon the psychosocial components of child health care. The American Board of Pediatrics, in 1951, and The Task Force on Pediatric Education, in 1978, identified the need to improve the pediatricians' educational training in the biosocial aspects of the practice of pediatrics. At least 35% of children seen in pediatric practice settings are estimated to have behavioral or developmental problems and, if everyday problems such as bedtime problems, excessive crying, and temper tantrums are included, that percentage would increase considerably (Christophersen & Barnard, 1978). As an example, Brazelton (1975) reports that over 85% of his pediatric practice involves giving advice, guidance, and counseling about psychological or developmental problems, while only 10–15% of his time is spent diagnosing and treating somatic problems. Clearly, the primary emphasis in pediatric practice is shifting from somatic disease to disease prevention and psychosocial and developmental problems.

EDUCATIONAL PREPARATION OF PHYSICIANS FOR IDENTIFYING AND MANAGING BIOSOCIAL PROBLEMS OF CHILDREN

Current Educational Programs

Changes within the educational and training programs in ambulatory pediatrics must occur in order to provide appropriate preparation for practitioners to identify and manage the biosocial problems now encountered in pediatric settings. The educational system in past years has ignored the pediatrician's function as a counselor and behavior therapist; as a result, the graduate practitioner is inadequately prepared to assess and manage the child presented to him with problems other than those of a somatic nature, that is, biosocial problems (Honig, Liebman, Malone, Koch, & Kaplan, 1976). From a survey study of 100 pediatricians practicing in the south Florida area, Toister and Worley (1976) reported that over one half of the 61 respondents stated that they had had no formal behavioral training during the last two years of medical school or during residency training (see Table I). Data from the survey of 1,600 members of the American Academy of Pediatrics conducted by The Task Force on Pediatric Education (1978) indicated that 54% of young pediatricians rated their residency as providing insufficient experience with psychosocial and behavioral problems. Of those surveyed, 36% stated that their training in the management of mental or emotional disorders was inadequate. Additionally, 37% of the respondents reported that their training in neurology and learning disabilities was inadequate. From their survey of 97 practicing pediatricians in five New England states, Dworkin, Shonkoff, Leviton, and Levine (1979) reported that 79% of the physicians surveyed rated their formal training in child development as inadequate. Almost 50% of the pediatricians surveyed regarded medical school as being of no value as a source of knowledge in developmental pediatrics. Only 30% of the respondents reported that residency training was highly valuable for acquiring knowledge in developmental pediatrics. As in the survey conducted by Toister and Worley (1976), the responses by the participants were independent of the number of years since medical school graduation.

Two Training Programs for Medical Students and Pediatric Residents

Medical education and residency training attempt, among other things, to teach the practitioner many clinical skills. There is some current empirical literature evaluating student practitioners' development of physical examination skills (Aloia & Jonas, 1976; Klachko & Reid, 1976; Leake, Barnard, & Christophersen, 1978) and history-taking and interviewing skills (Stillman, Sabers, & Redfield, 1977). However, an exhaustive search of the literature reveals no empirical literature on training programs for medical students and residents in child development or behavioral and psychosocial problems of childhood.

Two models for teaching child development and behavior to medical students and residents have been described in the literature. The first program, described by Honig *et al.* (1976), was a pediatric-psychiatric liaison program which developed between the staff pediatricians at the Children's Hospital of Philadelphia and the child psychiatrists at the Philadelphia Child Guidance Clinic. The authors did not incorporate any kind of experimental design into the program. The child psychiatrists met weekly with the staff pediatricians (who served as primary teachers for pediatric interns and residents) to discuss cases and problems encountered by the residents in the pediatric medical outpatient clinics. At the weekly meetings the psychiatrists taught the staff pedia-

Table I. Studies on Training and Practice of Behavioral Pediatrics by Pediatric Practitioners

References	Design	Purpose	Subject selection	Measurement
Dworkin, Shonkoff, Leviton, & Levine, 1979	Nonexperimental actuarial interview–survey	Survey of pediatricians in practice to assess the adequacy of their training in ambulatory pediatrics	Random selection; $n = 97$	Structured interview using questionnaire; interobservational reliability reported for pilot study only
Toister & Worley, 1976	Nonexperimental actuarial survey	Survey of pediatricians in practice to assess the adequacy of their training in ambulatory pediatrics	Nonrandom, experimenter selected; $n = 61$	Questionnaire (mailed); no reliability reported
Smith, 1978	Nonexperimental actuarial survey	The use of developmental screening tests by primary pediatricians	Nonrandom, experimenter selected; $n = 450$	Questionnaire; no reliability reported

tricians the principles and techniques for managing the problems encountered by the residents in the outpatient clinics. The staff pediatricians, in turn, taught the residents these principles and techniques. In the outpatient clinic setting, the residents presented a 30-minute conference before each clinic session. Topics for these conferences ranged from issues associated with normal psychosocial-sexual development to descriptions of specific clinical syndromes. Following the conferences, the interns and residents saw the families assigned to them in the clinic. During the time spent with families, the interns were given concentrated instruction and feedback on interviewing skills by the psychiatrist who accompanied them during the sessions with families. The residents received input on the clinical problems of their patients. Videotapes were made of the interns' and residents' interactions with their patients. The interns viewed these tapes individually with the staff pediatricians and psychiatrists and received further instruction and feedback. The residents' tapes were viewed by the entire group of interns, residents, and faculty in an attempt to emphasize therapeutic techniques and elicit peer-sharing feedback. In addition to the training procedures employed in the clinic setting, the pediatric director of the liaison program gave a series of 10 lectures to the residents rotating through the outpatient department. Topics for these lectures included growth and development, principles of interviewing, compliance, family interviewing, school avoidance, individuality in childhood, discipline, sleep problems, colic, and other common management problems in practice.

An informal evaluation of the residents' perceptions of the value of the training program revealed that most residents thought the program increased their abilities to manage psychological problems. Although 48% of the residents reported that they felt they had improved their interpersonal skills more significantly than their skills of identifying and managing psychological problems, no actual data were presented to support or refute the residents' personal reports.

The second model for teaching early child development was described by Shuman and Gross (1978). This model was used to teach psychological aspects of infant growth and development to groups of 8 to 10 medical students during a pediatric clinic rotation. The authors did not incorporate any kind of experimental design or data collection into the program. The goals of this teaching program were to demonstrate infant temperament, age-appropriate competencies of children attending the clinic, and the interplay between mothers and their children. Mother–infant pairs were randomly selected from the Pediatric Ambulatory Service at Hahnemann Medical College and Hospital, Philadelphia, Pennsylvania, or from child welfare agencies in the same city.

A group of medical students would observe, via a one-way mirror and intercom, a child development specialist conducting a 30–40 minute informal developmental assessment of a child and parent interview. A second child development specialist observed the interview with the group of students and described what was occurring with the child, the mother–child relationship, and the developmental assessment. After these observation sessions, the students were asked to evaluate the learning experience. All students thought the sessions were valuable and recommended that they remain a part of the pediatric rotation experience. However, there was no empirical evaluation to determine whether these students actually incorporated this information into their subsequent interviews and physical examinations. Like the model for teaching residents described by Honig *et al.* (1976) this model for teaching medical students about child development has not been empirically evaluated. Consequently, no meaningful conclusions can be drawn as to the actual effectiveness of these models in increasing the medical students' and residents' knowledge base and skills in identifying and managing biosocial problems of children.

Summary and Recommendations

As early as 1930 the need for pediatric practitioners to identify and manage biosocial problems of children and families was identified. However, survey studies of pediatricians indicate that the educational preparation to manage such problems is inadequate, or nonexistent, for most practitioners. Additionally, an exhaustive search of the literature reveals a void of empirically evaluated teaching programs designed to provide physicians with the knowledge base and management skills needed to care for children presenting to their practices with biosocial problems.

The educational preparation of the pediatrician to manage the biosocial problems of children must begin in medical school. A sound knowledge base of normal growth and development is as imperative for the practitioner who manages biosocial problems as is a sound knowledge base of anatomy, physiology, and pharmacology for the practitioner who manages somatic disease processes. A theory course on normal growth and development should be required for all medical students. Utilizing allied health professionals, for instance, child development specialists and child psychologists, to teach such a course would ensure a focus on the biosocial components of health rather than on the somatic complaints that are often related to biosocial problems. Additionally, a clinical practicum in patient care settings would be a necessary component of this course. Within the practicum setting, the students would observe pediatrician role models (as suggested by the 1978 Task Force on Pediatric Education) as they identified and treated biosocial problems. After a period of time, these students would, with supervision, begin managing children themselves. Such a practicum experience would enable medical students to increase their knowledge of and develop skills for managing biosocial problems in children. The theoretical course should precede the clinical practicum so that the students would have an understanding of the basic principles of growth and development before entering the practicum setting. Collaboration between the child development specialists and the clinical medical faculty would be necessary so that the medical students could begin to understand the relationship between the biosocial and medical components of ambulatory pediatrics. This collaboration of medical and psychological-behavioral professionals would also promote the development of the much-needed research in methods of educating physicians about the biosocial components of health as seen in ambulatory pediatric settings. Residency training would build upon this foundation acquired in medical school. The emphasis of the residency training would include further development and refinement of pragmatic management skills necessary for the practitioner in ambulatory pediatric settings. A continued collaborative relationship between child development specialists, child psychologists, and pediatric medical faculty would be necessary in order to provide a high-quality, efficient learning experience for the residents. Eventually, as the pediatric faculty demonstrated more skill in identifying and managing biosocial problems, they would conduct the bulk of the training, and the child development specialists and the child psychologists could assume a consultant role similar to that held by pediatric cardiologists, nephrologists, and neurologists. In the residency training programs, as in the medical student programs, it would be appropriate for the medical and psychological-behavioral professionals to evaluate empirically the methods used to develop and refine the residents' skills in identification and management of biosocial problems.

One method for refining residents' skills would employ tape recordings of physician–patient–parent interviews. Reviewing these tapes with the staff pediatrician and a behavioral child psychologist not only would provide feedback for refining interviewing skills but also could increase the residents' knowledge of growth and development and family dynamics and their impact on children, and could refine the residents' ability to distinguish between organic and behavioral problems. These tapes could also provide a data base for examining the *in vivo* effects of the training program.

The need exists for practicing pediatricians to increase their knowledge of biosocial problems of children. In addition to published literature, continuing education materials and workshops on specific biosocial problems such as discipline, compliance, enuresis, and car seat safety could be developed and dispersed by child health specialists. The 1978 Task Force on Pediatric Education also recommends the use of postgraduate minifellowships (offered by teaching hospitals and medical centers) as another way of improving the education of the practicing pediatrician. The practicing pediatrician would spend from a few days to a few weeks in a medical center pediatrics department, with supervision by the pediatric faculty, to acquire new skills and refine other skills. Pediatricians, behavioral psychologists, clinical nurse specialists, and continuing education specialists working collaboratively could not only provide knowledge to practitioners but also empirically evaluate such methods in relation to effectiveness and feasibility. This research is greatly needed. A more thorough discussion of physicians' assessment and management of biosocial problems follows in the next section of this paper.

RESEARCH ON PHYSICIAN ASSESSMENT AND MANAGEMENT OF THE BIOSOCIAL COMPONENTS OF AMBULATORY PEDIATRICS

Content of Well-Child Visits

The practitioners in ambulatory pediatrics are in key positions to influence directly the psychological, developmental, and behavioral status of children in their practices (Greenwood, 1973). However, predicting whether a particular child will have an emotional, behavioral, or developmental problem is very difficult, if not impossible. No one particular factor such as age, temperament, central nervous system dysfunction, or childrearing patterns can be singled out as a positive predictive factor. It appears that combinations of any of these factors, as well as other unknown factors, may be influential in the development of such problems (Chamberlin, 1974). Children most often at risk are those with chronic diseases and disabilities (Battle, 1975; Pless, Satterwhite, & VanVechten, 1976; Yule, 1977), children from families in which the parents are having marital problems, children from large families, and children whose mothers are depressed (Richman, 1977). Additionally, children perceived by their parents as difficult in infancy may be at risk if the parents' reactions to the problems of coping with the child negatively affect the parents' attitudes toward the child (Chamberlin, 1973). The

Table II. Studies on Content of Ambulatory Pediatric Visits

Reference	Patient population	Setting	Practitioner	Measurement	Content of visit with practitioner
Korsch, Negrete, Mercer, & Freemon, 1971	Pediatric patients receiving well-child care; 200 provider/patient contacts	Prepaid clinic; private practice; health department clinic	Nurse practitioner; nurse/physician teams; physicians	Parent interviews; audiotapes of parent/provider contact	27–36% of the content related to medical issues; 21–28% related to hygiene; 14–22% related to child's physical behavior; 11–14% related to personal problems; 3–10% related to social-emotional behavior. There was no significant difference between types of providers.
Foye, Chamberlin, & Charney, 1977	Pediatric patients 1 & 2 years of age; 64 provider/patient contacts	Private practice	Nurse practitioner; pediatricians	Previsit parent questionnaire; observation; audio tape recording	Nurses devoted an average of 13% of the visit to routine care, 60% to behavior and development, and 26% to medical issues. Physicians devoted an average of 16% of the visit to routine care, 47% to behavior and development, and 37% to medical issues. There were statistically significant differences between physicians and nurses in time spent discussing behavior/development issues and medical issues.

astute pediatrician can identify children who are at risk and, through early detection and intervention, may prevent exacerbation of problems and possible abuse of these children. However, in studies of physicians' behaviors (see Table II), it has been found that pediatric practitioners devote the majority of their time during well-child visits to routine care and medical issues and are apt to diagnose somatic problems more often than biosocial problems (Foye, Chamberlin, & Charney, 1977; Korsch *et al.*, 1971; Smith, 1978).

Table III provides details on four nonexperimental actuarial studies which have specifically investigated the content of well-child visits. Physicians spend most of their time during well-child visits dealing with routine care and medical issues. Time devoted to psychosocial and developmental issues ranged from 3% to 4%. In the Korsch *et al.* (1971) study, it is interesting to note that physicians in the private practice setting devoted more time (10%) to issues of social and emotional behavior and less time to medical issues (27%) than did the physicians in the public health clinics (4% and 37%, respectively). These private practice physicians were all board-certified pediatricians, whereas the physicians in the public health clinics varied in educational background. These studies provide objective assessments of the content of well-child visits and confirm that physicians devote a greater part of the well-child visit to medical concerns rather than to biosocial issues. Further, a survey of 121 primary care pediatricians revealed that only 63% of the respondents used developmental screening tests. Of those 63%, only 15–20% screen more than 10% of their patients with the tests (Smith, 1978). Clearly, physicians center the majority of their primary care efforts on issues of disease, even in well-child care settings.

Physician Recognition and Management of Biosocial Problems

Jenkins, Collins, and Andren (1978) and Jonsell (1977) suggested that somatic complaints are used by some parents and children as a way to make contact with the practitioners so that discussion of psychosocial and/or behavioral problems may occur. However, the sparse literature indicates that pediatric practitioners often do not acknowledge or initiate management of these problems.

Tables IV and V provide details on two studies designed to evaluate physicians' identification and management of biosocial problems. In the nonexperimental, actuarial study by Starfield and Borkowf (1969) (see Table IV), the data collected through parent questionnaires and chart audits indicate that the physicians identified only 42% of the psycho-

Table III. Research on Well-Child Visits

References	Design	Subject selection	Measurement	Followup
Foye, Chamberlin, & Charney, 1977	Nonexperimental actuarial	Random selection: $n = 5$; Nonrandom experimenter selected; $n = 9$	Parent questionnaire, tape recording of patient/physician visit; direct observation of patient/physician visit; reliability measurements on only 4 of the 64 tapes and reportedly indicated "good general agreement"	None
Starfield & Borkowf, 1969	Nonexperimental actuarial	Nonrandom experimenter selected; $n = 383$	Parent questionnaire; chart audit; reliability reported	None
Duff, Rowe, & Anderson, 1972	Nonexperimental actuarial	Random selection; $n = 25$	Direct observation of patient/physician visit and student/physician conferences; structured interview, audiotaped at clinic and home; chart review; no reliability reported	None
Korsche, Negrete, Mercer, & Freemon, 1971	Nonexperimental actuarial	Nonrandom, experimenter selected, $n = 450$	Tape recordings of patient/physician contact; interviews (semistructured); no reliability reported	None

Table IV. Study of Physician's Recognition of Behavior Complaints

Reference	Population	Setting	Measures	Number of behavioral complaints	Number of behavioral complaints recognized	Number of somatic complaints	Number of somatic complaints recognized	Outcome
Starfield & Borkowf, 1969	Pediatric population receiving consultant care; $n = 383$	Pediatric clinic	Parent questionnaires; chart audit	48	20	115	90	The difference between recognition of somatic and behavioral complaints is greater than can be attributed to chance.

Table V. Study of Physician's Contribution to Management of Somatic and Biosocial Problems

Reference	Population	Setting	Measurement	Types of problems	Number of patients with problems	Physician/medical student contribution to management
Duff, Rowe, & Anderson, 1972	Pediatric patients; $n = 25$	Pediatric clinic	Direct observation; structured interviews; audiotapes; chart audit	Physical	3	Physicians contributed to the management of all physical-only problems, 7 physical/biosocial problems (only 2 of the 7 were biosocial problems), and 1 biosocial-only problem.
				Physical/biosocial	13	
				Biosocial	9	

social problems of the patients and parents but recognized 78% of the somatic complaints. The physicians in this study had completed approved pediatric residencies and at the time of the study held a one-year fellowship in ambulatory pediatrics. Data for individual patients were not provided. Intraobserver reliability was 90% in determining whether the physicians were cognizant of the complaint. Data collected by direct observation of physician–parent interviews, structured interviews, audio tapes, and chart audits and reported in Table V support the findings by Starfield and Borkowf (Duff, Rowe, & Anderson, 1972). Physician–teachers and medical students contributed to the management of impetigo, congenital heart disease, allergic diarrhea, epistaxis, hernia, anemia, heart murmur, and speech disorder as well as two episodes of parent–child relationship problems. A total of 13 children in this study presented a combination of physical and psychosocial problems. The physicians and medical students contributed to the management of only one of these problems per child and in only two instances did they deal with the psychosocial problems. In the other seven children the somatic complaint received the physicians' and medical students' attention. There were a total of 46 psychosocial problems listed for the 25 patients. Among these were behavior disorders, family strife, retardation, understimulation, parent–child relationship problems, failure to thrive, encopresis, leg pains, speech disorders, infantilism, and maternal depression. The medical students were aware of all but 3 of these. However, they were dissatisfied with their disposition of only 25 of these patients' problems. For 8 of the remaining 21 psychosocial problems, the medical students reported having no opinion as to the value of introducing management measures. Interestingly, the parents of the patient population for this study were satisfied with the physician management of only 8 of their children's problems, 6 of which were physical problems. No reliability measures were reported for this study. The data collected in these two actuarial studies support the premise that physicians often do not identify biosocial problems; further, they often do not initiate treatment of these problems even when identified. Several reasons have been suggested to account for this state of affairs: (1) the physicians may feel inadequately equipped to deal with the problems; (2) they may not perceive these problems as within their purview of practice, (3) the management of such problems may not be rewarding in that improvement may be slow; (4) the physicians may have had unsatisfactory experiences in the past in attempting to manage behavior problems; (5) they may anticipate that the time required to treat these problems is too great in relation to the high volume of patients they see; and (6) they may be awaiting the development of adequate procedures for dealing with biosocial problems (Chamberlin, 1974; Starfield & Borkowf, 1969).

Summary and Recommendations

The challenge to practitioners in ambulatory pediatrics is to overcome those barriers that prevent them from focusing upon the biosocial components of their patients' health. Data collected through parent questionnaires, chart audits, audio tapes of patient–physician contacts, and direct observation demonstrate that pediatric practitioners in ambulatory pediatric settings often do not recognize biosocial problems. Further, these practitioners rarely initiate treatment of these problems, even if they are identified. A number of recommendations can be made.

The pediatric primary care providers, because of the nature of their profession, view biosocial problems of children as secondary to medical problems. However, in the ambulatory pediatric setting, the majority of children seen are essentially disease-free. In light of this fact, it appears that pediatric primary care practitioners must reorient themselves away from a strong disease emphasis and focus more on the biosocial aspects of children's health.

The process of reorientation must begin during the educational preparation of pediatric practitioners. Incorporating courses in behavioral pediatrics into the curricula for all preparation levels of practitioners is imperative. Vanderpool and Parcel (1979) described one such training activity for medical students. After a one-hour seminar on developmental characteristics of 5- and 8-year-olds, freshmen medical students attended a two-hour session in an elementary school to observe and interact with school-aged children. Each medical student observed for a short period of time in the classroom and then conducted a health education lesson with a small group of children. No empirical evaluation was done of this educational program; however, subjective analysis of the experience by medical students, course faculty, and school personnel indicated that it might be a valuable learning technique in behavioral pediatrics. Another program enabling medical students to develop an appreciation of issues in behavioral pediatrics was described by Chilton (1979). Third-year medical students selected a school and a school nurse from a list of volunteer school nurses in a metropolitan area. The students and school nurses were then given a list of questions to discuss with each other during a meeting. Topics from the list included the role of the nurse, the capabilities of the school nurse, the nature of conditions encountered by the nurse, and methods to improve coordination of physician and school health services. Students who participated in this program reportedly responded to questionnaires covering the issues discussed with the nurses in a manner more representative of practicing pediatricians than those students and residents who did not participate. Both of these strategies appear to be creative and promising in their attempts to reorient the education of medical students toward an emphasis on biosocial issues. Empirical evaluation of these two methods of teaching medical students about behavioral pediatric issues is imperative.

Orientation of pediatric residents toward an emphasis on issues in behavioral pediatrics is also needed. As we suggested earlier, collaboration between the pediatric faculty, residents, and pediatric psychologists in the clinical setting promises to be one effective way of accomplishing this goal. One way of insuring such a collaboration is to establish a behavioral clinic within the pediatric outpatient department. One such clinic was described by Sharrar and Linan (1980). The residents were assigned to the behavioral clinic one half-day a week. They were videotaped during interviews with each family and received feedback from the faculty pediatrician and child psychiatrist on their ability to identify the problems and the families' perceptions of them, on interviewing skills, and on intervention techniques. The clinic families were preselected for common problems that could be managed in 1 to 4 sessions. No empirical analysis of this learning program has been conducted; but, like those discussed earlier in this paper, the idea is interesting, educationally innovative, and worthy of further exploration and evaluation.

Reorientation of the practicing pediatrician toward an emphasis upon issues in behavioral pediatrics may be very challenging. In addition to the previously discussed suggestions of workshops, written materials, audiovisual materials, and minifellowships, changing the traditional format for continuing medical education programs might be helpful in meeting the needs of the practicing pediatric practitioner. For example, Caplan and Charney (1979) developed a longitudinal continuing medical education course. The format consisted of 12 weekly evening sessions on developmental pediatrics. Six topics were discussed; two sessions were devoted to each topic. The first session used a lecture format, the second was a practicum during which demonstrations with patients from the participants' practices illustrated the principles discussed during the lectures. No empirical evaluation was conducted of this program or its format; however, 96% of the 23 practicing pediatric practitioners reported satisfaction with the course. In light of such a high rate of satisfaction, this format deserves further development and empirical analysis. If such a format could be demonstrated to effect change in the behavior of pediatricians in their actual clinical practice, reorientation of these practitioners might not be an insurmountable task.

PHYSICIAN-ORIENTED RESEARCH ON INTERVENTION STRATEGIES FOR BIOSOCIAL PROBLEMS ENCOUNTERED IN AMBULATORY PEDIATRIC SETTINGS

This chapter has documented the existence of biosocial problems in children seen by pediatric practitioners in ambulatory settings. Parent concern about these problems in their children has also been demonstrated. Presently, data-based physician management of these problems is weak or nonexistent; however, as medical student and resident training programs and continuing education for practicing pediatric practitioners begin emphasizing these biosocial issues, physician management of these problems will improve.

Behavioral pediatrics is an area of specialization focusing upon children in medical pediatric settings who have nonpsychiatric clinical problems (Christophersen & Rapoff, 1980). Behavioral pediatrics differs from traditional child psychology not only in conceptualization, but also at the point of intervention and the manner of intervention (Walker, 1979). Behavioral pediatric intervention strategies are often introduced at the time of the well-child visit rather than during a visit to a mental health practitioner. Additionally, the manner of intervention is behaviorally oriented and short-term in nature as opposed to traditional child psychology. Because of the nature of behavioral pediatrics, further support exists for pediatric practitioners to employ behavioral intervention strategies to manage certain types of problems. However, the intervention strategies should be experimentally validated and feasible for use by pediatric practitioners. Five such pediatric psychology problem areas have been identified for management by pediatric practitioners in ambulatory settings: (1) car restraint seat use, (2) common behavior problems, (3) encopresis, (4) anticipatory guidance, and (5) childrearing practices (Christophersen & Rapoff, 1980). An extensive review of the literature has identified very few experimental studies concerned with these five areas; many of the studies could not be classified as experimentally sound research. Table VI describes the designs used for the studies reviewed in ambulatory pediatrics. The majority of the studies' designs would be classified as true experimental designs. Table VII surveys the literature reviewed.

Table VI. Designs Used in Ambulatory Research Reviewed

Design	Description	Comments
1. Experimental		
Multiple baseline across subjects	A treatment variable is applied sequentially to the same behavior across different but matched subjects sharing the same environmental conditions. Treatment phases for each subject can be conceptualized as separate A–B designs, with the length of baselines increased for each succeeding subject. (Mahoney, 1978)	The controlling effects of the treatment on each of the target behaviors is not directly demonstrated. However, the treatment effects are *inferred* strongly from comparison with untreated subjects.
Pretest–posttest control group	Subjects are randomly assigned to treatment and control groups and assessed before and after introduction of the treatment variable to determine the effects of the treatment intervention.	This design controls for all sources of internal invalidity but no sources of external invalidity. The control group enables the experimenter to have greater confidence that the treatment intervention contributed to the change. Generalization to other population samplings is weak.
Posttest only control group	Subjects are randomly assigned to experimental and control groups. Those in the experimental group are subjected to the treatment variable(s) and those in the control group are given "routine" treatment. The dependent variable(s) is measured in both groups upon termination of the treatment period.	This design controls for all sources of internal invalidity and the interaction of testing with the dependent variable(s). Generalization of findings is limited.

2. Preexperimental		
One-group pretest–posttest design	The dependent variable is measured before and after the experimental manipulation.	This design controls for only two sources of internal invalidity, i.e., selection and mortality, and no sources of external invalidity. It is impossible to conclude that the experimental manipulation itself effected the changed behavior. Generalization of findings cannot occur.
3. Nonexperimental		
Surveys	Questionnaires are used to elicit descriptive information.	Studies of this nature are important in the early developmental stages of the evaluation of issues in practice. They provide some information from which questions for empirical study arise. Inferences from the results are limited. Information is often subjective in nature.
Interviews	Another method for gathering descriptive information. Interviews may be structured or nonstructured.	
Case studies	Reports of individual cases in which a treatment intervention has been introduced. No objective preintervention data or posttreatment data are available.	
Actuarial	Direct observation of events to determine behavior that occurs.	Studies of this nature, though not experimental, provide more objective descriptive information than do surveys, interviews, or case studies.

Table VII. Review of Literature on Treatment Procedures for Biosocial Problems

References	Design	Purpose	Subject selection	Measurement	Follow-up
Christophersen, 1977	Multiple baseline across subjects	Assessment of car seat usage	Recruited via newspaper advertisement; $n = 11$	Direct observations with reliability	3 months
Christophersen & Gyulay, 1981	Multiple baseline across subjects	Assessment of effects of car seat use on behavior—used handout & interview	Recruited via newspaper advertisement; $n = 8$	Direct observations with reliability	3, 6, 12 months
Christophersen & Rainey, 1976	Nonexperimental case study	To evaluate effectiveness of treatment procedure for encopresis	Several "exemplary" cases; $n = 6$	Parents reports; no reliability	2–12 months
Barnard, 1977	Multiple baseline across subjects	To evaluate effectiveness of treatment procedure for encopresis	Sequential selection; $n = 4$	Direct observation with reliability; parent reports	None
Chamberlin, 1975	Nonexperimental actuarial descriptive study	To demonstrate the relationship between "positive contact" and child behavior patterns	Recruited and monitored for equal number of boys and girls; $n = 4$	Home interview; direct observations in the home; no reliability	None

Gutelius, Kirsch, MacDonald, Brooks, & McErlean, 1977	Posttest only control group design	To evaluate the effects of extensive child health supervision with emphasis on counseling and anticipatory guidance	Experimenter-selected; $n = 95$	Interview; standardized testing; observation; no reliability	None
Christophersen & Sullivan, in press	Random assignment to groups; posttest control group	To study the effectiveness of modeling on car seat usage	Sequentially selected; $n = 30$	Direct observations of car seat usage upon dismissal from hospital and 6-week checkup with reliability	6 weeks
Reisinger, Miller, Blatter, Wucher, & Williams, in press	Experimenter assignment to treatment groups; posttest control group	To study the effectiveness of physician counseling on use of smoke detectors in the home	Consecutively selected; $n = 240$	Home safety inspection; no reliability	None
Cullen, 1976	Posttest only control group with random assignment to control and experimental groups	To assess the effect of anticipatory guidance on children's behavior disorders	Unknown; $n = 246$ children (203 families)	Interview; standardized testing; no reliability	None

(Continued)

Table VII. (Continued)

References	Design	Purpose	Subject selection	Measurement	Follow-up
Koota, Kuritzkes, Grinjsztein, & Rubin, 1980	Nonexperimental actuarial descriptive report	To describe a parent education program in a private practice	Not reported	Increased number of patient visits	None
Chamberlin, Szumonski, & Zastowny, 1979	One-group pretest; posttest preexperimental design	To identify variables that might increase mother's knowledge of growth and development and promote optimal growth and development in children	Recruited from practices of 35 pediatricians; $n = 595$	Checklists; questionnaires; standardized testing; no reliability	None
Dershewitz & Ordakowski, 1980	True experimental pretest posttest control group design; random assignment to treatment group	To assess educational method in ambulatory pediatric setting	Sequentially selected; $n = 143$	Questionnaire; no reliability	1 and 6 weeks

Intervention Strategies

Car Restraint Seat Use

The procedures described by Christophersen (1977) and Christophersen and Gyulay (1981) for promoting car safety restraint seat usage have been judged feasible for use by the pediatric practitioner in an ambulatory setting. However, pediatric practitioners did not administer the intervention. In his 1977 study, Christophersen compared (by actual observations) the behavior of toddlers who usually rode in a restraint seat to the behavior of toddlers who usually rode unrestrained. The restraint seat toddlers behaved appropriately during over 95% of the observed intervals, whereas the unrestrained toddlers behaved appropriately less than 5% of the time. In the second study, Christophersen and Gyulay (1981) used a written protocol (see Appendix A) which instructed parents how to introduce the use of car seats and the importance of verbally attending to the children while in the automobile. They instructed parents on proper use of the car seats (i.e., fastening, releasing, and anchoring the top strap) and the need to show the children the advantages of a car seat. The parents were also instructed to talk more with the children during rides in the automobile. Measures of the children's appropriate behavior in the cars were assessed by direct observation during 15-minute automobile rides in both studies. When the car seats were used, children's inappropriate behaviors decreased significantly; these results were maintained at 12-month follow-up. Both studies used a multiple-baseline-across-subjects design, and demonstrated the validity of the procedure for increasing the use of car seats.

Common Behavior Problems

Common behavior problems such as temper tantrums and crying at bedtime and mealtime can be managed by the pediatric practitioner without referral to a mental health professional. However, research to validate the procedures for the ambulatory pediatric practitioner is virtually nonexistent. In an actuarial, nonexperimental, descriptive study of maternal use of positive contact, a positive correlation was found to exist between maternal use of positive contact and appropriate child behavior patterns (Chamberlin, 1975). Children whose mothers displayed more positive contact (read stories to or played with the children, hugged and kissed the children, took them on errands and family outings, and praised them) and children whose fathers played with them manifested significantly fewer hostile resistant behaviors than children whose mothers and fathers displayed little or no positive contact. No measures of reliability were reported in this study and no experimental manipulation occurred. Studies of this nature are needed, but intervention strategies and effective protocols must be created and evaluated if the practitioner is to have much impact on these problems.

Encopresis

Treatment regimens for encopresis have yet to be subjected to experimental analysis. In their 1976 publication, Christophersen and Rainey reported the results of their case studies on six encopretics. The protocol, originally suggested by Wright (1973), employed the use of glycerine suppositories and positive reinforcement in a daily routine until proper elimination occurred and was maintained. At the time of publication, the three children who had completed the program had not soiled since termination of the program. The other three children were still in treatment, but had reduced the number of soilings considerably. No measures of reliability on occurrence of soiling were conducted and no experimental manipulation was conducted. However, the establishment of a written protocol for use with families is a good beginning toward validation of this treatment procedure for encopresis.

Barnard (1977), in an unpublished thesis, conducted one of the few investigations of the treatment of encopresis that used a true experimental design (multiple baseline across subjects) and included reliable observation of the patients involved in the study. Results of this study indicate that the treatment strategy was effective in decreasing the incidence of encopresis. Although studies of this nature are difficult to conduct, they are essential to the development of a sound scientific base for the practice of behavioral pediatrics. The use of home observations with interobserver reliability checks represents a significant advance over previous studies that relied upon parental reports (cf. Pless, 1978).

Anticipatory Guidance

Anticipatory guidance is currently practiced to some extent by pediatricians (Brazelton, 1975; Christophersen, 1978). However, experimental validation of effective procedures has been subjected to only elementary analysis. In a longitudinal, posttest-only, control-group study (Cullen, 1976), the randomly assigned experimental group of parents of newborns received a total of twelve 20–30 minute interviews with the physician over a 5-year period. These interviews were spaced at 3-month intervals during the first year and 6-month intervals during the succeeding four years. The interviews consisted of discussions of age-appropriate developmental progress of the children and parent–child relationships. The control group did not receive these interviews. Cullen's data suggested that regular interviews with parents of preschool children did moderately reduce the incidence of behavioral disturbances in 6-year-old children. In a similarly designed study, expectant mothers in a low-income metropolitan area were randomly assigned to experimental and control groups (Gutelius, Kirsch, MacDonald, Brooks, & McErlean, 1977). The control group received prenatal care in prenatal clinics of the health department and infant care in the nearest well-baby clinics. Those in the experimental group were seen regularly by the same public health nurse during the remainder of the prenatal period, and newborn care was provided regularly by the same physician and nurse in a mobile coach parked in front of the patients' homes. These visits numbered at least nine, six, and four during the first, second, and third years of life, respectively. In addition, the nurse made one-hour visits between physician visits to promote cognitive stimulation for the infant. The data from this study suggest that the time spent with the mothers in the experimental group contributed to the significant differences between the two groups, with the experimental families displaying more positive outcomes in many areas, including infant development, childrearing practices, parental education, and parenting skills. Like Cullen's 1976 study, the study by Gutelius *et al.* (1977) suggests that time spent by health care providers in anticipatory guidance may have positively affected infants' development and behavior and promoted good relationships between parents and children as well. However, these two studies were not reported in a way that would allow replication. In addition, no reliability measures were taken in either study. Three other studies have evaluated the effectiveness of anticipatory guidance upon use of car seats (Christophersen & Sullivan, in press; Reisinger, Williams, Wells, John, Roberts, & Podgainy, 1981) and use of smoke detectors (Reisinger, Miller, Blatter, Wucher, & Williams, in press). These studies suggest that anticipatory guidance may effect changed behavior in the parents. However, the development of specific guidance protocols was not reported. A thorough search of the literature reveals that no attempt has been made to describe and define anticipatory guidance behaviorally.

Childrearing

As with anticipatory guidance, the state of experimental validation of protocols for teaching childrearing techniques in the ambulatory pediatric setting is in the elementary stages. A one-group pretest–posttest study was conducted to identify what variables might correlate highly with mothers' increased knowledge of child development and use of positive contact with their children (Chamberlin, Szumonski, & Zastowny, 1979). The data are not conclusive but suggest that the amount of teaching done by the physician is positively correlated with these two maternal characteristics. Another study was designed to evaluate the effectiveness of four teaching strategies (Dershewitz & Ordakowski, 1980). These strategies were verbal instruction by a registered nurse after the visit with the physician, a written fact sheet, a combination of these two, and verbal instructions by the physician only. This study utilized a pretest–posttest control group design with random assignment to the groups. The experimenters' data suggest that all methods of teaching specific information about fever control and diaper rash were effective and no method was superior.

Design

The literature describing research in ambulatory settings for car seat usage, encopresis, common behavior problems, anticipatory guidance, and child rearing practices is sparse. Many of those studies which have been conducted are uncontrolled studies with no measure of reliability, no replication, and no follow-up data.

The majority of these uncontrolled studies are actuarial and descriptive (Chamberlin, 1975; Koota, Kuritzkes, Grinjsztein, & Rubin, 1980). No attempt was made to assess a particular intervention strategy. The purpose of these studies was to describe those events which occur when certain variables are present. One study measured the effects of an experimental manipulation upon one group of subjects by employing a pre- and posttest design (Chamberlin *et al.*, 1979). One study was a case study of six encopretics with only anecdotal reports (Christophersen & Rainey, 1976).

The majority of the controlled studies reviewed have utilized the between-subject group design with random assignment to experimental and control groups. Multiple baseline designs were employed in three of the controlled studies (Barnard, 1977; Christophersen, 1977; Christophersen & Gyulay, 1981). In summary, most of the controlled and uncontrolled studies have used some type of between-group comparison design. These kinds of studies are needed in the early developmental stages of fields of clinical endeavors such as ambulatory pediatrics and behavioral pediatrics.

Reliability between Observers

Much of the literature on pediatric intervention strategies for anticipatory guidance, common behavior problems, and childrearing advice has relied upon interviews with parents, questionnaires, and standardized testing results. This low level of sophistication is consistent with the early development of the evaluation research being conducted. Future research in the area must depend upon direct observation, with appropriate interobserver reliability checks, as its data base.

Follow-Up Data

Follow-up data have not been reported routinely in the literature. This is another reflection of the state of empirical analysis of treatment regimens to be used by pediatric practitioners for problems of a biosocial nature. Follow-up data are important in assessing both the effectiveness over time of the intervention strategies employed, and the generalization of these strategies to similar problems. In both controlled and uncontrolled studies, follow-up data ranging from 6 weeks (Christophersen & Sullivan, in press; Dershewitz & Ordakowski, 1980) to 12 months (Christophersen & Gyulay, 1981; Christophersen & Rainey, 1976) have been reported. Over 50% of the studies reviewed included follow-up reports.

Summary and Recommendations

The majority of procedures for use in ambulatory pediatric settings for treatment of common behavior problems such as noncompliance, mealtime problems, bedtime problems, and tantrums have had no empirical evaluation. It has been suggested that children whose parents attend to them by reading to them, praising them, taking them on outings, and hugging and kissing them are less likely to be stubborn, to hit others, to be disobedient, and to throw temper tantrums than children whose parents do not do these things. However, for the pediatric practitioner who is attempting to manage common behavior problems no feasible intervention strategy has been empirically demonstrated to be valid. Research into anticipatory guidance and childrearing education as provided in the pediatric ambulatory setting is in an even earlier stage of development. Attempts have been made to demonstrate that anticipatory guidance has some value and that it is possible to teach childrearing skills in an ambulatory setting. However, content analysis, teaching-strategy evaluation, identification of and control for dependent and independent variables, and utilization of true and quasi-experimental designs are needed in further study of not only childrearing and anticipatory guidance, but also car seat use, common behavior problems, and encopresis. A thorough discussion of recommendations for future research in ambulatory pediatrics follows.

DIRECTIONS FOR FUTURE RESEARCH

Research in behavioral pediatrics as it pertains to the pediatrician must now become more experimental and less descriptive in nature. A number of recommendations for future research can be made. A few protocols of treatment for biosocial problems of children have been generated; however, experimental validation of these protocols has been inadequate. Additionally, other protocols for treatment must now be generated in order to insure the adoption of behavioral principles by pediatricians and parents in managing children with biosocial problems.

Protocols for Treatment

As we noted previously in this discussion, treatment protocols for car seat use and encopresis have been developed. Similar protocols are needed in other areas such as mealtime problems; temper tantrums; child rearing; and anticipatory guidance in growth and development, disease prevention, problem-solving, parenting, sexuality, and other areas related to providing health care to families in ambulatory pediatric settings. However, development of such protocols is only the beginning. Determining the feasibility of the protocols for use by pediatricians and validating the effectiveness of the protocols for treatment as well are imperative. Another pertinent research issue is determining the most effective method for dispensing the information in the protocol. The effectiveness of intervention strategies such as written handouts, audiovisual materials, and verbal instruction must be assessed for each treatment protocol. In addition, evaluation of treatment protocols and intervention strategies across settings, providers, patients, and time is needed to insure the generalizability and the reliability of the previous findings.

Replication

In addition to insuring generality and reliability of findings, replication of research conducted to evaluate treatment protocols and intervention strategies for problems in ambulatory settings will insure the identification and reporting of limitations and failures of these procedures (cf. Hersen & Barlow, 1976). The significance of such information is obvious. Therefore, investigators must be sure not only to specify what independent and dependent variables were assessed, but also to provide detailed descriptions of these, as well as methodology, measurement tools and methods, the sample population, reliability measurements, design, and calculation of data. The literature reviewed for this paper lacked such information which would allow for replication of the studies.

Experimental Design

In the early stages of evaluating a treatment protocol or intervention strategy, use of quasi-experimental designs and case studies is acceptable. Much of the research remaining to be done in the area of behavioral pediatrics may begin as case study reports and quasi-experimentally designed studies. However, once a treatment protocol has been developed and is thought to be feasible for use by pediatric practitioners and to be effective in producing the desired outcome, it must be subjected to controlled evaluation. Both single-subject and between-group designs would be valuable to this end. Advantages of single-subject designs in this research include (1) allowing investigators the liberty of adapting the treatment protocols and intervention strategies to the individual patient, (2) identifying sources of variability, and (3) enabling the investigators to build more powerful treatment packages (Hersen & Barlow, 1976). A series of single-subject designs also provides the information needed to explain why a treatment is or is not effective. At this point, the treatment protocols and intervention strategies can be submitted to group design studies. The advantages of group design studies include (1) demonstrating the magnitude of effect of the treatment protocols and intervention strategies, (2) demonstrating the effectiveness of a given treatment on a well-defined group (such as encopretics or children who have tantrums), and (3) providing comparison data for two or more treatment strategies for the same targeted behaviors. By employing both the single-subject and between-group designs sequentially for all treatment protocols and intervention strategies, investigators can better insure internal and external validity of the results.

CONCLUSIONS

Several conclusions can be made from this review of the behavioral pediatric literature as it pertains to the pediatric practitioner.

1. Education for pediatric practitioners in behavioral pediatrics is critically inadequate. Change in this area, although taking place, has been slow. Empirical analysis of methods currently employed to educate physicians in behavioral pediatrics is greatly needed.

2. Pediatric practitioners spend a majority of their time giving well-child care; however, they remain disease-oriented in their delivery of health care. Efforts to reorient the practicing physicians to identify and

manage biosocial problems must be undertaken and subjected to empirical analysis for effectiveness in promoting changed behavior in physicians.

3. Many problems in behavioral pediatrics could probably be managed by pediatricians. However, the scientific development of the field must make progress in demonstrating the feasibility of intervention by pediatric practitioners and in specifying and experimentally validating treatment protocols for these problems.

The relative containment of infectious disease and the emphasis upon disease prevention in the past few years has created a relatively medically healthy pediatric population. However, biosocial problems in children are not uncommon. Unfortunately, the pediatrician who is most likely to be sought out by parents and most likely to have the opportunity to identify and intervene in these problems is inadequately prepared to do so. Behavioral pediatric professionals must help meet this health care challenge by collaborating with medical educators and primary care providers to develop effective educational strategies and pragmatic intervention tools that will eliminate the void in care provided to children and their families who have biosocial problems.

Acknowledgments

The editorial assistance of Barbara Cochrane and Suzanne Gleeson is gratefully acknowledged.

APPENDIX

Using Automobile Car Seats: Guidelines for Parents*

Car riding can and should be a pleasant time for you and your child. This is an excellent time for pleasant conversation and teaching your child acceptable and appropriate behavior in the car. It is also the safest mode of travel, even on short trips, for your child.

1. Introduce the car seat to your child in a calm, matter-of-fact manner as a learning experience. Allow him to touch it and check it out.
2. Remind your child about the rules of behavior *nicely* before the first ride and in between rides.
3. Your first rides with the seat should be short practice ones to teach him the expected and acceptable behavior, *perhaps once around the block*. Point out interesting things that she/he can see. Make it a positive experience for both of you.

4. Praise him *often* for appropriate behaviors. (Example: "Mike, you are sitting so quietly in your seat. Mommy is proud of you. You are a good boy . . ."). This explanation teaches him the expected and appropriate behavior. Young children need specific directions. They cannot make the opposite connection of what is meant by "Quit that!" Catch him/her being good!!!!!! You cannot praise him too often!!!!
5. Include the child in pleasant conversation. (Example: "That was sure a good lunch . . . you really like hot dogs . . . you were a big help to me in the store . . . it'll be fun visiting grandma . . .").
6. This is also a good time to teach your child about his world. (Example: "Jon, see that *big, red* fire truck? Look at how *fast* it is going. What *do* firemen do? The light on the *top* is *red* . . . what else is red . . ."). This needs to be geared to the age of your child.
7. By your frequent praise, teaching and pleasant conversation, your child will remain interested and busy and not spend his time trying to get out of the seat. He will have your frequent attention.
8. Ignore yelling, screaming, and begging. The instant he is quiet, praise him for being quiet. You also should not yell, scream and beg. Remember, remain calm and matter-of-fact. Keep your child busy in conversation and observations of his world. Do not give in and let him out. This only teaches him that yelling, screaming, and begging will finally break mom or dad into letting him do what he wants. Show who's the boss!!!!
9. Older siblings should also be expected to behave appropriately. If the young child sees an older sibling climbing or hanging out the window, he will want to become a participant. The older sibling(s) should also be included in the conversation, praise and teaching.
10. Provide one or two toys that your child associates with quiet play such as: books, stuffed animals, dolls, etc. It may help to have special quiet riding toys that are played with only in the car. This decreases boredom. Remember, the young child's attention span is *very* short. Do not expect him to keep occupied for more than a couple of minutes or less, particularly at the beginning and depending upon his age. *Anticipating* this will *prevent* throwing of the toys, temper tantrums, crying or fussing.
11. Reward him with five to ten minutes of your time doing an activity that he likes to do immediately after the ride such as: reading a story, playing a game, helping prepare lunch, helping put away the groceries, etc. Do not get into the habit of buying your child favors or presents for his good behavior. He enjoys time *with you* and it's less expensive and more rewarding for both of you!!!! Remember, CATCH HIM BEING GOOD— AND PRAISE HIM OFTEN!!!!!

12. If your child even begins to try to release his seat belt or to climb out of the car seat, immediately tell him, "NO!" in a firm voice. On your first few trips, which should just be around the block, stop the car if you think that is necessary. Also, consider stating the rule once, clearly, "Do *not* take off your seat belt!", and discipline him/her if necessary.
13. Remember, without the praise and attention for good behavior in the car, your child will learn nothing from the training trips. The combination of praise and attention, with occasional discipline can and will teach the behavior you want in the car.

REFERENCES

Aloia, J. F., & Jonas, E. Skills in history-taking and physical examination. *Journal of Medical Education*, 1976, *51*, 410–415.

American Board of Pediatrics. A statement on training requirements in growth and development. *Pediatrics*, 1951, *7*, 430–432.

Anderson, J. E. Pediatrics and child psychology. *Journal of the American Medical Association*, 1930, *95*, 1015–1020.

Barnard, S. R. *An experimental analysis of a parent-mediated behavioral treatment program for encopresis.* Unpublished master's thesis, University of Kansas, 1977.

Battle, C. U. Chronic physical disease: Behavioral aspects. *Pediatric Clinics of North America*, 1975, *22*(3), 525–531.

Brazelton, T. B. Anticipatory guidance. *Pediatric Clinics of North America*, 1975, *22*(3), 533–544.

Caplan, S. E., & Charney, E. A longitudinal course for continuing medical education: A successful format in a community hospital. *Program and Abstracts of the 19th Annual Meeting of the Ambulatory Pediatric Association*, 1979, 47.

Chamberlin, R. O. Parental use of "positive contact" in child-rearing: Its relationship to child behavior and other variables. *Pediatrics*, 1975, *56*(5), 768–773.

Chamberlin, R. W. New knowledge in early child development: Its importance for pediatricians. *American Journal of Diseases of Children*, 1973, *126*, 585–587.

Chamberlin, R. W. Management of preschool behavior problems. *Pediatric Clinics of North America*, 1974, *21*(1), 33–47.

Chamberlin, R. W., Szumowski, E. K., & Zastowny, T. R. An evaluation of efforts to educate mothers about child development in pediatric office practices. *American Journal of Public Health*, 1979, *69*(9), 875–885.

Chilton, L. A. School health exposure for medical students. *Program and Abstracts of the 19th Annual Meeting of the Ambulatory Pediatric Association*, 1979, 49.

Christophersen, E. R. Children's behavior during automobile rides: Do car seats make a difference? *Pediatrics*, 1977, *60*(1), 69–74.

Christophersen, E. R. Behavioral pediatrics. *American Family Physician*, 1978, *17*(3), 134–139.

Christophersen, E. R., & Barnard, J. D. Management of behavior problems: A perspective for pediatricians. *Clinical Pediatrics*, 1978, *17*, 122–124.

Christophersen, E. R., & Gyulay, J. E. Parental compliance with car seat usage: A positive approach with long-term follow-up. *Journal of Pediatric Psychology*, 1981, *6*(3), 301–312.

Christophersen, E. R., & Rainey, S. K. Management of encopresis through a pediatric outpatient clinic. *Journal of Pediatric Psychology*, 1976, *4*(1), 38–41.

Christophersen, E. R., & Rapoff, M. A. Pediatric psychology: An appraisal. In B. B. Lahey & A. E. Kazdin (Eds.), *Advances in clinical child psychology* (Vol. 3). New York: Plenum Press, 1980.

Christophersen, E. R., & Sullivan, M. Increasing the protection of newborn infants in cars. *Pediatrics*, in press.

Cullen, K. J. A six-year controlled trial of prevention of children's behavior disorders. *Journal of Pediatrics*, 1976, *88*(4), 662–666.

Dershewitz, R. A., & Ordakowski, M. The evaluation of patient education in a pediatric outpatient setting. *Program and Abstracts of the 20th Annual Meeting of the Ambulatory Pediatric Association*, 1980, 41.

Duff, R. S., Rowe, D. S., & Anderson, F. P. Patient care and student learning in a pediatric clinic. *Pediatrics*, 1972, *50*(6), 839–846.

Dworkin, P. H., Shonkoff, J. P., Leviton, A., & Levine, M. D. Training in developmental pediatrics. *American Journal of Diseases of Children*, 1979, *33*, 709–712.

Foye, H., Chamberlin, R., & Charney, E. Content and emphasis of well-child visits. *American Journal of Diseases of Children*, 1977, *131*, 794–797.

Friedman, S. B. The challenge in behavioral pediatrics. *Journal of Pediatrics*, 1970, *77*(1), 172–173.

Greenwood, R. D. Psychiatry in pediatrics, emotional guidance in pediatric practice. *Journal of the Kansas Medical Society*, 1973, *74*, 220–223.

Gutelius, M. F., Kirsch, A. D., MacDonald, S., Brooks, M. R., & McErlean, T. Controlled study of child health supervision: Behavioral results. *Pediatrics*, 1977, *60*(3), 294–304.

Hersen, M., & Barlow, D. H. *Single case experimental designs: Strategies for studying behavior change*. New York: Pergamon Press, 1976.

Hoekelman, R. A. What constitutes adequate well-baby care? *Pediatrics*, 1975, *55*, 313–326.

Honig, P. J., Liebman, R., Malone, C., Koch, C. R., & Kaplan, S. Pediatric-psychiatric liaison as a model for teaching pediatric residents. *Journal of Medical Education*, 1976, *51*(11), 929–934.

Jenkins, G. H., Collins, C., & Andren, S. Developmental surveillance in general practice. *British Medical Journal*, 1978, *1*, 1537–1540.

Jonsell, R. Patients at a pediatric outpatient clinic. *Acta Paediatrica Scandinavica*, 1977, *66*, 723–728.

Klachko, D. M., & Reid, J. C. The effect on medical students of repetition of a physical examination routine. *Journal of Medical Education*, 1976, *51*, 670–672.

Koota, I. R., Kuritzkes, F. A., Grinjsztein, J. M., & Rubin, G. F. A model parent education and support program in a private pediatric office. *Program and Abstracts of the 20th Annual Meeting of the Ambulatory Pediatric Association*, 1980, 50.

Korsch, B. M., Negrete, V. F., Mercer, A. S., & Freemon, B. How comprehensive are well child visits? *American Journal of Diseases of Children*, 1971, *122*, 483–488.

Leake, H. C., Barnard, J. D., & Christophersen, E. R. Evaluation of pediatric resident performance during the well-child visit. *Journal of Medical Education*, 1978, *53*, 361–363.

Mahoney, M. J. Experimental methods and outcome evaluation. *Journal of Consulting and Clinical Psychology*, 1978, *46*(4), 660–672.

Pless, I. B. Accident prevention and health education: Back to the drawing board? *Pediatrics*, 1978, *62*(3), 431–435.

Pless, I. B., Satterwhite, B., & VanVechten, D. Chronic illness in childhood: A regional survey of care. *Pediatrics*, 1976, *58*(1), 37–46.

Reisinger, K. S., Miller, R. E., Blatter, M. M., Wucher, F. P., & Williams, A. F. The effect of pediatricians' counselling on the importance of smoke detectors upon subsequent detector installation. *American Journal of Diseases in Children,* in press.

Reisinger, K. S., Williams, A. F., Wells, J. R., John, C. E., Roberts, T. R., & Podgainy, H. J. The effect of pediatricians' counselling on proper infant restraint use in automobiles. *Pediatrics,* 1981, *67*(2), 201–206.

Richman, N. Behaviour problems in pre-school children: Family and social factors. *British Journal of Psychiatry,* 1977, *131,* 523–527.

Sharrar, W. G., & Linan, L. Teaching behavioral pediatrics. *Program and Abstracts of the 20th Annual Meeting of the Ambulatory Pediatric Association,* 1980, 75.

Shuman, B. J., & Gross, B. D. A model for teaching early child development. *Journal of Medical Education,* 1978, *53,* 497–499.

Smith, R. D. The use of developmental screening tests by primary care pediatricians. *Journal of Pediatrics,* 1978, *93*(3), 524–527.

Starfield, B., & Borkowf, S. Physicians' recognition of complaints made by parents about their children's health. *Pediatrics,* 1969, *43*(2), 168–172.

Stillman, P. L., Sabers, D. L., & Redfield, B. M. Use of trained mothers to teach interviewing skills to first-year medical students: A follow-up study. *Pediatrics,* 1977, *60*(2), 165–169.

Task Force on Pediatric Education. *The future of pediatric education.* Denver: Hirschfeld, 1978.

Toister, R. P., & Worley, L. M. Behavioral aspects of pediatric practice: A survey of practitioners. *Journal of Medical Education,* 1976, *51*(2), 1019–1020.

Vanderpool, N. A., & Parcel, G. S. Evaluation of an elementary school based training activity for first year medical students. *Program and Abstracts of the 19th Annual Meeting of the Ambulatory Pediatric Association,* 1979, 62.

Walker, C. E. Behavioral intervention in a pediatric setting. In J. R. McNamara (Ed.), *Behavioral approaches to medicine: Applications and analysis.* New York: Plenum Press, 1979.

Wright, L. Handling the encopretic child. *Professional Psychology,* 1973, *4*(2), 137–144.

Yule, W. The potential of behavioural treatment in preventing later childhood difficulties. *Behavioural Analysis and Modification,* 1977, *2*(1), 19–32.

V

HEALTH CARE BEHAVIORS, PREVENTION, AND THERAPEUTIC ADHERENCE

10

Pediatric Cardiology

Congenital Disorders and Preventive Cardiology

Thomas J. Coates and Bruce J. Masek

In contrast to adult cardiology, which is dominated by coronary artery disease and hypertension, pediatric cardiology has been concerned traditionally with the study and treatment of congenital disorders. Psychological interests include behavioral management of the child with congenital disorders and preparation of these children for diagnostic and surgical procedures. Two new areas are emerging as important for the pediatric cardiologist: early onset and treatment of elevated blood pressure and primary prevention of risk factors related to acquired cardiovascular disease.

CONGENITAL HEART DISEASE

Congenital heart disease occurs in 1% of all live births; 30,000 cases of congenital heart disease among children are found each year in the United States (Rowland, 1979). Table I presents the major categories of

THOMAS J. COATES • Johns Hopkins University School of Medicine, Baltimore, Maryland 21205. BRUCE J. MASEK • Children's Hospital Medical Center; Harvard Medical School, Boston, Massachusetts 02115. Preparation of this manuscript was supported by Young Investigator Grant No. 1-R23-H224297 to Thomas J. Coates, by Grant Nos. 1-R01-MH 35148, and by a Grant-In-Aid from the American Heart Association, Maryland Affiliate (Jean M. Kan, M.D., Principal Investigator).

Table I. Major Cardiac Malformations

Acyanotic lesions with increased blood flow to the lungs	
Defect causes blood to flow to the lungs in increased quantities; blood is properly oxygenated; pulmonary hypertension is a major problem.	
Ventricular septal defect (VSD)	opening in septum between left and right ventricles
Patent ductus arteriosus (PDA)	blood flow between aorta and pulmonary trunk; this is normal for embryos but normally closes after birth
Atrial septal defect (ASD)	opening between left and right atria
Obstructive lesions	
Generally in children present as obstruction of blood flow from the heart	
Coarctation of the aorta (COA)	narrowing of the descending aorta
Aortic stenosis (AOS)	obstruction of aortic outflow at or below aortic value
Aortic valvular stenosis (ARS)	orifice of aortic value is narrowed
Heart disease associated with left-to-right shunts	
Admixture lesions	a single chamber receives systemic and pulmonary venous returns
Transposition of the great arteries	anatomic reversal of aorta and pulmonary artery
Total anomalous pulmonary venous connection	pulmonary veins join vessel delivering blood to the right atrium
Persistent truncus arteriosus (PTA)	a single arterial blood vessel leaves the heart and delivers blood to pulmonary and systemic circulations
Cyanosis and diminished pulmonary blood flow	
Tetralogy of fallot (TOF)	ventricular septal defect, pulmonary stenosis, right ventrical hypertrophy
Tricuspal atresia (TA)	no communication between right atrium and right ventricle
Pulmonary atresia (LPA)	no direct blood flow from right artery to pulmonary artery

congenital heart disease. Surgical procedures developed during the last 20 to 30 years have increased the survival rate of these children. The basic inadequate oxygenation of the blood, resulting in cyanosis and pulmonary hypertension, are the pathophysiological consequences of congenital heart disease. This has the potential of causing central nervous system (CNS) damage and deterioration of cognitive functions. Congenital heart disorders are also significant in that they may be associated with early family trauma. The child typically must endure arduous medical procedures, and he or she may be regarded as special or different because of his or her unique medical history.

Management of the Infant Cardiac Patient

The New England Regional Infant Cardiac Program (Fyler, 1980) provides one example of improved medical management of the infant cardiac patient. The program is a voluntary association of all hospitals in New England to provide professional education, case identification and referral, transportation and subsidies, and continuing nursery nurse education. The program began in 1968 and continues to the present.

In the early years of this program, there was a 20% increase in the number of cases found, and there was a 50% decrease in infants who died from heart disease without ever reaching a hospital. There was also a consolidation of hospital services. Immediate survival following surgery improved regardless of anomaly. For example, mortality for infants with transposition of the great arteries was reduced from 47% in the first year to 36% in the years 1972 to 1974.

The Experience of Cure: Diagnosis and Surgery

There has emerged a new class of pediatric patient because surgical and medical management procedures have improved. Pediatric cardiac patients are unique in that they and their families have experienced a series of medical diagnoses and interventions for a vital organ. Correction of a cardiac defect usually involves diagnosis through cardiac catheterization, correction by surgery, and postsurgical care. In cardiac catheterization, the infant or child is admitted to the hospital, given a mild anaesthetic, and catheterized while awake. The procedure usually lasts 2 to 3 hours. Surgery, of course, involves more complete preparation and hospitalization.

In adults, anxiety surrounding fearful medical experiences (Janis, 1958) can lead to avoidance of necessary procedures, interference with the conduct of the procedure, and continuation of anxiety thereafter. Preoperative counseling and education with adults and children has improved cooperation and satisfaction (Wolfer & Visintainer, 1975), increased willingness to return to the hospital (Cassell, 1965), and reduced reported anxiety (Melamed & Siegel, 1975; Schmitt & Woolridge, 1973).

Young children undergoing medical interventions might be expected to experience trauma even greater than that afflicting adults. Their understanding of the procedures is less complete than that of adults; they may resent experiencing pain; the environment can be frightening; separation from parents may be hard to tolerate. Many studies to date have suggested significant psychosocial problems in children as a result of stressful medical procedures (e.g., Haller, Talbert, & Dombro, 1967; Petrillo, 1968). Aisenberg, Wolff, Rosenthal, and Nadas (1973) studied the reaction to cardiac catheterization of children 4 to 15 years of age. They reported that nearly all of the youngest, but less than half of the oldest, patients manifested negative behavioral and intrapsychic changes following catheterization. As with many studies in this area, the research is flawed. Ward nurses rated behavior, and intrapsychic states were assessed using projective techniques. Both assessments were biased, however, the former because the nurses knew of the surgical procedure and might expect negative behavior and the latter because projective tests are inherently unreliable (Anastasi, 1965).

Other studies have focused on counseling or education to ameliorate the effects of medical procedures (e.g., Petrillo, 1968; Blumgart & Korsch, 1964; Holt, 1970; Jolly, 1968). Although these studies form an important base for developing future procedures, they lack objective and unbiased data. The necessity is for objective and reliable descriptions of children's reactions to stressful medical procedures and careful evaluations of the efficacy of treatments for helping children to cope with the procedures more successfully.

In 1965, under the title, *The Psychological Responses of Children to Hospitalization and Illness: A Review of the Literature*, Vernon, Foley, Sipowicz, and Schulman (1965) reviewed over 200 articles and books dealing with the impact of hospitalization and medical procedures on the young. Only six studies provided empirical data; many of these studies used rating systems open to bias by the observers. Visintainer and Wolfer (1975), in reviewing the literature, found only two additional empirical studies from 1965 until 1975. Rie, Boverman, Grossman, and Ozoa (1968) tutored children with rheumatic illness about their disease or met with them in group psychotherapeutic meetings. Using clinical interviews, they found that those receiving psychotherapy reported less anxiety and had less concern about hospitalization both immediately and 19 months following treatment. Johnson and Stockdale (1975) used the Palmar Sweat Index to measure anxiety levels of hospitalized children. Children exposed to a puppet show designed to familarize them with hospital routines showed less anxiety by this index.

Visintainer and Wolfer (1975) (see also Wolfer & Visintainer, 1975) used blind ratings of behavioral upset and cooperation during medical procedures to evaluate the efficacy of various procedures for helping young children receiving medical care. Children given a combination of systematic education, rehearsal, and supportive care prior to each stressful experience showed less upset and more cooperation, and their parents showed less anxiety than those of children not receiving these treatments.

Two recent studies show promising progress in reducing stress in children and in documenting outcomes well. Peterson and Shigetomi (1981) taught 2.5 to 10.5 year old children admitted for tonsillectomy about hospitalization and how to use various coping strategies (relaxation, mental imagery, self-talk). A film modeling procedure was also used. Parent, nurse, and observer ratings of behavior, self-reports, and physiological measures (pulse, temperature, latency to void after surgery) were used to assess outcomes. Children trained in coping skills and receiving modeling consistently showed less anxiety across all rating and self-report measures than children receiving only the modeling or coping training.

Naylor, Coates, and Kan (1981) prepared 3- to 6-year-olds for cardiac catheterization by a booklet demonstrating the steps in catheterization and models of coping strategies, and by behavior rehearsal of the procedure and coping strategies. The control group received routine preparation and explanation from the pediatric cardiologist. Outcomes were assessed by direct observation of child behavior during the catheterization and by observations of child behavior at home. Parents were telephoned on specific days prior to and following the catheterization and asked to report occurrence or nonoccurrence of specific behavior. Children receiving the behavioral program showed less anxiety, aggression, and regression (e.g., wetting the bed) at home before and following the catheterization. During the catheterization, treatment subjects cried less, yelled for mother and complained of pain less of the time, and showed a lower rate of motor activity during the catheterization. Parents of treatment subjects were significantly less anxious before and after the catheterization, and child behavior during the catheterization was correlated with parent anxiety.

Psychosocial Adjustment and Cognitive Development of Children with Congenital Heart Disease

There are no definite conclusions that can be drawn about the psychosocial adjustment of children with congenital heart disease. Our review of the existing research literature suggests the following: (1) studies must use controls that do not differ in systematic and perhaps biasing ways from treatment cases; (2) studies must be designed to identify adaptive as well as maladaptive characteristics of children with cyanotic heart disease; (3) objective and unbiased measures of behavior and psychological functioning should be emphasized; and (4) environmental and family characteristics that contribute to adjustment or maladjustment should be identified.

Heart disease can affect the psychological adjustment of those it afflicts. This has been well documented in studies with adult patients (e.g., Lloyd & Cawley, 1978). Stern, Pascale, and Ackerman (1977), followed 68 patients with diagnosed myocardial infarctions for one year. Anxiety and depression were each present in 22% of the sample; 70% of these persons continued to report significant depressive symptomatology one year later. Taussig and her colleagues at Johns Hopkins Hospital documented partially the seemingly good adjustment of some of the patients treated there for congenital heart disease. Taussig, Crocetti, Eshaghpour, Keinonen, Yap, Bachman, Momberger, and Kirk (1971) followed 1,037 surgically treated patients with cyanotic heart disease over 15 years. About 20% of the respondents had achieved top professional ranks. Taussig, Kallman, Nagel, Baumgardner, Momberger, and Kirk (1975) concentrated on 441 patients alive at the beginning of the 15th postoperative year. A total of 8.1% had achieved professional status, 7.6% had received some graduate training, 19.8% had received bachelor degrees, and 13.7% had attended some college. Response bias (33% response rate in the first study and 45.6% in the second study) obviously precludes firm generalizations, but these data do suggest that at least some persons with congenital heart disease can pursue normal lives.

Children with congenital heart disease score lower on standardized tests of intellectual functioning than do children without disease. This retardation is particularly evident in the first three years of life and is more marked in cyanotic than in noncyanotic children (Rasof, Linde, & Dunn, 1967; Linde, Rasof, & Dunn, 1970).

Silbert, Wolff, Mayer, Rosenthal, and Nadas (1969) evaluated the intellectual functioning of 48 children with congenital heart disease between the ages of 4 and 8 years. Children with cyanotic heart disease had significantly lower IQ scores, poorer perceptual-motor task performance, and poorer gross motor coordination than did children with noncyanotic heart disease. Linde and his colleagues (1970) compared IQ scores of cyanotic and acyanotic children who either had surgery for repair of a cardiac defect or did not. The authors found no statistically significant differences between pretest and posttest IQ scores for the four groups. However, the postoperative increase in IQ for the cyanotic-operated group was significant.

It is difficult to interpret whether these findings are due to impaired cognitive or retarded motor functioning as a result of cyanotic heart disease. The developmental tests administered in the first years of life rely heavily on gross motor function. Because children with cardiac defects demonstrate impaired physical capacity, their performance on these tests may be a function of limited physical capacity rather than impaired intellectual ability. Older cyanotic children tend to have normal IQ scores on standard tests, although their results tend to fall in the lower end of the range of normal intelligence (Rasof *et al.*, 1967). Several factors could account for these differences: limited environmental experiences, limited social contacts, numerous hospitalizations, and deficits in school experience, to name a few. In addition, growth retardation is often present in children with congenital heart disease (Linde, Dunn, Schireson, & Rasof, 1967). As a group, children with cyanosis have physical retardation and delayed pubescence.

Emotional Development

The emotional development of children with congenital heart disease has not been well documented. Garson, Williams, and Redford (1974) compared 37 patients who had tetralogy of fallot (mean age = 19.1 years) with normal standards on the Cattell 16 PF Personality Inventory. The patients supposedly were more neurotic than norms on several dimensions: weak superego, dependent, overprotected, less well informed, operating on the basis of feeling and not thought, and lacking ambition. These data are suggestive, but limitations of the test (paper-and-pencil measures rarely correlate with behavior or social interactions) and the comparison group (college students used to derive normed standards) limit the interpretability of the findings. Linde, Rasof, Dunn, and Rabb (1966) compared emotional adjustment of children with cyanotic congenital heart disease (n = 98), acyanotic congenital heart disease (n = 100), normal siblings (n = 81), and normal children from a well-baby clinic (n = 40). The dependent measures were psychological ratings on a number of patient (general adjustment, physical dependence, anxiety, etc.) and parent (e.g., protectiveness, pampering, anxiety, guilt) variables. Children with cyanotic heart disease showed few differences from the other groups in degree of general adjustment and in degree of independence. The most striking finding from the later study is that poor psychological adjustment and anxiety in the child with heart disease were related more to maternal anxiety and pampering than to the child's degree of incapacity or severity of disease. These results must be interpreted cautiously, however. There were systematic age and IQ differences between cases and controls. The major dependent variables were subjective ratings; psychologists making these ratings were not blinded. The possibility of bias and also correlation among measures is obvious. Thus, it is unclear at present whether family life is disturbed or disrupted by congenital heart disease. One study suggests an affirmative response, while a second suggests a negative response.

A recent report (Lavigne & Ryan, 1979) used more objective measures to compare siblings of children treated in several specialty areas: pediatric hematology, cardiology, and plastic surgery. The adjustment measure was an objective paper-and-pencil test of children's emotional and behavioral problems. The siblings of patients were more likely to show symptoms of irritability, social withdrawal, fear and inhibition than normal controls.

Two factors could moderate the impact of the disease on the child and on the family. First, studies to date have not differentiated children in terms of residual medical difficulty. It seems quite plausible that children and families might differ in emotional, behavioral, and intellectual

Table II. Grouping of Congenital Disorders by Functional Criteria[a]

Group A.	Normal functional cardiac status
PDA	Postoperative (PO) with no residual defect
ASD	PO no residual defect, normal sinus rhythm (NSR)
VSD	PO no residual defect, NSR
VSD	Spontaneous closure
TOF	Corrected, no residual defect, NSR
Other	Anomalous subclavian artery, right aortic arch, dextrocardia and situs inversus, other defects of embryologic and possibly genetic significance but not affecting cardiac function
Group B.	Mild cardiac defect
VSD	Qp:Qs 2:1 with normal PA pressure, NSR unoperated or PO with residual shunt
PSV	RV–PA pressure difference 50 mm or less (unoperated or PO)
COA	PO normal blood pressure, arm to leg BP difference 20 mmHg
ECD	PO with mild residual mitral insufficiency
MIC	Congenital mitral insufficiency due to prolapse or other causes; mild or severe arrhythmia
TOF	PO mild residual pulmonic stenosis = insufficiency, = small VSD, = RBBB
Other	Bicuspid aortic valve (BVB) with mild insufficiency = LV-aortic pressure difference of 20 mm or less: small atrial or other shunts (PO or unoperated) with Qp:Qs 1.5:1 or less; peripheral pulmonic stenosis (PPS) mild.
Group C.	Moderate cardiac defect
VSD	Qp:Qs 2:1 and/or with aortic insufficiency, pulmonic stenosis or anomalous RV muscle bundle
PSV	RV–PA pressure difference 50 mm Hg (unoperated or PO)

Table II. (Continued)

COA	(Unoperated or PO) with BP differences arms and legs 30 mm Hg and/or arm BP greater than 150/95
TOF	PO with RV–PA pressure difference 50 mm Hg, grossly dilated or calcified outflow patch, aortic or tricuspid insufficiency or significant arrhythmias
ASV	Unoperated or PO with LV-aortic pressure difference 20 mm Hg = aortic insufficiency
TGV	Post-Mustard = residual shunts or arrhythmias
Other	Unoperated (or PO) PDA, ASD, AV canal, etc., requiring surgery or reoperation: aortic insufficiency, moderate. Arrhythmias: sick sinus syndrome, other. Cardiomegaly with C/T ration 60% = with any congenital lesion.
Group D.	Severe cardiac defect
Cyanotic	
TOF	Unoperated or with palliative shunt, PO open repair with homograft, prosthetic valve, or with severe arrhythmia
TGV	All not in Group C
SIV	All types with or without prior shunts
PROD	Eisenmenger reaction or primary pulmonary hypertension
Other	Truncus variants, pulmonary atresia with VSD, rate cyanotic lesion
Acyanotic	
Cardiomyopathy including hypertrophic subaortic stenosis (IHS) or following fibroelastosis or ligation anomalous left coronary artery, other	
LV outflow obstruction, e.g., subaortic tunnel or other unrelieved aortic stenosis with LV-aortic pressure difference of 80 mm Hg or greater	

[a]Adapted from Neill & Haroutunian (1977).

development and adjustment according to the degree to which medical complications continue beyond medical interventions. Specific medical diagnoses (as in Table I) appear to be less important in terms of psychological adjustment and cognitive and behavioral development than current functioning. Table II presents one classification of subjects based on current cardiac status. Group A represents those who presented with a variety of lesions but who now show normal functional cardiac status. Groups B and C demonstrate mild and moderate cardiac defect, and Group D demonstrates continuing severe cardiac defect. To date, no studies have examined the psychological, behavioral, or cognitive status of patients in terms of their current cardiac function. Rather, studies have focused on diagnostic classification regardless of outcome.

A second factor moderating outcomes might involve the reactions and coping skills of significant family members. Previous and current congenital heart disease may affect a child adversely only if parents and siblings react or behave in ways to promote lack of adjustment or fail to act in ways that will facilitate development. Not all families may be at risk for emotional and behavioral problems. Because the risk may be differential, it might be useful to study factors related to specific kinds of risks and develop interventions specifically for these families and problems.

Neurological Functioning

Complications of the central nervous system have been reported frequently in adult patients but not in children following cardiac surgery. Dysfunction reported in adults include abnormal neurological signs such as loss of deep tendon reflex; disorders of speech; facial nerve palsies; organic brain syndrome; and psychiatric disturbances such as delerium, confusion, and loss of memory.

In a study of 100 open-heart patients using extracorporeal circulation, Tufo, Ostfeld, and Shekelle (1970) found that *43% of all survivors developed neurological abnormalities* (visual field disturbances, snout and sucking reflexes, delerium, and confusion). Only 15% of the patients still showed signs of cerebral dysfunction upon discharge; the dysfunction was only transitory. Egerton, Egerton, and Kay (1963) reported that 10 of 16 open-heart patients developed moderate to severe neurological gross motor disturbances that disappeared by the fourth month. Kornfield, Zimberg, and Malm (1965) found that acute psychosis attributable to organic brain syndrome occurred in 39% of a sample of 99 adult patients. The authors reported few long-term complications but speculated that modification of the postoperative recovery environment could help to ameliorate the delirium and confusion experienced by these patients.

Neurological dysfunction is not a factor in postoperative recovery of children with congenital heart defects. Egerton and Kay (1964) reported that 35 of 36 children treated surgically for repair of a cardiac lesion experienced no neurological disturbance postoperatively. Kornfield *et al.* (1965) found no incidence of neurological complication postoperatively in their study of 20 children with congenital heart defects. Submaramian, Vlad, Fischer, and Cohen (1976) reported that 25 of 36 children were neurologically normal following cardiac surgery. Language delay, specific learning disability, and one case of temporal lobe seizure accounted for the remaining cases.

ADULT CARDIOVASCULAR DISEASE MAY BE PREVENTABLE

Cardiovascular disease is the leading cause of death and premature death in adults in the United States. In 1975, 52.5% of all deaths in the United States were attributed to diseases of the heart and blood vessels; 994,513 individuals died because of heart disease (American Heart Association, 1976). A number of behavior patterns are related to cardiovascular disease. Cigarette-smoking, elevated blood pressure, elevated levels of serum cholesterol—especially low density lipoproteins (LDL)—and diabetes increase risk for premature cardiovascular disease (Blackburn, 1974; Kuller, 1976; Kannel & McGee, 1979; Pooling Research Group, 1978). High levels of high density lipoproteins (HDL) decrease risk of cardiovascular disease (Miller, 1979). The Type A (coronary-prone) behavior pattern and elevated levels of serum triglyceride are important additional risk factors (Blackburn, 1974, 1978; Haynes, Feinleib, & Kannel, 1980).

Specific behavior patterns may be associated with increases or decreases in cardiovascular risk factors. Blood pressure, for example, is influenced by overweight, lack of physical activity, and excessive dietary salt intake (Task Force on Blood Pressure Control, 1977). Possible relationships between transient or chronic stress, Type A behavior and blood pressure have also been documented (Shapiro & Surwit, 1976; Coates, Parker, & Kolodner, 1982; Voors, Sklov, Wolf, Hunter, & Berenson, 1982). Increased HDL is influenced by increases in physical activity, reductions in dietary saturated fat, increases in dietary fiber, and decreases in weight (Blackburn, 1974, 1978).

If cardiovascular risk factors are related to specific behavior patterns, then it may be possible to prevent cardiovascular disease if people can be helped to refrain from smoking, maintain ideal weight, increase consumption of dietary fiber, decrease consumption of dietary cholesterol, saturated fat, salt, and sugar, increase physical activity, reduce stress, and modify coronary-prone behavior.

The degree to which children and adolescents should be encouraged to change common life patterns remains controversial. Some assert that all American young people are at increased risk for cardiovascular disease relative to children in other countries and should be encouraged to change eating, exercise, smoking, and stress-related habits. Kannel and Dawber (1972) stated quite strongly that "atherosclerosis, although uncommonly observed before age 40, has its onset in infancy and childhood." Although acquired cardiovascular diseases occurs in adults, it is clear that it does not develop overnight. Presumably, starting young will help people begin and maintain healthful living, and working with the young can be prudent and beneficial.

Elevated Blood Pressure

Primary ("essential") hypertension is a major health problem among adults, affecting approximately 20% of the population over 18 years of age (Kannel, 1971; Stamler, Schoenberger, Shekelle, & Stamler, 1974). Hypertension is the most common cause of stroke and a major risk factor for coronary heart disease (Blumenthal, 1978). Recent studies have demonstrated that elevated blood pressure in juveniles (age 10–18) usually is not secondary but rather is primary (Kilcoyne, 1978; Voors, Webber, & Berenson, 1978). Important strides have been made in the past 10 years in describing the prevalence and epidemiology of juvenile blood pressure. Studies of the etiology, pathophysiology, natural history, and clinical management of juvenile primary hypertension are beginning (Torok, 1979).

The norms published by the Task Force Report on Blood Pressure Control in Children (1977) are used currently by most investigators and clinicians to evaluate blood pressure status in young persons. These norms are based on the combined data from the Muscatine Study, the Mayo Clinic, and Miami, Florida. Pressures were determined using a single casual blood pressure. By contrast, the data in the Bogalusa Heart Study (Voors, Foster, Frerichs, Webber, & Berenson, 1976) were obtained using nine readings taken over 90 minutes. The readings were taken at least 5 minutes apart by three trained observers. Table III presents mean blood pressures for different ages from these studies. Neither set of norms can necessarily be regarded as correct, but a choice between them has important implications for treatment. These different methods of blood pressure reading yield widely different average blood pressures and cutoff points for determining who is to be considered hypertensive. Which set should be used for determination of persons in need of treatment remains an empirical question, possibly to be determined by the degree of hypertension observed in children in different blood pressure strata.

Measurement and Evaluation of Juvenile Hypertension

The estimation of human blood pressure is subject both to human and instrumental inaccuracies. Casual observation of blood pressure is often confounded by (1) physiologic variables such as distended bladder or full stomach (resulting in a rise in blood pressure); (2) environmental

Table III. Mean Blood Pressures from Different Studies

		Age					
		5		10		15	
		Systolic	Diastolic	Systolic	Diastolic	Systolic	Diastolic
Bogalusa (Voors *et al.*, 1976)	M F	97.6	62.8	98.6	61.3	108.8	67.6
Muscatine (Lauer *et al.*, 1975)	M F	94.4 91.8	63.1 61.3	109.3 108.6	72.8 72.3	122.3 116.3	78.4 78.3
Miami (Task Force Report, 1977)	M F	99.8 97.3	63.9 64.8	— —	— —	— —	— —
Chicago (Miller & Shekelle, 1976)[a]	M F	— —	— —	— —	— —	126.1 119.7	68.1 68.1
National Center for Health Statistics (1973, 1977)[b]	M F	105.5[c] 105.7	65.0 65.3	110.9 113.3	66.8 68.3	133.8 127.9	75.8 75.3

[a]Data given for white males and females only; subjects include 15- and 16-year-old students.
[b]White subjects only.
[c]6-year-olds.

variables such as time of day, ambient temperature, and exposure to cold; and (3) patient variables such as state of relaxation, expectations, and prior experience with blood pressure measurement.

Most blood pressure measurements are made using the auscultatory method. The competency of the observer is the greatest single problem in hypertension screening. Highly trained observers can vary up to 16 mmHg systolic and 45 mmHg diastolic in reading a prerecorded blood pressure (Rose, Holland, & Crowley, 1969). Sources of observer error that have been well documented include: (1) auditory acuity, (2) bias toward over-or underestimation, (3) preference for 0 or 5 as the terminal digit, (4) reaction time differences, (5) rate of inflation and deflation of the cuff, and (6) proper cuff size and placement (Maxwell, 1974). Choice of diastolic Korotkoff phase, that is, Phase IV (muffling of the sounds), or Phase V (complete disappearance of the sounds), introduces further variability in reported pressures.

The implications are obvious for epidemiological and clinical studies. Rigid protocols must be established for training observers and for checking their accuracy continuously. Without quality controls such as those used in the Bogalusa Study (Berenson, 1980), believability of results declines considerably.

Semi-automated devices using the Doppler principle or transduction of pulsitile blood flow have been developed to control for observer error.

These devices do not provide a perfect solution. They are calibrated against an independent observer, and precision of blood pressure measurement is also dependent on proper cuff size and placement. Cost is a major disadvantage to their widespread use at present.

The evaluation of high blood pressure in young people requires repeated measurement within a single session and repeated observations made on successive examination. Voors *et al.* (1978) recommend that three blood pressure readings be taken monthly for 3 or 4 serial examinations to determine a consistent percentile status. This point appears to be particularly important because of the extreme lability of blood pressure in children (Loggie, 1977; Torok, 1979).

Epidemiology

The prevalence of juvenile hypertension based on an age- and sex-adjusted percentile criterion is estimated at between 3% and 10% (Dube, Kapoor, Ratner, & Tunick, 1975; Voors, Foster, Frerichs, Webber, & Berenson, 1976). Prevalence estimates based on fixed blood pressure level (usually greater than 140 mmHg for systolic and greater than 90 mmHg for diastolic pressure) are only slightly higher (Kilcoyne, 1978). Data on blood pressure measurements from studies of normal children vary markedly, however. Again, differences in blood pressure measurement appear to be the chief reason for the variability (Voors *et al.*, 1978). The percentage of juvenile hypertension diagnosed as primary has ranged widely from 6% (Still & Cotton, 1967) to 91% (Londe, Bourgoyne, Robson, & Goldring, 1971) and averages 26% (Londe, 1978), although the percent attributed to primary hypertension has been increasing in recent years.

The issue of tracking is especially important in studying blood pressure in adolescents. Juvenile blood pressures stay at a relatively fixed point on the age- and sex-adjusted percentile distribution curve throughout childhood and into adult life (Voors, Webber, & Berenson, 1979). In the Bogalusa Heart Study, blood pressures were collected on 3,800 children over 4 years. Correlation coefficients from year 1 to year 2 ranged from 0.62 to 0.73 systolic and from 0.41 to 0.54 diastolic. Only a slight drop in the correlation was noted from year 1 to year 4. By the fourth assessment, 33% of those who were initially in the top decile remained there; 60% of those high in 3 successive years were high in year 4. Up to 71% systolic and 45% of diastolic variability in year 4 blood pressure measurements were accounted for by previous blood pressure and changes in anthropometric measures (e.g., changes in height and weight). Although these multiple correlations are high, there still remain questions regarding why some subjects who are initially high regress toward normalcy while others do not.

Because the mortality rate increases with duration of illness, vigorous intervention programs at an early age are warranted. Longitudinal studies on adolescents indicate that blood pressure levels found at the initial examination are related closely to hypertension in adulthood (Paffenbarger, Thorne, & Wing, 1968) and further support the need for early intervention before arterial wall or renal complications develop.

Pathophysiology

The exact causes of primary hypertension have not yet been established firmly. Numerous factors regulate blood pressure. Autonomic nervous system regulation of peripheral vascular resistance, heart rate, and cardiac output is one control mechanism. The renin-angiotensin system, with its capacity to constrict peripheral arterioles, appears to be another important mechanism. However, understanding of this system is far from complete (Kowarski, Edwin, Akesode, Pietrowski, & Hamilton, 1978). Adrenal corticosteroids may also exert an important role in the development of hypertension, but research is just beginning to elaborate humoral control mechanisms (Hamilton, Zadik, Edwin, Hamilton, & Kowarski, 1979). The kidneys play a major role, more central than previously thought, in long-term control of blood pressure. Because the kidneys can decrease and increase blood volume as needed, they set baseline blood pressure. The autonomic nervous system, by stimulating the kidneys, can set the baseline at a higher level, as can angiotensin. The kidney in effect functions as a servomechanism for controlling pressure, and other factors influence the level at which the kidney mechanism operates.

Hemodynamic studies suggest that abnormalities in the myocardium may be involved in the pathogenesis of juvenile primary hypertension. Zahka, Neill, Kidd, Cutiletta, and Cutiletta 1981 have demonstrated a high incidence of left ventricular hypertrophy in juveniles with mild hypertension. Heart rate, cardiac output, and stroke volume of the hypertensive subjects were similar to normotensive controls, which suggests that hemodynamic circulatory state is not primarily involved. They hypothesized, on the basis of animal models, that increased muscle mass in the left ventricle produces greater pressure as it pushes blood through the circulatory system.

Hemodynamic response to stress has also been postulated as important in the etiology of hypertension and as a key variable in predicting essential hypertension later in life. It has been hypothesized that hypertension develops in young people who respond to environmental or emotional stress by high cardiac output. When faced with a challenging stimulus, the organism responds with increased output of epinephrine and norepinephrine. The catecholamines, together with direct sympathetic stimulation, cause a rise in peripheral resistance and cardiac output. Acute increases in blood pressure result. It is hypothesized that fixed hypertension occurs with progressive changes in anatomical structure and with blood pressure regulatory mechanisms.

The Coronary-Prone Behavior Pattern

The Western Collaborative Group Study established that the Type A or coronary-prone behavior pattern is associated with increased risk of cardiovascular disease (Rosenman, Brand, Sholtz, & Friedman, 1976). Type A subjects exhibited 2.37 times the rate of new coronary heart disease that was observed in their Type B counterparts. After adjustment for the four traditional risk factors (age, cholesterol, systolic blood pressure, smoking), the approximate relative risk for Type A men was 1.97. Jenkins, Zyzanski, and Rosenman (1976) reported that Type A behavior was associated with increased risk of reinfarction among persons already having clinical coronary disease. The Type A behavior pattern has also been associated with increased severity of atherosclerosis in two separate studies as confirmed by coronary angiography (Zyzanski, Jenkins, Ryan, Flessas, & Everist, 1976; Blumenthal, Williams, King, Schanberg, & Thompson, 1978).

Finally, Haynes *et al.*, (1980), in an 8-year prospective study, established that Type A behavior as measured by the Framingham Type A scale predicted coronary heart disease, myocardial infarction, and angina among men and women.

Matthews (1978) has been especially prominent in investigating developmental antecedents of Type A behavior in children and adolescents. The Matthews Youth Test for Health (MYTH; Matthews, 1976) contains 17 5-point rating scales of children's competitiveness, impatience, and aggression as rated by classroom teachers. The test is reliable (interrate reliability = .83; Cronbach's alpha = .90), and data on validity are currently being collected and assessed. Matthews and Angulo (1980) tested the construct validity of the MYTH in a subsample of children who were challenged to win a car race against an experimenter. Type A children won the race against a female experimenter but not against a male experimenter by a wider margin than did Type B children. Type A children also demonstrated aggression against a doll earlier and were more impatient than Type B children throughout the session. Matthews and Krantz (1976) reported that the hard-driving and competive components of the Type A pattern have a modest genetic component. Matthews, Glass, and Richins (1977) found that mothers gave fewer positive evaluations of task performance to Type A children than to Type B children, and that Type A children were pushed to try harder more often than Type B children. In a third study, Matthews (1977) found that Type A children, relative to Type B children, elicited more positive evaluations and positive pushes (e.g., "You did so well. Why not try for five?"), from Type B caregivers. Other studies have reported modest similarities between parents and children (Bortner, Rosenman, & Friedman, 1970), suggesting a modeling effect. The data suggest a reciprocal effect. Children will become more competitive when urged to behave this way and when reinforced for competitiveness. At the same time, competitive and impatient children elicit a greater frequency of positive evaluations and urges to perform better than the less competitive children. Matthews (1978) also drew from the literature on achievement motivation, aggression, and time urgency to suggest how behavioral clusters could be transmitted through modeling and reinforcement.

Coates, Parker, and Kolodner (1982) studied the relation of traditional factors (body weight, age, parental hypertension) and Type A in predicting baseline blood pressure and blood pressure response to stress in black and white male adolescents. Each subject completed the Bortner Type A Scale (Spiga & Peterson, 1981) adapted for adolescents and then was escorted into an experimenter room where an automated blood pressure cuff was attached to the left arm and instructions were conveyed. Blood pressure was monitored every 3 minutes during a 15-minute baseline period. Following this, each subject spend 15 minutes working with the Alluisi Performance Battery (Emurian, Emurian, & Brady, 1978). This performance battery requires subjects to complete 5 tasks simultaneously: probability monitoring, horizontal addition and subtraction, matching to sample histograms, detecting when a stationary signal moves, and detecting when a moving signal becomes stationary.

Table IV presents multiple regression equations showing predictors of baseline systolic and diastolic blood pressure. Ponderosity, age, and parental hypertension were important predictors of systolic and diastolic blood pressure along with Type A behavior. Table V presents equations using changes in blood pressure as the dependent variable. Behavioral factors entered most strongly in predicting relative and absolute change. Performance on the Alluisi was negatively related to blood pressure increases, while Type A was positively related to blood pressure increases.

Table IV. Dependent Variable: Baseline Blood Pressure

Independent variable	Beta	R
Systolic[a]		
Weight/height^2	.410	.47
Parental hypertension	.298	.55
Age	.228	.58
Job involvement (JAS)	.212	.62
Type A (Bortner)	.149	.64
Diastolic[b]		
Weight/height^2	.608	.56
Type A (Bortner)	.286	.64
Hard-driving (JAS)	.307	.67
Age	.275	.69
Type A (JAS)	−.295	.71

[a] $R^2 = .41$; $F = 4.62$ ($p < .01$).
[b] $R^2 = .50$; $F = 7.05$ ($p < .001$).

Blood Lipids

The Muscatine epidemiologic study (Lauer, Connor, Leaverton, Reiter, & Clarke, 1975) has characterized both the cross-sectional and longitudinal characteristics of total serum cholesterol. Serum cholesterols are similar for children at all ages with the mean cholesterol being 182 mg/dl. Twenty-four percent of children had total cholesterol levels greater than 200 mg/dl, 9% were greater than 220 mg/dl, and 3% were greater than 240 mg/dl. Tracking of serum cholesterol was quite high. In Muscatine, 1,502 school-age children have been measured three or four times over 6 years. Correlations between repeated cholesterol levels were .60 (Clarke, Schrott, Leaverton, Connor, & Lauer, 1978).

The Bogalusa study has provided some data on cholesterol fractions and racial differences. Black children have greater total serum cholesterol and lower triglyceride levels than do white children (Frerichs, Srinivasan, Webber, & Berenson, 1976). Blacks have higher levels of HDL cholesterol; this accounts for the racial difference observed in total cholesterol. Between the ages of 5 and 14 years, black children have 5% greater levels of total cholesterol, 14% greater levels of HDL cholesterol, approximately equal levels of LDL cholesterol, and 18% lower levels of VLDL cholesterol (Srinivasan, Frerichs, Webber, & Berenson, 1976). If HDL is indeed protective, blacks may have a metabolic protective edge for risk of cardiovascular disease as early as 4 to 5 years of age.

Table V. Change in Blood Pressure

Independent variable	Beta	*R*
Change in systolic blood pressure[a]		
Bortner Type A	.132	.36
Alluisi points	−.107	.53
Weight/height^2	3.13	.57
Parental hypertension	−.313	.61
Relative change in systolic blood pressure[b]		
Bortner Type A	.372	.36
Alluisi points	−.404	.53
Weight/height^2	.254	.57
Parental hypertension	−.214	.60
Age	.078	.61
Relative change in diastolic blood pressure[c]		
Weight/height^2	−5.20	.32
Age	.115	.38
Race	.195	.39

[a] $R^2 = .38$; $F = 4.93$ ($p < .01$).
[b] $R^2 = .381$; $F = 4.06$ ($p < .01$).
[c] $R^2 = .16$; $F = 3.62$ ($p < .05$).

Smoking

Adoption of the smoking habit is an adolescent phenomenon; few take up the habit after age 25. Data from the most recent National Institute of Education survey of adolescents show that young adolescents are still taking up the smoking habit (Green, 1979). Table VI presents the most recent estimates of regular adolescent smokers by age. Smoking among adolescent males has decreased in recent years, whereas smoking among adolescent females has remained at levels observed in previous surveys (Green, 1980).

Some data on racial differences in smoking onset have emerged

Table VI. Percentage of Adolescents Reporting Regular Smoking[a]

	12–14 yrs	15–16 yrs	17–18 yrs
Males	5.2%	13.5%	19.3%
Females	4.3%	11.8%	20.2%

[a] Adapted from Green (1979).

from the Bogalusa study. Hunter, Webber, and Berenson (1980) administered a smoking questionnaire to 3,014 children ages 8 to 17. Venipuncture for assessment of blood lipids also permitted drawing a sample for assessment of plasma thiocyanate. Using self-reported data, white boys adopted smoking earlier (32% of 14- to 15-year-olds) than the other racial/sex groups. White girls caught up to males by 16 to 17 years (41% adopted). Black boys and girls lagged behind in taking up the smoking habit and never did smoke as much as whites.

Predisposing Factors

Obesity

The Muscatine Study reported that obese children were overrepresented in the upper ends of the distributions of blood lipids and blood pressure. Lauer *et al.* (1975) reported that 17.8% of those whose skinfold thicknesses exceeded the 90th percentile were at or above the 90th percentile in serum cholesterol; 23.5% were at or above the 90th percentile for triglyceride. Clarke *et al.* (1978) reported relationships between lipoprotein fractions and ponderosity. Obese boys showed significantly higher total and LDL cholesterol and slightly lower HDL cholesterol. Obese girls showed no increase in total cholesterol but had significantly lower HDL and higher LDL/HDL ratios.

In the Bogalusa study, white children showed higher levels of serum triglyceride and pre-beta lipoproteins and lower levels of alpha lipoprotein than did blacks (Srinivasan *et al.*, 1976). White children also showed greater adiposity than black children (Voors, Webber, Harsha, Frank, & Berenson, in press).

Frerichs, Webber, Voors, Srinivasan, and Berenson (1979) selected a stratified random sample of 278 children aged 7 to 15 years from the Bogalusa population. A questionnaire on habitual physical activity found a negative correlation between reported exercise and obesity, confirming previous observations that obese children are less active than nonobese children (Waxman & Stunkard, 1980; Bullen, Reid, & Mayer, 1964). White children showed a stronger relationship between adiposity and serum triglycerides and the level of the glucose tolerance curve. The investigators hypothesized that fat transport in whites may differ from that observed in blacks.

Epidemiological studies have found consistently that increases in blood pressure are correlated with increases in weight. These relationships hold in black and white children and across the entire range of blood pressures and age groups (Dube *et al.*, 1975; Holland & Beresford, 1975; Miller & Shekelle, 1976; Voors *et al.*, 1976). In the Muscatine study, for example, students in the upper decile of relative weight were overrepresented in the upper end of the blood pressure distribution: 28.6% had systolic pressures above the 90th percentile and 28.4% had diastolic pressures above the 90th percentile. In the Bogalusa study, correlations between body weight and systolic/diastolic blood pressures were .54/.48. The ponderosity index consistently entered first in a stepwise multiple regression equation in predicting systolic and diastolic blood pressure among all age groups (Voors *et al.*, 1978).

Parents

In the 1976–1977 screen of the Bogalusa Heart Study, 80% (n = 4,074) of the eligible students participated in the study. Parental history was obtained in the consent form preceding the examination. The children of parents with a history of heart attack were significantly heavier and had significantly higher pre-beta lipoprotein concentrations than did children of parents without history of heart attack. Although not statistically significant, there was a trend for children of parents who had had heart attacks to have thicker triceps skinfold, higher blood pressures, higher total cholesterol, triglyceride and beta-lipoprotein, and lower alpha-lipoprotein values. Children whose parents had a history of high blood pressure had significantly greater weight and skinfold thickness than children whose parents did not have a history of high blood pressure. These children already had significantly higher systolic and diastolic blood pressure levels.

The well-known existence of familial clustering of blood pressure among parents and siblings appears to be stable over time and is present at all ages (Biron, Mongeau, & Bertrand, 1975). Twin and adoptee studies have established the role of heredity in accounting for significant proportions of the variance in basal blood pressure levels (Feinleib, Garrison, Borhani, Rosenman, & Christian, 1975). Coates, Parker, and Kolodner (in press) found that both parental hypertension and overweight contributed to variance in baseline blood pressures of black and white male adolescents.

It is a commonly documented phenomenon that children are more likely to take up the smoking habit if parents smoke (Coates, Stephen, Perry, & Schwartz, 1980). In fact, parental smoking appears to be more strongly related to smoking onset than peer factors, even though the latter is given primary attention in the common literature (Green, 1980).

Diet

The role of diet in increasing risk factors for disease remains quite controversial (National Academy of Sciences, 1980). During one total school population screening in the Bogalusa Heart Study, 24-hour dietary recall data were collected on 50% of the 10-year-olds (Frank, Berenson, Schilling, & Moore, 1977; Frank, Voors, Schilling, & Berenson, 1977). Snack foods accounted for 34% of total calories. School lunches and dietary intakes were high in fat (40%), high in calories (refined sugar accounted for 50% of carbohydrates), and high in cholesterol. Dietary intakes were correlated with serum lipids. Moreover, when groups at each end of the serum cholesterol distributions were compared, there were significant differences between them in dietary intake of saturated fat and cholesterol.

Diet has never been related to interindividual differences in obesity. In fact, it has been asserted that the obese and nonobese differ in physical activity and not in customary food habits (Huenemann, Hampton, Behnke, Schapiro, & Mitchell, 1974). Waxman and Stunkard (1980) conducted behavioral observations on obese children and on a nonobese sibling at home and a friend at school. Documented for the first time, possibly because direct behavioral observations were used, were food-intake differences between the obese and nonobese child. Both physical activity and caloric intake appear to be responsible at least for the maintenance of obesity, at least in young children.

PROMISING TRENDS FOR THE PREVENTION OF CARDIOVASCULAR DISEASE

Although helping adolescents to live differently may hold great promise for promoting health, building programs that are effective may be more of a challenge with young persons than with adults. Children and adolescents may have no immediate health hazard to prompt consideration of change. It cannot be assumed that a healthful life-style will persist once it is begun.

Reducing Elevated Blood Pressure

A collaborative effort between the Division of Pediatric Cardiology, the Division of Behavioral Biology, and the Division of Health Education at the Johns Hopkins Medical Institution is in progress. Three interventions are being evaluated: medical management, health education, and biofeedback/relaxation.

In an initial study of health education and a psychophysiological approach, 20 adolescents, (10 males and 10 females), ages 12–18, were referred from the Johns Hopkins Pediatric Blood Pressure Center. Eligibility for inclusion in the study was based on clinic observations of systolic or diastolic pressure levels above the 90th percentile for sex and age using the norms from the Task Force Report on Blood Pressure Control (1977).

The experimental design of the treatment program called for 10 weekly visits over a three-month period. This was reduced to 6 biweekly visits for subjects from outlying areas. During each session, subjects were asked to engage in Bensonian meditative relaxation. They were instructed to practice meditation regularly at home. Before, during, and following each 10-minute meditation period, several measures of physiological activity were recorded. Blood pressure was recorded every 5 minutes using the Dinamap Automated Blood Pressure Recorder. Pulse transit time was monitored continuously from the R-wave of an EKG signal and the radial artery pressure pulse. Heart rate was monitored continuously from the EKG, and digital skin temperature was monitored from a thermistor taped to an index finger. A computer system recorded averages of each measure over successive one-minute intervals. The subject was provided with feedback for changes in blood pressure and/or skin temperature during the meditation periods. In addition, these subjects were provided with a cuff and stethoscope, trained to take their own blood pressure, and requested to keep daily records of home pressure levels in a notebook. Finally, each subject was requested to minimize use of table salt and to record the occurrence of salting in a notebook.

The data from 14 subjects who completed the protocol showed that meditation was associated with physiological effects. These effects varied both within and between subjects, primarily as a function of initial levels of physiological arousal. Small but significant increases in skin temperature were associated with meditation. Decreases in blood pressure (mean = 4/2 mmHg) were observed during meditation, with or without accompanying changes in heart rate. Pulse transit time measures did not correlate significantly with blood pressure variations.

Arterial blood pressure levels decreased across sessions in 11 of 14 subjects (mean = 9.2 ± 0.7 mmHg; $p < .01$). That these effects were not due simply to habituation to the clinic environment was suggested by the fact that there was a significant negative correlation between magnitude of blood pressure decrease and body weight ($r = -.61$). Two of the 3 most obese subjects showed no blood pressure decreases. (No such correlations were observed between blood pressure changes and sex or age.) Compliance in recording home blood pressure levels and salt usage was variable and generally related to socioeconomic level. Follow-up studies are in progress, combining psychophysiological procedures with systematic behavior modification of salt and food intake.

Coates (1980) developed a variation of relaxation therapies for elevated blood pressure termed *stress management training*. Adolescents with elevated blood pressure first are taught progressive muscle relaxation and are urged to practice it at home once a day. Following this, they are trained in identifying behavioral, cognitive, and physiological reactions to stress. Finally, they are taught several strategies for reducing stress when it occurs.

Data evaluating the program are quite preliminary, but an example from one case study illustrates the techniques and outcomes (see Figure 1). The patient was a 14-year-old white female whose clinic blood pressures were consistently above the 90th percentile (using the norms from

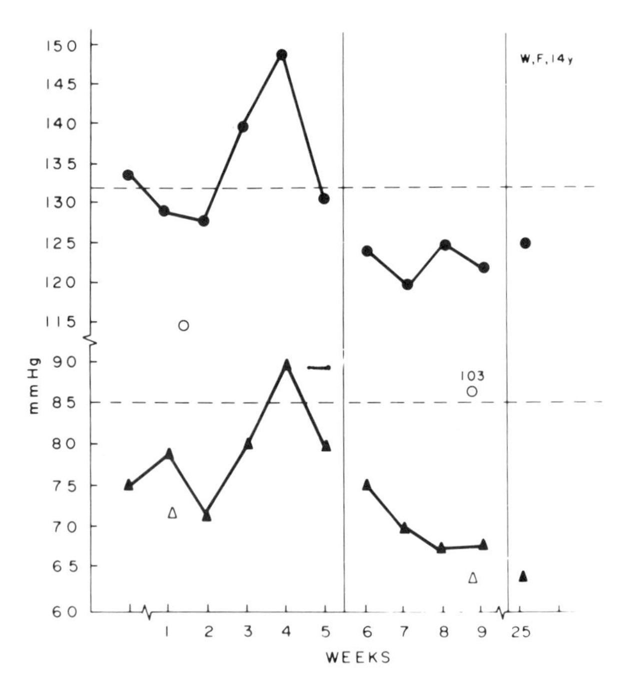

Figure 1. Average blood pressures (systolic and diastolic) for white female, 14 years of age, at baseline (weeks 1–5) during treatment (weeks 6–9), and at follow-up (week 25). Each data point represents the average of 3 successive blood pressures taken 5 minutes apart.

the Task Force on Pediatric Blood Pressure, 1977). She was referred to us by a school nurse who took her blood pressure when the patient complained of frequent headaches. Baseline systolic and diastolic pressures showed considerable variability, the former hovering around the 90th percentile. Training was associated with lowering of blood pressure, and the lowered levels were maintained at follow-up.

We also attempted to evaluate the program using automated blood pressures taken every $7\frac{1}{2}$ minutes in the natural environment (Horan, Kennedy, & Padgett, 1981). Results before and after treatment for one subject are presented in Figures 2 and 3. Both systolic and diastolic pressures were lowered, together with variances associated with each. Distributions of both systolic and diastolic pressures were shifted to the lower end of the distributions for both variables.

One further line of study merits attention. In view of the results of Coates, Parker, and Kolodner (1982), training students to increase competency in specific tasks (e.g., math) or to modify Type A characteristics might prove beneficial in promoting acute and chronic blood pressure decreases.

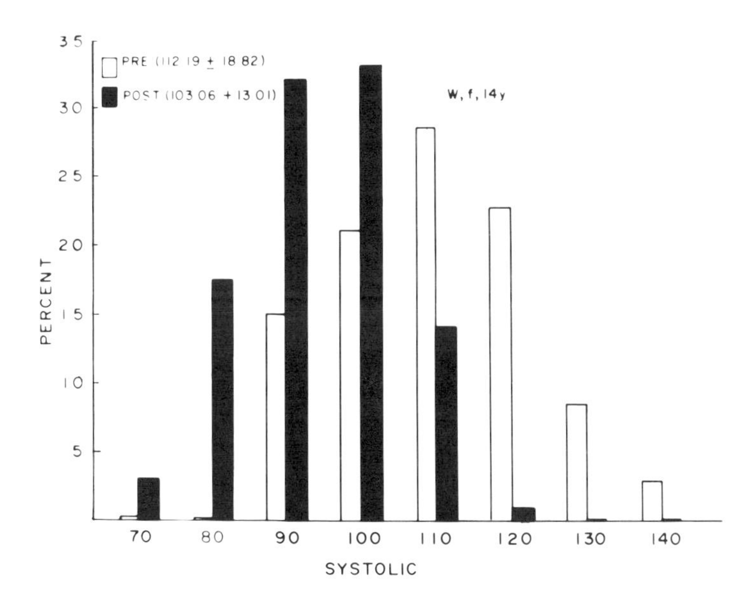

Figure 2. Results of 24-hour blood pressure monitoring presented in terms of percentage of readings within each blood pressure range. Data for pre and for post represent readings for one 24-hour period.

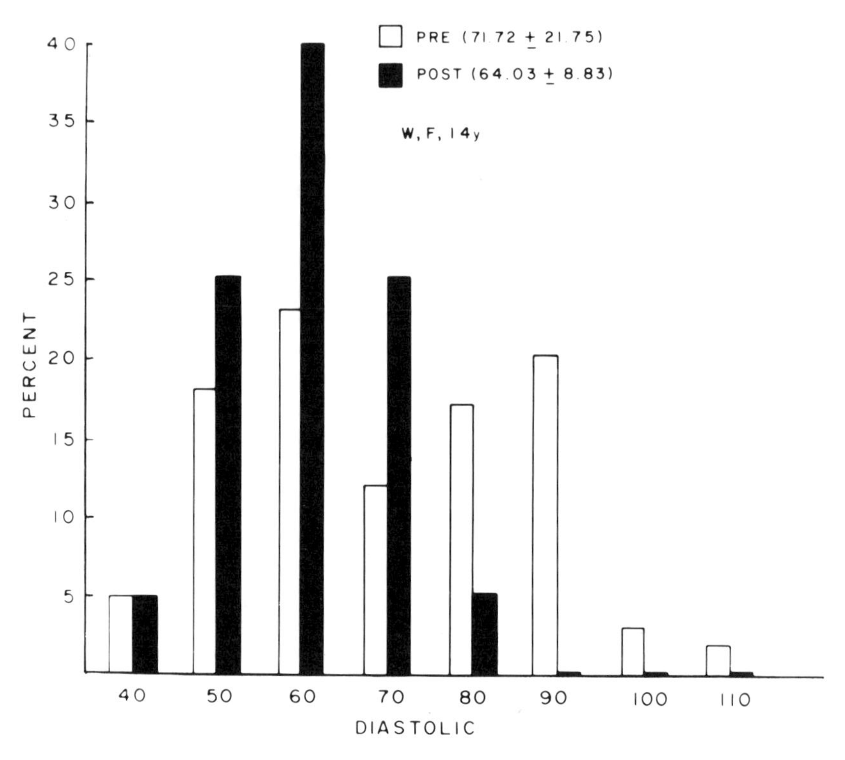

Figure 3. Results of 24-hour blood pressure monitoring presented in terms of percentage of readings within each blood pressure range. Data for pre and for post represent readings for one 24-hour period.

Medication Management

There is as yet no consensus on the most appropriate course of medical management of juvenile hypertension. Children and adolescents with borderline or mild elevations in blood pressure typically do not require medication (Sinaiko & Mirkin, 1978). In these cases, the Task Force on Blood Pressure Control (1977) recommends annual reevaluation, counseling on weight control and salt intake, and encouragement toward physical activity. Antihypertensive drug therapy is recommended in individuals with persistent elevations in diastolic blood pressure above 90 mmHg if they are 7½ years of age and in certain individuals 13–18 years of age. Medication is uniformly recommended for elevations above 100 mmHg for all individuals from 7 to 18 years of age.

Medication regimens are implemented in a "stepped-care" approach. Thiazide diuretics are usually selected first. The antihypertensive effects are related to decreases in plasma volume and extracellular fluid. The only known side effect of thiazide therapy in children is a decrease in serum potassium which is treated with potassium supplements.

If the child's blood pressure does not respond to thiazide, then a beta-blocking agent or methyldopa is recommended *as a supplement*. The mechanism of action of beta-blocking agents in lowering blood pressure is poorly understood. Suppression of the renin-angiotensin system and action on beta-andrenergic receptors located in blood pressure regulatory mechanisms in the brain have been proposed (Sinaiko & Mirken, 1978). These agents should not be used if asthma is diagnosed, and bradycardia is an infrequent complication of therapy.

Methyldopa competes with dopa in the synthesis of norepinephrine, and stores of norepinephrine are depleted. However, it is thought that methyldopa acts on the central vasomotor control center in lowering blood pressure (Day & Roach, 1974). Side effects include sedation, hepatitis, and hemolysis.

Should this course prove ineffective, then a vasodilator such as hydralazine is usually added to the regimen. Vasodilators act by reducing vascular wall tension (Sinaiko & Mirken, 1978). The high incidence of side effects limits their usefulness. Regularly seen are headache, flushness, tachycardia, salt and water retention, and increased cardiac output.

It is also possible that aggressive antihypertensive therapy will not lower the child's blood pressure to an acceptable level and hospitalization may be required.

Important issues related to antihypertensive drug therapy are the potential effects on the central nervous system. Long-term deleterious effects on the development of various organs and systems must also be evaluated to validate the continued use of these medications. Of more immediate concern to the physician is the predicted high rate of nonadherence to medical regimens that potentially can undermine therapy.

Smoking Prevention and Cessation

One of the most influential approaches to the prevention of smoking is based on the work of Evans (1976), which was inspired by McGuire's (1964) concept of psychological inoculation. An analogy is drawn between social pressures to adopt smoking and germs that infect the body. "Infection" might be prevented if adolescents can be exposed to a weak dose of those "germs" to develop "antibodies" (i.e., skills for resisting pressure to smoke).

Evans and his colleagues (Evans, Rozelle, Mittlemark, Hansen, Bane, & Havis, 1978) translated these concepts into a usable classroom strategy. The package used in their study involved specific information on the social pressures to smoke, focused discussions, feedback, and monitoring. A total of 750 seventh-grade students participated in the program. Students receiving all components of the program (those given the feedback alone and those who were monitored for smoking but not given feedback) were smoking at significantly lower rates than those students in the control group at the end of the program. Over 18% of the control group but only 10% of the treatment groups had begun smoking by the end of the program. The powerful effect of monitoring, and the relatively short intervention phase (10 weeks) are noteworthy.

Evans, Rozelle, Maxwell, Raines, Dill, Guthrie, Henderson, and Hill (in press) evaluated the program in 13 junior high schools. At the end of the eighth and ninth grade, experimental group subjects smoked infrequently and intended to smoke less frequently than did control group subjects.

McAlister, Perry, and Maccoby (1979) expanded Evans' work by employing older students as peer leaders to deliver the intervention. Active involvement in identifying and resisting pressures to smoke through role-playing, contests, and small group discussions were expected to enhance student learning of pressure resistance skills. Peer leaders taught these social skills in teams during a 7-session program with an entire seventh-grade class ($n = 336$) in California.

In the first reported results of this intervention (McAlister *et al.*, 1979), there was one treatment school and two control schools. About 2% of the students in the treatment and control schools reported smoking during the previous week at the beginning of the school year. By the end of the school year, 9.9% of the students in the control schools, but only 5.6% of the students in the treatment school, reported smoking during the previous week.

During the second year of the intervention, the treatment group received two follow-up treatments focusing on the immediate physiological effects of smoking. When surveyed the following June, 7.1% of the treatment subjects, 18.8% of students in one control school, and 21.0% of the students in the second control school, reported smoking in the previous week. The largest increase in onset occurred during the spring (McAlister, Perry, Killen, Telch, & Slinkard, 1980). Obviously, smoking prevention requires continual skills training and reinforcement to maintain treatment effects, particularly during the years in which people are most vulnerable to the tobacco habit.

The same principles were expanded further in projects at the University of Minnesota (Murray, Johnson, Luepker, Pechacek, Jacobs, & Hurd, 1979). Four schools received a variety of treatments and five monitoring sessions during the intervention year (1977–1978) and follow-up year (1978–1979). Seventh-grade students in one school were monitored for smoking but received no instruction. The students in a second school were shown films similar to those designed by Evans' group. In the third school, same-age peers (opinion leaders) were trained to direct intervention sessions similar to those in the second school.

The Minnesota group used self-reports and saliva thiocyanate measures to assess the amount of smoking in each seventh-grade group. Smokers were classified as regular (at least once per month), experimental (trying once or twice), or nonsmokers (never having smoked). Students in the monitoring-only school were much more likely than students in the two schools with interventions to begin smoking; extra components appeared to enhance the efficacy of the total program.

Reducing Overweight

Reducing overweight can be effective in lowering blood pressure and in modifying blood lipids (increasing HDL and decreasing LDL cholesterol) (Coates, Jeffery, Slinkard, Killen, & Danaher, in press). However, reducing overweight in children and adolescents is not easy to accomplish. Coates and Thoresen (1980) made three recommendations for the treatment of obesity in young persons, based on theoretical understandings of change processes and advances documented in promising experimental studies.

1. The distinction between *learning, practice,* and *performance* in the natural environment is often forgotten. Bandura (1977) has repeatedly emphasized the difference between learning and performance; programs typically attend to the former without recognizing the central importance of the latter. In Bandura's reciprocal interaction model (Bandura 1977, 1978), modeling and performance are influential in teaching new skills. The external environment, as filtered through the person's belief system, is central in influencing current behavior. The environment both cues and consequates behavior performed by the person.

When translated into weight-loss programs, the social learning model implies not only that effective teaching strategies (e.g., vicarious and participant modeling) be used, but also that environments be constructed to promote and support ongoing weight loss. We cannot teach people about weight loss in a classroom setting and expect them to apply those concepts and skills to lose weight (Jeffery & Coates, 1978).

2. Weight loss is facilitated by highly structured programs. These

kinds of programs provide the supports necessary for achieving and maintaining behavior changes.

3. Programs should place emphasis on and reward weight loss instead of habit change. There is no magical set of weight loss strategies applicable to or needed by everyone (Coates & Thoresen, 1981). People can lose weight in a variety of ways. They must be successful in their weight loss endeavors, identify strategies that work for them, and be encouraged to continue practicing those strategies for maintenance of weight loss.

Other recommendations can be gleaned from promising empirical studies in the behavioral treatment of obesity.

4. Weight loss early in a program might be especially important (Jeffery, Wing, & Stunkard, 1978). Gross, Wheeler, and Hess (1976) treated 10 obese adolescent girls in 10 weekly sessions. At the end of 10 weeks, 4 subjects had gained or maintained, 3 subjects had lost from 2 to 8 pounds, and 3 had lost more than 15 pounds. At a 27-week follow-up, the following results were documented: (1) of those maintaining or gaining immediately following treatment, 3 continued to gain while 1 lost 10 pounds; (2) the losses of those losing 2 to 8 pounds during treatment ranged from 6 to 14½ pounds; and (3) of those losing 15 or more pounds, the losses now range from 21 to 40 pounds. Continued success could be predicted from weight losses during the program.

5. Wing and Jeffery (1979) noted in their comparison of treatments that studies using *intensive motivational procedures* achieved the best weight losses. Heyden, de Maria, Barbee, and Morris (1973) placed 8 adolescents (11 to 17 years of age, 4 black females, 3 white females, 1 white male, 145 to 259 pounds) on a 700-calorie diet, fasting for 1 or 2 days per week, sodium restriction, and vitamin and protein supplements. Mothers and children saw the physicians frequently for encouragement and direction. These subjects lost an average of 40.2 pounds in 7 months. Losses ranged from 10 pounds (in the subject originally weighing 145 pounds) to 74 pounds. Only one subject was unsuccessful. She lost 25 pounds in 3 months but then regained it at 6 months. These impressive outcomes were replicated later with 7 other adolescents.

Coates, Jeffery, Slinkard, Killen, and Danaher (in press) combined frequent contact and contingency contracts in a weight loss program for adolescents. Thirty-six adolescents (13–17 years old, 9–100% overweight) were taught basic weight loss skills in ten one-hour sessions using videotape, modeling, role-playing, group discussion, and reading. Overweight students were taught problem-solving skills to analyze their individual eating patterns; they then devised and tried out possible solutions for problems identified. All subjects were required to deposit the equivalent of 15 weeks of their allowance or 50% of their estimated earnings from part-time employment. Subjects in the weight-loss reward groups (Groups 1 and 2) received deposit refunds back for achieving weight-loss goals of at least one pound per week. In the habit change group (Groups 3 and 4), subjects received refunds for keeping caloric intake below an individually established goal level. Subjects in the daily contact groups (Groups 1 and 2) came in each morning or afternoon to receive refunds for meeting weight loss or habit change goals. Weekly contact subjects (Groups 3 and 4) visited once per week to earn refunds. The daily-contact weight-loss group (Group 1) was the only group to show significant weight changes from baseline to weight at 15 weeks, and from baseline to weight at follow-up. The same pattern was observed in analyzing pounds above ideal weight.

While all of these results are hopeful, they are not overwhelming in terms of the clinical significance of the findings. Additional laboratory work and field studies are essential. Especially needed are programs of research that deviate from the models currently being used so that we can advance beyond present impotence.

Nutrition Behavior Change

Nutrition educators are aware of the need to motivate behavior change in students and to disseminate knowledge and to evaluate outcomes so that progress is possible (Whitehead, 1957). Although the importance of documented behavior change is acknowledged, the principles of behavior change are often not employed systematically and evaluations are frequently weak.

Go (1976) completed a content analysis of 90 curriculum guides designed for kindergarten through the 12th grade. The major conclusion of the analysis pointed to the inadequacies of the guides: (1) concepts frequently were underdeveloped and often were not related to lesson objectives and learning activities; (2) cognitive and affective learning objectives were stressed while behavioral objectives were not; (3) because practice was downplayed, students would not be given appropriate opportunities to practice critical skills; (4) the programs were strongly teacher-dominated rather than being oriented toward student activity and improvement; and (5) most of the guides provided no guidelines for evaluation, and none of the guides was accompanied by data attesting to their efficacy. Only with careful and objective evaluation will it be possible to determine which approaches are promising and which should be disseminated.

Coates, Jeffery, and Slinkard (1981) developed a cardiovascular nutrition and activity change program using principles of social learning theory and they sought to evaluate the program by using direct observations of eating and physical activity at school.

The program was developed to teach concepts and to promote significant behavior change using 12 45-minute class sessions over a four-week period. Six nutrition classes were followed by six exercise classes. Three elements of this program appeared to be most critical: behavioral commitment, feedback and incentives, and family involvement. Studies at this (or any) age often have difficulty translating general principles into specific behavior changes. Daily goal sheets, on which students made a written commitment to substitute specific "heart healthy" foods for foods normally in their lunches and to engage in "heart healthy" playground activities, appeared to be essential in promoting behavior change. Students knew precisely what had to be done to meet program goals and to earn available social rewards.

The program was implemented first in two elementary schools adjacent to Stanford University in two successive years. In each year, three 4th-grade classes in one school and three 5th-grade classes in a second school were involved. A time series multiple baseline design was employed: lessons in the 5th grade were lagged behind lessons in the 4th grade classes.

Direct observation, paper-and-pencil measures, and phone interviews to parents were used to evaluate the programs. While the students were eating lunch, observers approached each student individually and wrote down the contents of the lunch. Following lunch, individual students were observed for a one-minute period. At each 5-second interval, the observer made a judgement regarding the student's activity level (sitting, standing, walking, climbing, running, standing still but moving upper trunk), according to preestablished operational definitions.

Foods in lunches were classified as heart healthy (target food items in the lessons) or not heart healthy. At both schools, the average daily number of heart healthy foods in students' lunches increased during the program, remained high following the program, and maintained above baseline levels when follow-up data were collected. The postprogram data were collected at the end of the spring term, while the follow-up data were collected in the fall following summer vacation. Students also increased in knowledge and in reported performance for heart healthy foods and activities after the program. These results were maintained at follow-up.

Beginning Early and Maximizing Prevention

Obesity among infants is common (Hutchinson-Smith, 1970; Taitz, 1971). Shukla, Forsyth, Anderson, and Marwah (1972) reported that 16.75 of the infants studied were obese, and 27.7% were between 10% and 20% above standard weight. Prevention may be the best treatment; focusing effort on preventing obesity during infancy and other high-risk periods would seem especially useful.

Piscano, Lichter, Ritter, and Siegel (1978) placed 80 infants born in 1970 on a prudent diet at 3 months of age (or when they reached 13 pounds if they did not weigh that much at 3 months). The diet emphasized low saturated fat, low salt, low sugar, and low cholesterol. Typical items included fresh fruit cooked without sugar, fresh vegetables, meats (white/red ratio of 2:1), skimmed milk and cheeses, plain yogurt, and natural gelatin. The comparison group was 50 newborns, born between 1964 and 1970, fed conventional diets. At 3 months of age (before the diet had started) 37.5% of control females and 35% of the control males were overweight (weight percentiles exceeded height percentiles by more than one standard deviation) as compared to 29.4% of diet females and 21% of diet males. By age 3 the findings were dramatic: 16.7% of control females and 34.8% of control males were overweight as compared to 0.0% of diet females and 2.56% of diet males.

Primary prevention may also be effective in promoting low levels of serum cholesterol and in increasing levels of physical activity. Friedman and Goldberg (1976) placed infants on a low saturated fat and low cholesterol diet from birth. When compared to a matched control group at 3 years, significant differences between the groups were found in percentile height, weight, head circumference, skinfold thickness, total serum protein, hemoglobin, number of sick visits, or number of failures on the Denver Developmental Screening Test. Total serum cholesterol levels between the two groups were significantly different (145.1 ± 1.4, $p < .05$) for the low and regular diet subjects respectively.

Rose and Mayer (1968) studied activity and caloric intake in relation to relative weight in 31 4- to 6-month-old infants living in normal home conditions. Activity level rather than size was related to caloric intake, with the more active infants consuming more calories. Calories consumed and weight gained were not correlated, whereas activity and total calories were correlated positively ($r = .47$, $p < .01$) and triceps skinfold was negatively correlated with activity ($r = -.53$, $p < .01$) and with calories eaten per kilogram of body weight ($r = -.71$, $p < .01$). Thus, as the infants became more overweight, they consumed less and also were less active. Data on caloric intake are from dietary records and thus may suffer possible reporting biases. Replications would be useful, with more sensitive measures of caloric intake.

CONCLUSIONS

Space has permitted only a cursory review of current activity in behavioral pediatrics selected for cardiovascular disorders. Both the treatment of congenital disorders and the prevention of acquired disorders are vital; it is important in both areas that psychologists not become part of the problem. Psychological approaches must expand thinking and conceptualization rather than accept those provided by medicine. Behavioral pediatrics of cardiovascular disorders must represent a new science and not merely the addition of "behavior" to existing thought and practice.

REFERENCES

Aisenberg, R. B., Wolff, P. N., Rosenthal, A., & Nadas, P. Psychological impact of cardiac catheterization. *Pediatrics*, 1973, *51*:1051–1059.

American Heart Association. *Heart facts*. Dallas: Aita, 1976.

Anastasi, A. *Psychological testing*. New York: Holt, 1965.

Bandura, A. *Social learning theory*. Englewood Cliffs, N.J.: Prentice-Hall, 1977.

Bandura, A. The self-system in reciprocal determinism. *American Psychologist*, 1978, *33*, 344–358.

Berenson, G. S. *Cardiovascular risk factors in children: The early natural history of atherosclerosis*. New York: Oxford, 1980.

Biron, P., Mongeau, J. G., & Bertrand, D. Familial aggregation of blood pressure in adopted and natural children. In O. Paul (Ed.), *Epidemiology and control of hypertension*. New York: Stratton, 1975.

Blackburn, H. Progress in the epidemiology and prevention of coronary heart disease. In P. Yu & J. F. Goodwin (Eds.), *Progress in cardiology*. New York: Lea & Feibiger, 1974.

Blackburn, H. Physical activity and cardiovascular health: The epidemiological evidence. In G. Landry & P. Urban (Eds.), *Physical activity and human well being*. Miami: Symposia Specialists, 1978.

Blumenthal, J. A., Williams, R. B., King, Y., Schanberg, A. M., & Thompson, L. W. Type A behavior pattern and coronary atherosclerosis. *Circulation*, 1978, *58*, 634–639.

Blumenthal, S. Implications for the future study of hypertension. *Pediatric Clinics of North America*, 1978, *25*, 183–186.

Blumgart, E., & Korsch, B. M.: Pediatric recreation: An approach to meeting the emotional needs of hospitalized children. *Pediatrics*, 1964, *34*, 133–136.

Bortner, R. W., Rosenman, R. N., & Friedman, M. Familial similarity in pattern A behavior. *Journal of Chronic Diseases*, 1970, *20*, 525–533.

Bullen, B., Reid, R. B., & Mayer, J. Physical activity of obese and non-obese girls appraised by motion picture sampling. *American Journal of Clinical Nutrition*, 1964, *14*, 211–223.

Cassell, S. Effect of brief puppet therapy upon the emotional responses of children undergoing cardiac catheterizations. *Journal of Consulting Psychology*, 1965, *29*, 1–8.

Clark, W. R., Schrott, H., Leaverton, P. E., Connor, W., & Lauer, R. L. Tracking of blood lipids and blood pressures in school age children: The Muscatine study. *Circulation*, 1978, *58*, 626–634.

Coates, T. J. Stress management for adolescent hypertension. Presented at the Symposium on Hypertension in the Young, Johns Hopkins University, 1980.

Coates, T. J., & Thoresen, C. E. Obesity in children and adolescents: The problem belongs to everyone. In B. Lahey & A. E. Kazdin (Eds.), *Advances in clinical child psychology (Vol. 2)*. New York: Plenum Press, 1980.

Coates, T. J., & Thoresen, C. E. Behavior and weight changes in three obese adolescents. *Behavior Therapy*, 1981, *12*, 383–389.

Coates, T. J., Stephen, M., Perry, C., & Schwartz, B. *Precursors of cigarette smoking in adolescents*. Unpublished manuscript, John Hopkins School of Medicine, 1980.

Coates, T. J., Jeffery, R. W., & Slinkard, L. A. The heart health program: Introducing and maintaining nutrition changes among elementary school children. *American Journal of Public Health*, 1981, *71*, 15–23.

Coates, T. J., Jeffery, R. W., Slinkard, L. A., Killen, J. D., & Danaher, B. G. Monetary incentives and frequency of contact in weight loss, blood pressure reduction, and lipid change in adolescents. *Behavior Therapy*, in press.

Coates, T. J., Parker, F. C., & Kolodner, K. Stress and heart disease: Does blood pressure reactivity offer a link? In T. J. Coates, A. C. Peterson, & C. Perry (Eds.), *Adolescent health: Crossing the barriers*. New York: Academic Press, 1982.

Day, M. D., & Roach, A. G. Central alpha and beta adrenoreceptors modifying arterial blood pressure and heart rate in conscious cats. *British Journal of Pharmacology*, 1974, *51*, 325–333.

Dube, S. K., Kapoor, S., Ratner, H., & Tunick, F. L. Blood pressure studies in black children. *American Journal of Diseases of Children*, 1975, *129*, 1177–1180.

Egerton, N., Egerton, W. S., & Kay, J. H. Neurologic changes following profound hypothermia. *Annals of Surgery*, 1963, *157*, 366–374.

Egerton, N., & Kay, J. H. Psychological disturbances associated with open heart surgery. *British Journal of Psychiatry*, 1964, *110*, 433–439.

Emurian, H. H., Emurian, C. J., & Brady, J. V. Effects of a pairing contingency on behavior in a three-person programmed environment. *Journal of the Experimental Analysis of Behavior*, 1978, *29*, 319–329.

Evans, R. I. Smoking in children: Developing a social psychological strategy of deterrance. *Preventive Medicine*, 1976, *5*, 122–127.

Evans, R. I., Rozelle, R. M., Mittlemark, M. B., Hansen, W. R., Bane, A. L., & Havis, J. Deterring the onset of smoking in children: Knowledge of immediate physiological effects and coping with peer pressure, media pressure, and parent modeling. *Journal of Applied Social Psychology*, 1978, *7*, 126–135.

Evans, R. I., Rozelle, R. M., Maxwell, S. E., Raines, B. E., Dill, C. A., Guthrie, T. J., Henderson, A. H., & Hill, L. C. Social modeling films to deter smoking in adolescents: Results of a three-year field investigation. *Journal of Applied Psychology*, in press.

Feinleib, M., Garrison, R., Borhani, N., Rosenman, R., & Christian, J. Studies of hypertension in twins. In O. Paul (Ed.), *Epidemiology and control of hypertension*. New York: Grune & Stratton, 1975.

Frank, G. C., Berenson, G. S., Schilling, P. E., & Moore, M. L. Adapting the 24-hour dietary recall for epidemiologic studies of school children. *Journal of the American Dietetic Association*, 1977, *71*, 26–31.

Frank, G. C., Voors, A. W., Schilling, P. E., & Berenson, G. S. Dietary studies of rural school children in a cardiovascular survey. *Journal of the American Dietetic Association*, 1977, *71*, 31–35.

Frerichs, R. R., Srinivasan, S. R., Webber, L. S., & Berenson, G. S. Serum cholesterol and triglycerides in 3446 children from a biracial community: The Bogalusa Heart Study. *Circulation*, 1976, *54*, 302–309.

Frerichs, R. R., Webber, L. S., Voors, A. W., Srinivasan, S. R., & Berenson, G. S. Cardiovascular disease risk factor variables in children at two successive years: The Bogalusa Heart Study. *Journal of Chronic Diseases*, 1979, *32*, 251–262.

Friedman G., & Goldberg, S. J. An evaluation of the safety of a low-saturated fat, low-cholesterol diet beginning in infancy. *Pediatrics*, 1976, *58*, 655–652.

Fyler, D. C. Report of the New England Regional Infant Cardiac Program. *Pediatrics*, 1980, *65*, 377–461.

Garson, A., Williams, R. B., & Redford, T. Long term follow up of patients with Tetralogy of Fallot: Physical health and psychopathology. *Journal of Pediatrics*, 1974, *85*, 429–433.

Go, C. *An analysis of selected curriculum guides used for nutrition education in grades K–12*. Berkeley, Calif.: Society for Nutrition Education, 1976.

Green, D. W. *Teen-age smoking: Immediate and long-term patterns*. Washington, D.C.: U. S. Government Printing Office, 1979.

Green, D. W. Teen-age smoking behavior. In F. R. Scarpetti & S. K. Daterman (Eds.), *Drugs and youth culture*. Beverly Hills, Calif.: Sage, 1980.

Gross, M. A., Wheeler, M., & Hess, K. The treatment of obesity in adolescents using behavioral self-control. *Clinical Pediatrics*, 1976, *15*, 920–924.

Haller, J. A., Talbert, J. L., & Dombro, R. H. *The hospitalized child and his family*. Baltimore Johns Hopkins University Press, 1967.

Hamilton, B. P., Zadik, Z., Edwin, C. M., Hamilton, J. H., & Kowarski, A. A. Effect of adrenal suppression with dexamethasone in essential hypertension. *Journal of Clinical Endocrinology and Metabolism*, 1979, *21*, 423–428.

Haynes, S. G., Feinleib, M., & Kannel, W. B. The relationship of psychosocial factors to coronary heart disease in the Framingham study. *American Journal of Epidemiology*, 1980, *111*, 37–58.

Heyden, S., de Maria, W., Barbee, S., & Morris, M. Weight reduction in adolescents. *Nutrition and Metabolism*, 1973, *15*, 295–304.

Holland, W. W., & Beresford, S. A. A. Factors influencing blood pressure in children. In O. Paul (Ed.), *Epidemiology and control of hypertension*. New York: Grune & Stratton, 1975.

Holt, J. R. Play PRN in pediatric nursing. *Nursing Forum*, 1970, *9*, 288–309.

Horan, M. J., Kennedy, H. L., & Padgett, N. E. Do borderline hypertensive patients have labile blood pressure? *Annals of Internal Medicine*, 1981, *94*, 466–468.

Huenemann, R. L., Hampton, M. C., Behnke, A. R., Schapiro, L. R., & Mitchell, B. W. *Teen-age nutrition and physique*. Springfield, Ill.: Charles C Thomas, 1974.

Hunter, S. MacD., Webber, L. S., & Berenson, G. S. Cigaretts smoking and tobacco usage behavior in children and adolescents: The Bogalusa Heart Study. *Preventive Medicine*, 1980, *6*, 701–712.

Hutchinson-Smith, B. The relationship between the weight of an infant and lower respiratory infection. *Medical Officer*, 1970, *123*, 257.

Janis, I. L. *Psychological stress*. New York: Wiley, 1958.

Jeffery, R. W., & Coates, T. J. Why aren't they losing weight? *Behavior Therapy*, 1978, *9*, 856–860.

Jeffery, R. W., Wing, R. R., & Stunkard, A. J. Behavioral treatment of obesity: The state of the art in 1976. *Behavior Therapy*, 1978, *9*, 189–199.

Jenkins, C. D., Zyzanski, S. J., & Rosenman, R. H. Risk of new myocardial infarction in middle-aged men with manifest coronary heart disease. *Circulation*, 1976, *53*, 342–347.

Johnson, P. A., & Stockdale, D. F. Effect of puppet therapy on palmar sweating of hospitalized children. *Johns Hopkins Medical Journal*, 1975, *137*, 1–5.

Jolly, R. H. Play and the sick child. *Lancet*, 1968, *2*, 1286–1287.

Kannel, W. B. The origins and evils of essential hypertension. *New England Journal of Medicine*, 1971, *284*, 444–445.

Kannel, W. B., & Dawber, T. R. Atherosclerosis as a pediatric problem. *Journal of Pediatrics*, 1972, *8*, 544–559.

Kannel, W. B., & McGee, D. C. Diabetes and cardiovascular risk factors: The Framingham study. *Circulation*, 1979, *59*, 8–12.

Kilcoyne, M. Natural history of hypertension in adolescence. *Pediatric Clinics of North America*, 1978, *25*, 47–53.

Kornfield, D. S., Zimberg, S., & Malm, J. R. Psychiatric complications of open-heart surgery. *New England Journal of Medicine*, 1965, *273*, 287–292.

Kowarski, A. A., Edwin, C. M., Akesode, A. P., Pietrowski, L. S., & Hamilton, B. P. The integrated concentration of plasma renin activity and aldosterone in essential hypertension. *The Johns Hopkins Medical Journal*, 1978, *142*, 35–41.

Kuller, L. H. Epidemiology of cardiovascular diseases: Current perspectives. *American Journal of Epidemiology*, 1976, *104*, 425–456.

Lauer, R. M., Connor, W. B., Leaverton, P. E., Reiter, M. A., & Clarke, W. R. Coronary heart disease factors in school children: The Muscatine study. *Journal of Pediatrics*, 1975, *86*, 697–706.

Lavigne, J. V., & Ryan, M. Psychologic adjustment of siblings of children with chronic illness. *Pediatrics*, 1979, *63*, 616–627.

Linde, L. M., Rasof, B., Dunn, O. J., & Rabb, E. Attitudinal factors in congenital heart disease. *Pediatrics*, 1966, *38*, 92–101.

Linde, L. M., Dunn, O. J., Schireson, R., & Rasof, B. Growth in children with congenital heart disease. *Journal of Pediatrics*, 1967, *70*, 413–419.

Linde, L. M., Rasof, B., & Dunn, O. J. Longitudinal studies of intellectual and behavioral development in children with congenital heart disease. *Acta Paediatrica Scandinavica*, 1970, *59*, 169–176.

Lloyd, G. G., & Cawley, R. H. Psychiatric morbidity in men one week after acute myocardial infarction. *British Medical Journal*, 1978, *2*, 1453–1455.

Loggie, J. M. H. Prevalence of hypertension and distribution of causes. In M. I. New & L. S. Levine (Eds.), *Juvenile hypertension*. New York: Raven Press, 1977.

Londe, S. Causes of hypertension in the young. *Pediatric Clinics of North America*, 1978, *25*, 55–65.

Londe, S., Bourgoyne, J. J., Robson, A. M., & Goldring, D. Hypertension in apparently normal children. *Journal of Pediatrics*, 1971, *78*, 569–577.

Matthews, K. A. *Mother–child interaction and determinants of Type-A behavior*. Unpublished doctoral dissertation, University of Texas, Austin, 1976.

Matthews, K. A. Caregiver–child interactions and the Type-A coronary prone behavior patterns. *Child Development*, 1977, *48*, 1752–1756.

Matthews, K. A. Assessment and developmental antecedents of the coronary-prone behavior pattern in children. In T. M. Dembroski, S. M. Weiss, J. L. Shields, S. G. Haynes, & M. Feinleib (Eds.), *Coronary-prone behavior*. New York: Springer, 1978.

Matthews, K. A., & Angulo, J. Measurement of the Type A behavior: impatience, anger, and aggression. *Child Development*, 1980, *51*, 466–476.

Matthews, K. A., & Krantz, D. S. Resemblance of twins and their parents in Pattern-A behavior. *Psychosomatic Medicine*, 1976, *28*, 140–144.

Matthews, K. A., Glass, D. C., & Richins, M. The mother–son observation study. In D. C. Glass (Ed.), *Behavior patterns, stress, and coronary disease*. Hillsdale, N.J.: Erlbaum, 1977.

Maxwell, M. H. A functional approach to screening. In *The hypertension handbook*. West Point, Pa.: Merck, Sharp, & Dohme, 1974.

McAlister, A., Perry, C., & Maccoby, N. Adolescent smoking: Onset and prevention. *Pediatrics*, 1979, *63*, 650–658.

McAlister, A., Perry, C. P., Killen, J. D., Telch, M., & Slinkard, L. A. Pilot study of smoking, alcohol, and drug abuse prevention. *American Journal of Public Health*, 1980, *70*, 719–721.

McGuire, W. Inducing resistance to persuasion: Some contemporary approaches. In L. Berkowitz (Ed.), *Advances in experimental social psychology* (Vol. 1). New York: Academic Press, 1964.

Melamed, B. G., & Siegel, L. J. Reduction of anxiety in children facing hospitalization and surgery by use of filmed modeling. *Journal of Consulting and Clinical Psychology*, 1975, *43*, 511–521.

Miller, N. E. The evidence for antiatherogenicity of high density lipoproteins in men. *Lipids*, 1979, *13*, 914–919.

Miller, R. A., & Shekelle, R. B. Blood pressure in tenth grade students. *Circulation*, 1976, *54*, 993–1000.

Murray, D., Johnson, C. A., Luepker, R., Pechacek, R., Jacobs, D., & Hurd, P. *Social factors in the prevention of smoking in seventh grade students*. Paper presented at the meeting of the American Psychological Association, New York, 1979.

National Academy of Sciences Food and Nutrition Board. *Toward healthful diets*. Washington, D. C.: National Research Council, 1980.

National Center for Health Statistics. *Blood pressures of youths 6–11 years*. Publication (HRA) 74–1617, Vital and Health Statistics, Series 11, No. 135. Washington, D. C.: U. S. Government Printing Office, 1973.

National Center for Health Statistics. *Blood pressures of youths 12–17 years*. Publication (HRA) 77–1645, Vital and Health Statistics, Series 11, No. 163. Washington, D. C.: U. S. Government Printing Office, 1977.

Naylor, D., Coates, T. J. & Kan, J. J., Reducing stress in pediatric cardiac catheterization. *Circulation*, in press (abstract).

Neill, C. A., & Haroutunian, L. M. The adolescent and young adult with congenital heart diseases. In A. J. Moss, F. H. Adams, & G. C. Emmanouilides (Eds.), *Atherosclerosis in heart disease in infants, children, and adolescents*. Baltimore: Williams & Wilkins, 1977.

Paffenbarger, R. S., Thorne, M. C., & Wing, A. L. Chronic diseases in former college students: VIII. Characteristics of youths predisposing to hypertension in later years. *American Journal of Epidemiology*, 1968, *88*, 25–32.

Peterson, L., & Shigetomi, C. The use of coping techniques to minimize anxiety in hospitalized children. *Behavior Therapy*, 1981, *12*, 1–14.

Petrillo, M.: Preventing hospital trauma in pediatric patients. *American Journal of Nursing*, 1968, *68*, 1469–1473.

Piscano, J. C., Lichter, H., Ritter, J., & Siegal, A. P. An attempt at the prevention of obesity. *Pediatrics*, 1978, *61*, 360–364.

Pooling Project Research Group. *Relationship of blood pressure, serum cholesterol, smoking habit, relative weight, and ECG abnormalities to incidence of major coronary events*. Dallas: American Heart Association Monograph, 1978.

Rasof, B., Linde, L. M., & Dunn, O. J. Intellectual development in children with congenital heart disease. *Child Development*, 1967, *38*, 1043–1053.

Rie, H. E., Boverman, H., Grossman, B., & Ozoa, N. Immediate and long-term effects of intervention early in prolonged hospitalization. *Pediatrics*, 1968, *41*, 755–765.

Rose, G. A., Holland, W. W., & Crowley, E. A. A sphygmomanometer for the epidemiologist. *Lancet*, 1964, *1*, 296–298.

Rose, H. E., & Mayer, J. Activity, caloric intake and the energy balance of infants. *Pediatrics*, 1968, *41*, 18–29.

Rosenman, R., Brand, R. J., Sholtz, R. I., & Friedman, M. Multivariate prediction of coronary heart disease during 8.5 year follow-up in the Western Collaborative Group Study. *American Journal of Cardiology*, 1976, *37*, 902–910.

Rowland, T. W. The pediatrician and congenital heart disease—1979. *Pediatrics*, 1979, *64*, 180–186.

Schmitt, F. F., & Wollridge, P. J. Psychological preparation of surgical patients. *Nursing Research*, 1973, *22*, 108–116.

Shapiro, D., & Surwit, R. J. Learned control of physiological function and disease. In H. Leitenberg (Ed.), *Handbook of behavior modification and behavior therapy*. Englewood Cliffs, N.J.: Prentice-Hall, 1976.

Shukla, A., Forsyth, H. A., Anderson, C. M., & Marwah, S. M. Infantile overnutrition in the first years of life: A field study in Dudley, Worcestershire. *British Medical Journal*, 1972, *4*, 507–515.

Silbert, A., Wolff, P. H., Mayer, B., Rosenthal, A., & Nadas, A. S. Cyanotic heart disease and psychological development. *Pediatrics*, 1969, *43*, 192–200.

Sinaiko, A. R., Mirkin, B. L. Clinical pharmacology of antihypertensive drugs in children. *Pediatric Clinics of North America*, 1978, *25*, 137–157.

Spiga, R., & Peterson, A. C. *The coronary-prone behavior pattern in early adolescence*. Paper presented at the meetings of the American Educational Research Association, Los Angeles, 1981.

Srinivasan, S. R., Frerichs, R. R., Webber, L. S., & Berenson, G. S. Serum lipoprotein profile in children from a biracial community. The Bogalusa Heart Study. *Circulation*, 1976, *54*, 309–318.

Stamler, J., Schoenberger, J. A., Shekelle, R. B., & Stamler, R. Hypertension: The problem and the challenge. In *The hypertension handbook*. West Point, Pa.: Merck, Sharp & Dohme, 1974.

Stern, M. J., Pascale, L., & Ackerman, A. Life adjustment postmyocardial infarction. *Archives of Internal Medicine*, 1977, *137*, 1680–1685.

Still, J. L., & Cotton, D. Severe hypertension in young persons. *Archives of Diseases in Childhood*, 1967, *42*, 34–49.

Submaramian, S., Vlad, P., Fischer, L., & Cohen, M. E. Sequelae of profound hypothermia and cardiocirculatory arrest in the corrective treatment of congenital heart disease in infants and small children. In B. S. L. Kidd & R. D. Rowe (Eds.), *The child with congenital heart disease after surgery*. New York: Futura, 1976.

Taitz, L. S. Infantile overnutrition among artificially fed infants in the Sheffield region. *British Medical Journal*, 1971, *1*(2), 315–323.

Task Force on Blood Pressure Control. Report: *Pediatrics*, 1977, *59*, 797–820.

Taussig, H. B., Crocetti, A., Eshaghpour, E., Keikonen, R., Yap, K. J., Bachman, D., Momberger, N., & Kirk, H. Long time observations on the Blalock Taussig operations. III: Common complications. *Johns Hopkins Medical Journal*, 1971, *129*, 274–289.

Taussig, H. B., Kallman, C. H., Nagel, D., Baumgardner, R., Momberger, N., & Kirk, H. Long time observations on the Blalock Taussig with a tetralogy of Fallot. *Johns Hopkins Medical Journal*, 1975, *137*, 13–19.

Torok, E. The beginning of hypertension: Studies in childhood and adolescence. In F.

Gross & T. Strasser (Eds.), *Mild hypertension: History and management*. Chicago: Year Book Medical Publishers, 1979.

Tufo, H. M., Ostfeld, A. M., & Shekelle, R. Central nervous system dysfunction following open heart surgery. *The Journal of the American Medical Association*, 1970, *212*, 1333–1340.

Vernon, D. T. A., Foley, J. M., Sipowicz, R. R., & Schulman, J. L. *The psychological responses of children to hospitalization and illness: A review of the literature*. Springfield, Ill.: Charles C Thomas, 1965.

Visintainer, M. A., & Wolfer, J. A. Psychological preparation for surgical pediatric patients: The effect of children's and parents' stress responses and adjustment. *Pediatrics*, 1975, *56*, 187–202.

Voors, A. W., Foster, T. A., Frerichs, R. R., Webber, L. S., & Berenson, G. S. Studies of blood pressure in children ages 5–14 in a total biracial community: The Bogalusa Heart Study. *Circulation*, 1976, *54*, 319–327.

Voors, A. W., Webber, L. S., & Berenson, G. S. Epidemiology of essential hypertension in youth: Implications for clinical practice. *Pediatric Clinics of North America*, 1978, *25*, 15–27.

Voors, A. W., Webber, L. S., & Berenson, G. S. Time course studies of blood pressure in children: The Bogalusa Heart Study. *American Journal of Epidemiology*, 1979, *109*, 320–327.

Voors, A. E., Sklov, M., Wolf, T., Hunter, S., & Berenson, G. S. Cardiovascular risk factors in children and coronary-related to behavior. In T. J. Coates, A. C. Petersen, & C. Perry (Eds.), *Adolescent health: Crossing the barriers*. New York: Academic Press, 1982.

Voors, A. W., Webber, L. J., Harsha, J. W., Frank, G. C., & Berenson, G. S. Obesity: Determinants and consequences in children of an entire biracial population: The Bogalusa Heart Study. *Journal of Chronic Diseases and Therapeutic Research*, in press.

Waxman, M., & Stunkard, A. J. Caloric intake and expenditure of obese boys. *Journal of Pediatrics*, 1980, *96*, 187–193.

Whitehead, F. E. Nutrition education for children in the U. S. since 1900: Part II. *Journal of the American Dietetic Association*, 1957, *33*, 885–890.

Wing, R. R., & Jeffery, R. W. Outpatient treatments of obesity: A comparison of methodology and clinical results. *International Journal of Obesity*, 1979, *3*, 261–279.

Wolfer, J. A., & Visintainer, M. A. Pediatric surgical patients' and parents' stress responses and adjustment. *Nursing Research*, 1975, *24*, 244–255.

Zahka, K. G., Neill, C. A., Kidd, L., Cutiletta, M. A., & Cutiletta, A. F. Cardiac involvement in adolescent hypertension: Echocardiographic determination of myocardial hypertrophy. *Hypertension*, 1981, *3*, 664–668.

Zyzanski, S. J., Jenkins, C. D., Ryan, T. J., Flessas, A., & Everist, M. Psychological correlates of coronary and angiographic findings. *Archives of Internal Medicine*, 1976, *136*, 1234–1237.

11

Therapeutic Adherence

Bruce J. Masek and William R. Jankel

It is our intent in this chapter to examine the problem of therapeutic nonadherence in pediatrics. Research with adult populations will only be considered in the absence of illustrative examples from the pediatric literature. We will discuss methodological issues including definitional and measurement difficulties that have hampered progress in developing effective solutions for nonadherence. The research that has attempted to identify the causes of nonadherence and treatments for nonadherence will also be reviewed. The remainder of the chapter will focus on a review by disease entity of the pediatric research in the area that is noteworthy for innovation and/or objectivity in its approach to the assessment and treatment of nonadherence. We have chosen to review the majority of pediatric research by disease entity for the sake of convenience and to identify and integrate with medical management of the disease the salient variables from an applied behavior analysis perspective relevant to assessing and treating nonadherence. Finally we offer some guidelines for future inquiries in this important area of behavioral pediatrics research.

THE MAGNITUDE OF THE PROBLEM

It is generally acknowledged that a major problem facing the health care system is nonadherence by patients to prescribed courses of treat-

BRUCE J. MASEK • Children's Hospital Medical Center; Harvard Medical School, Boston, Massachusetts 02115. WILLIAM R. JANKEL • John F. Kennedy Institute; Johns Hopkins University School of Medicine, Baltimore, Maryland 21205.

ment. As evidence of this concern, the number of published articles in the area of therapeutic adherence was 245 prior to 1974, but by 1977 this number had grown to 853 (Haynes, 1979). However, when attempting to estimate the magnitude of the problem, reviewers of this research have produced widely varying figures with questionable generalizability due to a number of methodological problems that have plagued research in this area (Sackett & Snow, 1979). Figures for adult nonadherence, for example, range from 7% to 92% (Davis, 1966; Marston, 1970). Estimates of nonadherence for pediatric populations have proved even more difficult to ascertain but are generally reported to be less than the adult figures. Despite the difficulties in providing reliable estimates of adherence rates with various therapeutic regimens and in various populations, evidence continues to accumulate to suggest that patient nonadherence is widespread and costly to the health care system (Masek, Epstein, & Russo, 1981). Real progress in understanding the magnitude of the problem will require considerable refinement in the definition and assessment of nonadherence (Gordis, 1979).

THE PROBLEM OF DEFINITION

Definitions of nonadherence are as varied as the diagnoses, measurement systems, and types of therapeutic regimens investigated in each study. Typically, these definitions lack precision in terms of the behavior necessary and sufficient to achieve the desired therapeutic effect which, in many instances, is poorly understood as well. For example, Haynes (1979) has suggested that the definition be nonjudgmental in terms of attribution of blame for adherence failures to the parents, physician, or environmental conditions and that it focus solely on the correlation of a patient's behavior with medical advice. Ruley (1978) defined adherence as "the extent to which the life style of a patient is changed to correspond to a therapeutic regimen including the necessity of taking medications, returning for follow-up medical care, or variations in diet or exercise." Although these definitions may have utilitarian value at present, progress in the field will depend, in large part, on definitions that incorporate aspects of: (1) specific behavior involved in therapeutic adherence, (2) expected biological response in the patient, and (3) the distribution of these variables in relation to therapeutic outcome. We are a long way from developing this type of definition and consequently will continue to focus on the effect of patient nonadherence in terms of the failure to achieve a treatment goal. This is currently the case in acute conditions, where the failure to complete a course of antibiotics may result in recurrent infections (Daschner & Marget, 1975); in chronic conditions including seizures (Freiman & Buchanan, 1978) and hypertension (Sackett & Haynes, 1976), where control may be compromised by irregular administration of the medications; or with more general health care issues, such as iron supplements during pregnancy, tuberculosis screenings, vaccinations, and appointment-keeping. (Blackwell, 1973; Melchoir, 1965; Porter, 1969).

MEASUREMENT OF ADHERENCE

While the phenomenon of nonadherence with therapeutic regimens has been extensively investigated, procedures for assessing and improving adherence have not been rigorously researched. This situation, fortunately, appears to be changing (Rapoff & Christophersen, 1980). Current concepts in the measurement of adherence favor two approaches: bioassay of some substance reflective of the medical regimen under investigation and behaviorally-based observational methods. The problems associated with the use of tablet counts of unused medication, physician estimates of adherence, treatment outcome, and patient report of estimates of adherence are numerous (Masek, 1982; Rapoff & Christophersen, 1980). The use of these techniques for research purposes should be avoided unless they are adjunctive to a more objective assessment. However, in clinical settings, these procedures may be all that are available or feasible. In this situation, Rapoff and Christophersen (1980) offer some suggestions as to how to minimize sources of bias inherent in their use.

Considered to be the most objective assessment of medication adherence (and certain dietary restrictions) is the bioassay of urine, blood, saliva, or stool for the presence of the medication, metabolite, or a pharmacologically inert substance that has been incorporated into the medication. For example, Eney and Goldstein (1976) monitored theophylline levels of asthmatic children from saliva samples analyzed by gas chromatography. Bergman and Werner (1963) tested urine samples for the presence of penicillin in children with streptococcal infections using a cultured microorganism, *Sarcina lutea*, highly sensitive to pencillin. A filter paper disk was dipped into the urine sample and then placed on the culture plate and growth inhibition (indicative of adherence) was assessed approximately 12 hours later. Seizure clinics routinely draw blood from patients for serum anticonvulsant level determinations, a procedure which can readily be incorporated into a research program to test the efficacy of various strategies to improve adherence (e.g., Lund,

Jorgensen, & Kuhl, 1964). Incorporating a tracer into the medication, such as riboflavin which is easily detected in the urine (Hobby & Deuschle, 1959) or phenazopyridine which colors the urine bright orange (Epstein & Masek, 1978), has also been used successfully to estimate patient adherence.

Bioassay techniques are the standard by which other methods of assessing adherence are compared. Their principle advantage lies in the fact that they do not rely on patient self-report, which is the least reliable gauge of therapeutic adherence. The disadvantages of using this type of measure are several and relate to issues of practicality, reliability, and validity.

Practically, some substances do not presently lend themselves to laboratory analysis and require sophisticated pharmacological techniques such as radioimmunoassay (e.g., Weintraub, Au, & Lasagna, 1973) to be detected in the blood or urine. Cost can be a major deterrent to employing bioassay techniques routinely for adherence screening. For example, a single Dilantin blood level determination can cost $30.00. Finally, patients may not always cooperate with the procedure used to obtain the sample for bioassay, particularly if they are subjected to the procedure at frequent intervals.

Of more immediate concern to both the clinician and the researcher is the issue of the reliability of the bioassay procedure used to estimate adherence. Several investigators have shown that the results of drug bioassays can significantly vary from laboratory to laboratory using blood obtained from the same sample (Brett, 1977; Richens, 1975). This research points out the need to check independently the reliability of the laboratory results when conducting drug adherence research using bioassay estimates.

Assuming that the bioassay procedures are reliable still presents problems in interpreting the results in terms of the degree of adherence. Soutter and Kennedy (1974) point out that optimum therapeutic ranges for many drugs are routinely difficult to specify. Individual differences in metabolic rate and the timing of the bioassay in relation to the dosage schedule (particularly for drugs with short metabolic half-lives) are additional factors that can influence drug levels in the blood or urine. If the patient is taking other drugs which can induce or inhibit metabolism of the drug under investigation, further variability in the adherence estimate can be introduced. Finally, patients themselves can manipulate adherence estimates relying on bioassay data by inadvertently omitting or intentionally "loading" one or more doses of the drug just prior to testing. Thus, a patient who is generally adherent may be classified as nonadherent and vice versa.

Despite these objections to bioassay procedures, they are likely to be central to future studies of procedures to improve therapeutic adherence for two reasons. First, a carefully developed research protocol can circumvent most of these problems. For example, unscheduled sampling of the blood or urine can reduce the probability of obtaining "manipulated" data (Haynes, Sackett, Gibson, Taylor, Hackett, Roberts, & Johnson, 1976). Bioassay procedures that are less invasive such as salivary assay for theophylline (Eney & Goldstein, 1976) or nicotine (Evans, 1976) is another development that should lead to increased patient cooperation with repeated testing for adherence. Second, future research should focus on the correlation of a bioassay and a behavioral assessment of adherence with therapeutic outcome to broaden our understanding of the consequences of various degrees of nonadherence with therapeutic outcome. An illustration of how this second point might be developed within the context of a seizure clinic will be discussed later in this chapter.

Behaviorally-based observational methods provide another objective assessment of adherence that is only beginning to be used in clinical medicine (Russo, Bird, & Masek, 1980). Participant observers have been used to test the efficacy of an automated tablet-dispensing device (Azrin & Powell, 1969) and to assess the reliability of diet adherence data collected by a young diabetic (Lowe & Lutzker, 1979). Direct observation of a patient's behavior is often obtrusive or not practical, but combined with biochemical or physical evidence (e.g., weight or increased respiratory flow), it could be used initially or intermittently to begin to evaluate the relationship between therapeutic adherence and outcome.

IMPROVING ADHERENCE

Controlled studies investigating procedures to improve adherence that have employed an objective assessment of adherence are rarely reported in the research literature. Generally, the interventions tested in these studies have been designed from the health education or behavioral perspective. Collectively, this research has yielded some preliminary but promising methods for improving adherence.

Adherence with short-term therapy such as antibiotics for uncomplicated infections can be achieved if the physician, nurse, or pharmacist provides more detailed written or verbal information about the medication, dosage schedule, and importance of adherence (Colcher & Bass, 1972; Linkewich, Catalano, & Flack, 1974; Sharpe & Mikeal, 1974). However, the results with increased patient education for long-term therapy such as antihypertensive medications have been disappointing (Mc-

Kenney, Slining, Henderson, Devins, & Barr, 1973; Sackett, Haynes, Gibson, Hackett, Taylor, Roberts, & Johnson, 1975).

Providing patients with feedback (direct or by mail) regarding the results of blood level analyses of their medication has proved to be another effective short-term approach to establishing adherence with the therapeutic regimen (Gundert-Remy, Remy, & Weber, 1976; Lund *et al.*, 1964; Sherwin, Robb, & Lechter, 1973). This procedure may also be useful in maintaining adherence to long-term therapies; however, 12 weeks was the longest period of treatment in these studies, and follow-up was not conducted.

Gordis and Markowitz (1971) demonstrated that the comprehensive care approach characterized by increased accessibility to the same doctor produced no difference in the adherence levels of pediatric rheumatic heart patients maintained on long-term penicillin therapy when compared to a similar group seen infrequently in a speciality clinic. This study, while not strictly an example of a behavioral or health educational strategy, suggests that increased attention from the health care establishment (cf. Culliton, 1978) is a cost-inefficient approach for most therapeutic adherence problems.

Behavioral approaches have proved to be the most promising in solving therapeutic adherence problems, particularly those more complicated (i.e., with multiple behavioral components) or refractory to other forms of treatment.

Token economies (Ayllon & Azrin, 1968) have been successfully used to increase and maintain adherence to complex therapeutic regimens. For example, Lowe and Lutzker (1979) increased adherence to a foot-care program, special diet, and urine testing for sugar in a 9-year-old diabetic to 100% of the criteria established for each therapy using a memo and point system. Magrab and Papadopoulou (1977) used a token economy to effect adherence to dietary and fluid intake restrictions in children requiring hemodialysis.

Response cost procedures employing loss of monies deposited with the therapist contingent on nonadherence with Antabuse therapy (Bigelow, Strickler, Liebson, & Griffiths, 1976) and a vitamin C regimen (Epstein & Masek, 1978) have proved to be effective interventions for problematic medication adherence. In the former study, regular ingestion of Antabuse set the occasion for partial refunds of monies deposited with the therapist. This effected a significant reduction in alcohol consumption. In the latter study, subjects who normally took no medications of any kind adhered to a nontherapeutically prescribed vitamin C regimen rather than forfeit deposited monies.

Self-monitoring of the behavior requisite of adherence appears to be of limited usefulness, probably best employed as one of several components in a treatment program to improve adherence (Carnahan & Nugent, 1975; Epstein & Masek, 1978; Haynes *et al.*, 1976).

These early results of behavioral approaches to the problem of therapeutic nonadherence are encouraging, but suggest that considerably more research into developing effective procedures must be done. None of these studies begins to address cost-effectiveness issues, nor is follow-up usually conducted to assess maintenance of treatment effects. Furthermore, the feasibility of implementing most, if not all, of these procedures on a routine basis within the context of clinical medicine is doubtful. This is not to undermine the ingenuity and significance of these early attempts, but rather to suggest that practically implemented procedures are likely to be more enthusiastically received by the medical establishment. Future research should address this issue more directly, unless the clinical researcher is presented with unusually difficult adherence problems requiring more elaborate and costly interventions.

DETERMINANTS OF NONADHERENCE

A considerable amount of research effort has been devoted to the examination of factors which may be related to or characterize the nonadherent patient (Blackwell, 1976; Davidoff, 1976; Haynes, Taylor, & Sackett, 1979). The focus of these studies has been on demographic and psychological attributes, the type of illness, and the proposed therapeutic intervention. In the first exhaustive literature review by Sackett and Haynes (1976) of 246 studies of adherence to therapeutic regimens, they reported that only 4 of 193 factors showed a positive correlation with nonadherence. These factors included race, complexity of treatment, length of treatment, and the belief in the effectiveness of the prescribed treatment. Intelligence, side effects, and severity of illness were not predictive of adherence, although these findings are open to question since the majority of studies had failed to use objective measures of adherence. A study by Epstein and Masek (1978), which included several objective measures of adherence of adults to a vitamin C regimen, reported that only age and self-report of nonprescription drug use discriminated the adherent and nonadherent groups. Mattar and Yaffe (1974) reported that adherence among children is superior to that of adults, suggesting that perhaps parents feel more responsible for the health of their children than for their own. However, their data revealed that no specific factors such as age, sex, race, ethnic background, number of children in the family, parents' educational level, income, employment, or marital status had a direct bearing on adherence behavior.

Social behaviors, on the other hand, of either parents or children may be related to adherence. In both acute (Becker, Drachman, & Kirscht, 1972; Charney, Bynum, Eldridge, Frank, MacWhinney, McNab, Scheiner, Sumpter, & Iker, 1967; Francis, Korsch, & Morris, 1969; Gabrielson, Levin, & Ellison, 1967) or chronic (Elling, Whittemore, & Green, 1960; Gordis, Markowitz, & Lilienfeld, 1969; Heinzelmann, 1962) pediatric illnesses, the most consistent predictor of adherence was the patient's or parent's estimation of the seriousness of the illness. Previous experience with a severe illness was also more likely to increase adherence. The lack of experience with severe illness or absence of physical discomfort might lead patients to feel that their condition was less severe than was the case and might therefore contribute to nonadherence. Another social factor that has been implicated in relation to pediatric nonadherence in particular is family instability (Alpert, 1964; Macdonald, Hagberg, & Grossman, 1963). At least one study (Gray, Kesler, & Moody, 1966) reported that behaving in accordance with the expectations of friends or relatives may be a strong motivator for adherence with medical recommendations.

Specific characteristics of a disease have been thought to be related to adherence, although very few have actually been correlated with pediatric illness. Positive correlations have been related to hospitalization. As the degree of disability increases, the adherence to treatment also increases. When there appears to be a lack of symptomology, adherence decreases (Blackwell, 1976).

PEDIATRIC RESEARCH IN NONADHERENCE

Most examples of nonadherence from the pediatric literature involve oral medications and can occur at any point in the behavioral sequence once the prescription has been written. This includes failure to get the prescription filled, failure to take the medication or omitting doses, or discontinuing the course of treatment prematurely. Occasionally, the therapeutic outcome may be unaffected by the prescribed medications and nonadherence may therefore not adversely affect health. In most cases, however, nonadherence or partial adherence leads to extended illness, complications, and further cost. Subsequent discussion will attempt to analyze the reasons and proposed and potential solutions for these aspects of therapeutic nonadherence.

Otitis Media

Charney *et al.* (1967) reported that nonadherence with a penicillin regimen increased dramatically when the mother perceived the illness to be less serious than it was. Nonadherence rates of 48% for the "less serious" illness versus 28% for the "more serious" illness were reported. They further reported no relationship between adherence and duration of symptoms, the physician's estimate of the seriousness of the illness, the type of penicillin used, whether other medications were being used concurrently, and whether throat cultures were performed. More general variables such as age, sex, number of children in the family, any other illnesses currently in the family, the educational level of the parents, or the recency of a previous illness in the child were also unrelated to level of adherence. In a study involving 245 patients, Mohler, Wallin, and Dreyfus (1955) reported that 34% of 84 patients admitted taking less than the prescribed amount. The reasons given were as follows:

Felt well	37%	(31)
Carelessness	27%	(23)
Lack of money	17%	(14)
Child refused	11%	(9)
Instructions misunderstood	8%	(7)

They found that the adherence rate for children was approximately 68% as compared to 50% for an adult population with similar diagnoses.

Thus, in the case of middle ear inflammation, there is data to indicate that the perception of the seriousness of the illness by the mother is the most significant factor related to adherence. This suggests that increased education and instruction should be a routine preventive intervention for nonadherence in the pediatrician's office. A telephone follow-up timed to coincide with the end of treatment could also serve as a relatively inexpensive means of assessment of and consequence for adherence.

Rheumatic Fever

Studies by Elling, Whittemore, and Green (1960) and Gordis, Markowitz, and Lilienfeld (1969) have examined factors related to adherence with treatment for rheumatic fever, focusing upon social characteristics. They defined six factors which appeared to be strong predictors of the likelihood of nonadherence. These include: (1) being female, (2) being adolescent, (3) having several brothers and sisters, (4) not having been hospitalized for an acute episode, (5) having no restrictions imposed on general activities, and (6) being unaccompanied by a parent for their clinic visit. Gordis *et al.* found that if four of these factors were present, the incidence of nonadherence rose to approximately 90%. This study unfortunately does not readily point to practical solutions when this situation is detected.

In contrast, a study by Korsch, Gozzi, and Francis (1968) examined factors related to adherence from patients at a hospital emergency room that offer practical guidelines for the health care provider. In this condition, when the family is under stress, the quality of the physician–parent relationship appeared to affect the adherence outcome. The authors reported that in this situation children are likely to receive their medications if: (1) the mother feels that the physician has praised her as a mother; (2) the physician has established a friendly, easy communication with the mother; (3) the mother understands why the child needs the medications; (4) the physician respects the mother; and (5) the physician has been honest and open. Thus, the ability to communicate effectively appears to be more important than the supposed authority of the physician in motivating the appropriate adherence behaviors in parents. However, the parent–child relationship, after the parents have been effectively informed, remains a critical area of investigation if the child is to be considered adherent or nonadherent to a particular treatment.

Asthma

The broad range of frequency and severity of asthmatic episodes both across patients and in the same patient over time, as well as variations in a patient's ability to metabolize antiasthmatic pharmacological agents, makes it difficult to prescribe medication effectively. This difficulty is compounded when either the patient is nonadherent or there is no adequate method for determining serum levels of the medications. An innovative approach for one antiasthmatic medication was proposed by Eney and Goldstein (1976), who compared salivary and serum levels of theophylline and reported a nearly linear relationship ($r = .97$). Thus, measuring salivary theophylline provides a noninvasive technique which measures theophylline levels independently of other xanthines (coffee, tea, cocoa, etc.). In two studies involving a total of 90 ambulatory chronic asthmatic children, they found that when the patients knew that they were being monitored, adherence, defined as reaching the effective therapeutic serum level, rose from 11% to 41%. The percentage of total nonadherence dropped from 23% to slightly more than 6%. They attributed the continuing nonadherence in these patients to the absence of close parental supervision, the taste of the medication, the necessity of administration during school hours, or a decreased motivation to take medications when they were less severely stressed. These speculations parallel those of Elling *et al.* (1960) and Gordis *et al.* (1969), who reported nonadherence rising to 90% if these factors were present in the outpatient setting.

Cystic Fibrosis

Meyers, Dolan, and Mueller (1975) investigated the relationship between the degree of adherence to an antibiotic regimen and the severity of cystic fibrosis in children. Cystic fibrosis is almost always fatal before the 21st year of life. In this study, 61 children were followed over a two-month interval. On the basis of urine samples, 80% of the children were defined as completely adherent and 20% as either intermediate or poor. While other studies of adherence in chronic illnesses have noted that adherence generally decreases with illness duration and the complexity of medical treatment, the high degree of adherence in this study further supports the position that disease severity and the consequences of nonadherence are principal components in determining adherence.

Encopresis

The literature on encopresis reveals it to be a common childhood dysfunction which up until recently has resisted most forms of intervention, including psychotherapy (Easson, 1960; Pinkerton, 1958); suppositories, enemas, and mineral oil (Davidson, Kugler, & Bauer, 1963; Levine & Bakow, 1976); and bowel retraining (Levine & Bakow, 1976).

Levine (1975) had previously shown the heterogeneity of the encopretic population, and consequently Levine and Bakow (1976) pursued the development of a treatment program with broad applicability. In a study of 127 children between the ages of 4 and 16 years, they focused on a home-based approach that required substantial cooperation from both parents and child for several months.

The active phase of treatment began with a disimpaction procedure using hyperphosphate (Fleet's) enemas and bisacodyl (Dulcolax) suppositories in order to begin retraining with a well-cleaned-out bowel. Following this procedure, the children were required to take light mineral oil (30cc) twice each day and were told they must do so for at least six months. In addition, they were further instructed to take two multiple vitamins per day between oil doses to minimize potential malabsorption problems. Both parents and children were trained to recognize stool retention signs and to use supplemental oral laxatives if necessary.

Training consisted of habituating the children to the toilet by requiring that they sit on the toilet with adequate foot support for at least ten minutes, twice each day. Children under 8 were additionally reinforced by the use of a token economy consisting of a star chart in which each star represented a sitting episode and stars could be exchanged for other goods. Follow-up contacts were made at three weeks after beginning and then every four to eight weeks depending on the presence of symptoms. Results at one year revealed that 8% were unchanged, 14% showed some improvement, 27% showed marked improvement, and 51% were in remission.

What this study illustrates is that a close monitoring procedure coupled with reinforcement technology can result in substantial symptomatic improvement in a disorder which is difficult to manage. The more severe cases of incontinence and constipation, and children who were having additional difficulty in meeting the performance demands placed on their general or academic behavior, were most likely to resist management. Actual treatment failure was attributed to lack of adherence, the severity of the problem, and lack of immediate improvement.

Diabetes

Insulin-dependent diabetics require the most widespread alteration of daily habits, extending beyond the correct taking of medications at the appropriate times to balanced diets which avoid sugars, regular urine-testing for glycosuria, and attention to activity levels. Consequently, the incidence of nonadherence with all or part of the therapeutic regimen is high and difficult to treat. For example, Gabriele and Marble (1948) reported a 61% failure to follow dietary planning in children at a summer camp. Simonds (1975) also studied a population of summer camp attendees and reported that 43% of the diabetic children disregarded their therapeutic regimen due to either emotional or behavioral factors. The children in this study used false positive urine tests (sugar in the urine when in fact there was none) to receive additional attention or false negative reports to avoid parental confrontations. A second sample of diabetics revealed that 33% were falsifying urine test data to achieve desired consequences—extra sweets, increased activity time, avoidance of punishment.

The therapeutic regimen in diabetes requires the regular commission of several complex procedures and there are many occasions for nonadherence. As Simonds (1979) pointed out, these occasions include: (1)

dietary behavior, such as skipping meals, overeating, eating sweets, and improper food exchange; (2) *urine-testing*, including not testing and reporting false positives or negatives; (3) *injection behavior*, including the use of inappropriate equipment, skipping injections, injections at the wrong time, or inaccurate dosages; (4) *activity behavior*, including not adjusting dosages and diet in accord with activity levels or being excessively restrictive; and (5) *appointment behavior*, including missing appointments or not bringing test results.

A study by Lowe and Lutzker (1979) employed a multiple baseline design to investigate the effects of memos and a point system to increase adherence in a juvenile diabetic. The indexes of improvement were foot care, dieting, and urine-testing. Briefly, they reported that adherence to the diet increased to a high level and was maintained during the memo condition, while urine-testing and foot care reached 100% and were maintained by the introduction of a point system. Follow-up demonstrated a maintenance effect of 100% adherence.

This study not only demonstrates the utility of Simonds's (1979) analysis and the effectiveness of a behavioral intervention in an insulin-independent diabetic, but exemplifies a cost-effective approach to which the health care system is most likely to be responsive in the future. Additional research in each of the areas suggested by Simonds using similar behavior management techniques is clearly indicated.

Urinary Tract Infections

In one of the first studies of adherence in children with recurrent urinary tract infections (UTI), Daschner and Marget (1975) reported a significant relationship between nonadherence to an antibiotic regimen and incidence of recurrent bacteriuria despite long-term therapy. Long-term therapy has been suggested as better than short-term treatment in UTI (Normand & Smellie, 1965; Pryles, Wherrett, & McCarthy, 1964; Stansfeld & Webb, 1954), probably partially in response to pervasive nonadherence in the early clinical experience with short-term therapy. However, even poorer adherence to long-term therapy by mothers and children has since been documented by Becker *et al.* (1972).

Perhaps the most important data from the Daschner and Marget study is that even irregular use of antibiotics reduced the infection rate. Since the typical prescription requires 4 doses per day, the above study suggests that a simplified regimen might be more adaptable to a child's and/or parent's daily routine, resulting in higher rates of adherence and decreased infections. In a test of this approach, Caloza, Fields, and Bernfeld (1978) studied 548 patients with urinary tract infections under double-blind conditions. They found that 1 gm of cephradine orally twice a day provided serum and urine levels which were similar to those of 500 mg 4 times a day. The clinical responses were nearly identical for the two regimens, and the authors suggested that a twice-per-day regimen would provide equivalent protection to a 4-per-day regimen for this drug. This position is also supported by Breese, Disney, Talpey, Green, and Tobin (1974), and Derrick and Dillon (1973), and Shapera, Hable, and Matsen (1973).

Hemodialysis

As part of the overall management of children with renal failure, dietary management plays a critical role. Children who fail to adhere to their diets are at risk for heart failure, hypertension, hyperkalemia, azotemia, or bone disease. It is particularly important to provide adequate nutrients and calories to maintain growth (Holliday, 1972), but since fluid and electrolyte allowances are based on body weight, strict dietary control is necessary. Wilson, Potter, and Holliday (1973) have commented on the numerous adjustments an adolescent must face during hemodialysis which may seriously affect treatment outcome and lead to dietary abuse.

Armed with the knowledge that behavioral strategies have been applied successfully to other dietary problems (Stuart & Davis, 1972; Stunkard, 1972), Magrab and Papadopoulou (1977) examined the effects of a behavioral intervention (token economy) on dietary adherence in four children undergoing hemodialysis. They found that a token program resulted in significant positive dietary changes, better adherence to fluid-intake restrictions, and a reduction of intersession weight gains of 45%. The amount of weight fluctuation which is detrimental to proper management also decreased. The withdrawal of the token contingency led to weight gains and fluctuations between dialysis sessions. The very rapid change which resulted from the institution of the behavioral contingency and the apparent motivating nature of the tokens led the authors to conclude that token systems may be an effective way to manage dietary adherence in children.

Seizures

Because the majority of epileptics are seen in an outpatient setting, like diabetics, there is a growing recognition that inadequate seizure control is primarily a function of nonadherence to the medical regimen.

The tendency for long-term epileptic patients to be nonadherent has been discussed by Kutt (1974). In adults, Buchanan, Shuenyane, Mashigo, Mtangai, and Unterhalter (1977) reported a rate of 50%, as did Mucklow and Dollery (1978). In the pediatric population it is estimated at 33% (Freiman & Buchanan, 1978).

In the Freiman and Buchanan study, a comparison between medication adherence in a hospital and in an outpatient setting was conducted. In both groups a rate of dosage omissions of over 50% was reported, with the midday dose being the most commonly omitted. They reported that side effects of phenytoin, phenobarbital, or primidone did not contribute to omissions. Patients who denied omitting doses had a higher level of seizure activity, and 11 of 75 subjects had no detectable phenytoin levels.

Lund (1973) suggested that over 70% of epileptic patients would be seizure-free if the plasma level of phenytoin was in the range of 10–20 μ/ml. At least two groups (Bochner, Hooper, Sutherland, Eadie, & Tyler, 1974; Cook, Anderson, Poole, Lesser, & O'Tuama, 1975) have demonstrated that saliva levels are an accurate measure of plasma levels and could therefore be used to assess therapeutic levels and adherence. For example, Mucklow and Dollery (1978) found that only 32% of the hospital group and 19% of the outpatient group had saliva concentrations of anticonvulsants in the therapeutic range. Since the dosages provided were sufficient to produce the appropriate plasma level, and additional examinations ruled out rapid drug elimination as the reason for low salivary levels, the authors concluded, albeit tentatively, that poor adherence was the major cause.

In general, adherence is inversely correlated with the complexity of the regimen, and a strategy of fewer daily doses, a simple behavioral manipulation, should be investigated further, particularly since Cocks, Critchley, Hayward, Owen, Mawer, and Woodcock (1975) and Reynolds, Chadwick, and Galbraith (1976) have shown that the amount of phenytoin necessary to achieve seizure control and therapeutic plasma levels can be obtained by ingesting a single larger dose without risking increased side effects.

The reasons for nonadherence with anticonvulsant regimens in this disorder remain elusive. Initial poor prognosis, prolonged multidrug treatment, drug interactions, and side effects all have a role (Reynolds, 1978). One could also speculate that the cyclic nature of the seizures may lead patients to omit doses and thus lower plasma levels during interseizure intervals, making them responsible for less effective control of epilepsy than the currently available anticonvulsant drug armamentarium would appear to warrant.

SUMMARY: SOME GUIDELINES FOR FUTURE RESEARCH

As we pointed out earlier, considerably more research must be conducted before a basic understanding of the variables predictive of therapeutic nonadherence, effective procedures to measure and improve adherence, and the relationship of therapeutic adherence to health and illness can be realized. By way of summary, a guideline for the conduction of a medication adherence study is given in the hope that the methodological shortcomings of earlier research will be avoided and research in this important area stimulated.

There are a number of effective anticonvulsant drugs currently available for the treatment of epilepsy. However, as we have noted, nonadherence to long-term therapy in pediatric populations is a significant problem (Freiman & Buchanan, 1978). An outpatient seizure clinic at a tertiary-care hospital provides a useful setting for evaluating assessment and treatment methodologies relating to therapeutic nonadherence.

In this context, single-subject research designs are excellent tools for procedure development and validation. For example, a multiple baseline design specifying two groups of 6–8 subjects allows for sophisticated hypothesis-testing without committing large amounts of resources. Treatment is implemented systematically after varying baseline length across the two groups. Subjects are selected by type of seizure disorder, medication regimen, and previous objective evidence of nonadherence. Typically, Dilantin, Zarontin, and Tegretol are the most problematic medications from an adherence standpoint.

During baseline, a reliable monitoring system for seizure frequency is established. Self-monitoring of seizure behavior with reliability checks by a participant observer has proved to be a fairly objective assessment of seizure rate which provides important information about the correlation between stated adherence, objective evidence of adherence, and therapeutic outcome. A multiple assessment of adherence is proposed, using self-report through a medication diary and bioassay for the drug(s) in the blood. Each patient is informed that in addition to the regular obtaining of blood samples during clinic visits, occasional (once or twice during the study) unannounced home visits to obtain blood will be made.

Patients are seen at the beginning of the study and at 3-week intervals thereafter so that the effects of the treatment on adherence behavior can be evaluated more closely. Blood is drawn at each clinic visit, and patients receive state lottery tickets for keeping their appointments.

Data on seizure frequency and medication ingestion are mailed in weekly, and adherence with this procedure nets the patient a chance in a "clinic lottery" held in conjunction with clinic visits. The clinic lottery serves two functions. One function is to reinforce adherence with the data collection system. The second function is as a treatment for nonadherence. Objective evidence of adherence (i.e., blood levels of the drug) above the lower therapeutic limit results in the earning of additional clinic lottery chances. Prizes consist of inexpensive goods and services such as gift certificates. The probability of winning can be manipulated to be high at first and then lowered at a later point. The expense of the prizes might be secured through donations from businesses, charitable organizations, and research foundations. The additional initial start-up expenses are mainly linked to the cost of professional time (usually a clinic coordinator) to organize the program.

A follow-up study in which half of the subjects are maintained on the program and half are returned to baseline conditions is conducted to assess maintenance of treatment effects once the experimental contingencies are removed. A cost–benefit analysis of the program would be essential to determine whether the procedures would be appropriate for routine application in seizure clinics.

This proposal offers an outline of minimally acceptable design criteria for a therapeutic adherence study. The actual treatment proposed is relatively unimportant compared to the assessment and research design issues discussed. Other questions were left unaddressed but must be considered before a study is actually conducted, for instance, the reliability of the proposed objective assessment of adherence. Conducting methodologically sound and clinically useful therapeutic adherence research is a challenging task but a highly accessible one, particularly for those investigators working in larger medical centers.

What appears to be clear from the preceding review and discussion is that until the parameters which define adherent behavior are better understood and the assessment technology becomes more accessible, the efficient use of scores of existing medical treatments will not yet be realized. In view of over 10 years of concentrated research efforts into the problem of nonadherence with medical regimens, it is somewhat puzzling that so little has been accomplished. This observation may reflect the fact that nonadherence is a multifactorial social and behavioral problem making research efforts extremely difficult. On the other hand, what is probably needed is a reordering of priorities in the research effort, moving away from the descriptive to the procedure-testing approach described above.

REFERENCES

Alpert, J. J. Broken appointments. *Pediatrics*, 1964, *34*, 127–132.

Ayllon, T., & Azrin, N. H. *The token economy*. New York: Appleton-Century- Crofts, 1968.

Azrin, N. H., & Powell, J. Behavioral engineering: The use of response priming to improve prescribed self-medication. *Journal of Applied Behavior Analysis*, 1969, *2*, 39–42.

Becker, M. H., Drachman, R. H., & Kirscht, J. P. Predicting mothers' compliance with pediatric medical regimen. *Journal of Pediatrics*, 1972, *81*, 843–854.

Bergman, A. R., & Werner, R. J. Failure to receive penicillin by mouth. *New England Journal of Medicine*, 1963, *268*, 1334–1338.

Bigelow, G., Strickler, D., Liebson, I., & Griffiths, R. Maintaining disulfiram ingestion among outpatient alcoholics: A security deposit contingency contracting procedure. *Behaviour Research and Therapy*, 1976, *14*, 378–381.

Blackwell, B. Drug therapy: Patient compliance. *New England Journal of Medicine*, 1973, *289*, 249–252.

Blackwell, B. Treatment adherence in hypertension. *American Journal of Pharmacy*, 1976, *148*, 75–85.

Bochner, F., Hooper, W. D., Sutherland, J. M., Eadie, M. J., & Tyler, J. H. Diphenylhydantoin concentration in saliva. *Archives of Neurology*, 1974, *31*, 57–59.

Breese, B. B., Disney, F. A., Talpey, W. P., Green, J. L., & Tobin, J. Streptococcal infections in children: Comparison of the therapeutic effectiveness of erythromycin administered three times daily. *American Journal of Diseases of Childhood*, 1974, *128*, 457–461.

Brett, E. Implications of measuring anticonvulsant blood levels in epilepsy. *Developmental Medicine and Child Neurology*, 1977, *19*, 245–251.

Buchanan, N., Shuenyane, E., Mashigo, S., Mtangai, P., & Unterhalter, B. Factors influencing drug compliance in ambulatory black urban patients. *South African Medical Journal*, 1979, *55*, 368–373.

Caloza, D. L., Fields, L. A., & Bernfeld, G. E. Discontinuous antibiotic therapy for urinary tract infections. *Scandinavian Journal of Infectious Diseases*, 1978, *10*, 75–78.

Carnahan, J. E., & Nugent, C. A. The effects of self-monitoring by patients on the control of hypertension. *The American Journal of the Medical Sciences*, 1975, *269*, 69–73.

Charney, E., Bynum, R., Eldredge, D., Frank, D., MacWhinney, J. B., McNab, M., Scheiner, A., Sumpter, E., & Iker, H. How well do patients take oral penicillin? *Pediatrics*, 1967, *40*, 188–195.

Cocks, D. A., Critchley, E. M., Hayward, H. W., Owen, V., Mawer, G. E., & Woodcock, B. G. Control of epilepsy with a single dose of phenytoin sodium. *British Journal of Clinical Pharmacology*, 1975, *2*, 449–453.

Colcher, I. S., & Bass, J. W. Penicillin treatment of streptococcal pharyngitis: A comparison of schedules and the role of specific counseling. *Journal of the American Medical Association*, 1972, *222*, 657–659.

Cook, C. E., Anderson, E., Poole, W. K., Lesser, P., & O'Tuama, L. Phenytoin and phenobarbital concentrations in saliva and plasma measured by radioimmunoassay. *Clinical Pharmacology and Therapeutics*, 1975, *18*, 742–747.

Culliton, B. J. Health care economics: The high cost of getting well. *Science*, 1978, *200*, 883–885.

Daschner, F., & Marget, W. Treatment of recurrent urinary tract infection in children. *Acta Pediatrica Scandinavica*, 1975, *64*, 105–108.

Davidoff, F. Compliance with antihypertensive therapy: The last link in the chain. *Connecticut Medicine*, 1976, *40*, 378–383.

Davidson, M., Kugler, M. M., & Bauer, C. H. Diagnosis and management in children with severe and protracted constipation and obstipation. *Journal of Pediatrics*, 1963, *62*, 261.

Davis, M. S. Variations in patients' compliance with doctor's orders: Analysis of congruence between survey responses and results of empirical investigations. *Journal of Medical Education*, 1966, *41*, 1037–1048.

Derrick, C. W., & Dillon, H. C. Streptococcal pharyngitis: Comparisons of therapeutic agents b.i.d. and q.i.d. *Clinical Research*, 1973, *21*, 120–126.

Easson, W. M. Encopresis—Psychogenic soiling. *Journal of the Canadian Medical Association*, 1960, *82*, 624.

Elling, R., Whittemore, R., & Green, M. Patient participation in a pediatric program. *Journal of Health and Human Behavior*, 1960, *1*, 183–191.

Eney, R. D., & Goldstein, M. D. Compliance of chronic asthmatics with oral administration of theophylline as measured by serum and salivary levels. *Pediatrics*, 1976, *57*, 513–517.

Epstein, L. H., & Masek, B. J. Behavioral control of medicine compliance. *Journal of Applied Behavior Analysis*, 1978, *11*, 1–9.

Evans, R. I. Smoking in children: Developing a social psychological strategy of deterrence. *Preventive Medicine*, 1976, *5*, 122–127.

Francis, V., Korsch, B., & Morris, M. Gaps in doctor–patient communication: Patient's response to medical advice. *New England Journal of Medicine*, 1969, *280*, 535–540.

Freeman, J. Personal communication, June 1978.

Freiman, J., & Buchanan, N. Drug compliance and therapeutic considerations in 75 black epileptic children. *Central African Journal of Medicine*, 1978, *24*, 136–140.

Gabriele, A. J., & Marble, A. Experiences with 116 juvenile campers in a new summer camp for diabetic boys. *American Journal of Medical Science*, 1948, *218*, 161–171.

Gabrielson, I. W., Levin, L. S., & Ellison, M. D. Factors affecting school health follow-up. *American Journal of Public Health*, 1967, *57*, 48–59.

Gordis, L. Conceptual and methodological problems in measuring patient compliance. In R. B. Haynes, D. W. Taylor, & D. L. Sackett (Eds.), *Compliance in health care*. Baltimore: Johns Hopkins University Press, 1979.

Gordis, L., & Markowitz, M. Evaluation of the effectiveness of comprehensive and continuous pediatric care. *Pediatrics*, 1971, *48*, 766–776.

Gordis, L., Markowitz, M., & Lilienfeld, A. M. Why patients don't follow medical advice: A study of children on long-term antistreptococcal prophylaxis. *Journal of Pediatrics*, 1969, *75*, 957–968.

Gray, R. M., Kesler, J. P., & Moody, P. M. The effects of social class and friends' expectations on oral polio vaccination participation. *American Journal of Public Health*, 1966, *56*, 2028–2032.

Gundert-Remy, U., Remy, C., & Weber, E. Serum digoxin levels in patients of a general practice in Germany. *European Journal of Clinical Pharmacy*, 1976, *10*, 97–100.

Haynes, R. B. Determinants of compliance: The disease and the mechanics of treatment. In R. B. Haynes, D. W. Taylor, & D. L. Sackett (Eds.), *Compliance in health care*. Baltimore: Johns Hopkins University Press, 1979.

Haynes, R. B., Sackett, D. L., Gibson, E. S., Taylor, D. W., Hackett, B. C., Roberts, R. S., & Johnson, A. L. Improvement of medication compliance in uncontrolled hypertension. *Lancet*, 1976, *1*, 1265–1268.

Haynes, R. B., Taylor, D. W., & Sackett, D. L. (Eds.). *Compliance in health care*. Baltimore: Johns Hopkins University Press, 1979.

Heinzelmann, F. Factors in prophylaxis behavior in treating rheumatic fever: An exploratory study. *Journal of Health and Human Behavior*, 1962, *3*, 73–81.

Hobby, G. K., & Deuschle, K. W. The use of riboflavin as an indicator of isoniazid ingestion in self-medicated patients. *American Review of Respiratory Diseases*, 1959, *80*, 415–423.

Holliday, M. A. Calorie deficiency in children with uremia: Effect upon growth. *Pediatrics*, 1972, *50*, 590.

Korsch, B. M., Gozzi, E. K., & Francis, V. Gaps in doctor–patient communication: I. Doctor–patient interaction and patient satisfaction. *Pediatrics*, 1968, *42*, 855–869.

Kutt, H. The use of blood levels of antiepileptic drugs in clinical practice. *Pediatrics*, 1974, *53*, 557–559.

Levine, M. D. Children with encopresis: A descriptive analysis. *Pediatrics*, 1975, *56*, 412–417.

Levine, M. D., & Bakow, H. Children with encopresis: A study of treatment outcome. *Pediatrics*, 1976, *58*, 845–852.

Linkewich, J. A., Catalano, R. B., & Flack, H. L. The effect of packaging and instruction on outpatient compliance with medication regimens. *Drug Intelligence and Clinical Pharmacy*, 1974, *8*, 10–15.

Lowe, K., & Lutzker, J. R. Increasing compliance to a medical regimen with a juvenile diabetic. *Behavior Therapy*, 1979, *10*, 57–64.

Lund, L. Effects of phenytoin in patients with epilepsy in relation to its concentration in the plasma. In D. S. Davies & B. N. C. Prichard (Eds.), *Biological effects of drugs in relation to their plasma concentration*. London: Macmillan, 1973.

Lund, M., Jorgensen, R. S., & Kuhl, V. Serum diphenylhydantoin (Dilantin) in ambulant patients with epilepsy. *Epilepsia*, 1964, *5*, 51–58.

Macdonald, M. E., Hagberg, K. L., & Grossman, B. J. Social factors in relation to participation in follow-up care of rheumatic fever. *Journal of Pediatrics*, 1963, *62*, 503–513.

Magrab, P. R., & Papadopoulou, Z. L. The effect of a token economy on dietary compliance for children on hemodialysis. *Journal of Applied Behavior Analysis*, 1977, *10*, 573–578.

Marston, M. V., Compliance with medical regimens: A review of literature. *Nursing Research*, 1970, *19*, 312–323.

Masek, B. J. Compliance and medicine. In D. M. Doleys, R. L. Meredith, & A. R. Ciminero (Eds.), *Behavioral medicine: Assessment and treatment strategies*. New York: Plenum Press, 1982.

Masek, B. J., Epstein, L. H., & Russo, D. C. Behavioral perspectives in preventive medicine. In S. M. Turner, K. S. Calhoun, & H. E. Adams (Eds.), *Handbook of clinical behavior therapy*. New York: Wiley, 1981.

Mattar, M. E., & Yaffe, S. J. Compliance of pediatric patients with therapeutic regimens. *Postgraduate Medicine*, 1974, *56*, 181–188.

McKenney, J. M., Slining, J. M., Henderson, H. R., Devins, D., & Barr, M. The effect of clinical pharmacy services on patients with essential hypertension. *Circulation*, 1973, *48*, 1104–1111.

Melchoir, J. C. The clinical use of serum determinants of phenytoin and phenobarbital in children. *Developmental Medicine and Child Neurology*, 1965, *7*, 387–391.

Meyers, A., Dolan, T. F., & Mueller, D. Compliance and self-medication in cystic fibrosis. *American Journal of Diseases of Children*, 1975, *129*, 1011–1013.

Mohler, D. N., Wallin, D. G., & Dreyfus, E. G. Studies in the home treatment of streptococcal pneumonia. *New England Journal of Medicine*, 1955, *252*, 1116–1118.

Mucklow, J. C., & Dollery, C. T. Compliance with anticonvulsive therapy in a hospital clinic and in the community. *British Journal of Clinical Pharmacology*, 1978, *6*, 75–79.

Normand, J. C. S., & Smellie, J. M. Prolonged maintenance chemotherapy in the management of urinary infection in children. *British Medical Journal*, 1965, *1*, 1023.

Pinkerton, P. Psychogenic megacolon in children: The implications of bowel negativism. *Archives of Disease in Childhood*, 1958, *33*, 371.

Porter, A. M. W. Drug defaulting in a general practice. *British Medical Journal*, 1969, *1*, 218–222.
Pryles, C. V., Wherrett, B. A., & McCarthy, J. M. Urinary tract infections in infants and children: Long term prospective study: Interim report on results of six weeks chemotherapy. *American Journal of Diseases of Children*, 1964, *108*, 1–13.
Rapoff, M. A., & Christophersen, E. R. *Compliance with medical regimens: A review and evaluation*. Paper presented at the Banff International Conference on Behavioral Medicine, Banff, Alberta, Canada, March 1980.
Reynolds, E. H. Drug treatment of epilepsy. *Lancet*, 1978, *3*, 721–725.
Reynolds, E. H., Chadwick, D., & Galbraith, A. W. One drug (phenytoin) in the treatment of epilepsy. *Lancet*, 1976, *1*, 923.
Richens, A. Quality control of drug estimations. *Acta Neurologica Scandinavica*, 1975 (Supplement), *60*, 81–84.
Ruley, E. J. Compliance in young hypertensive patients. *Pediatric Clinics of North America*, 1978, *25*, 175–182.
Russo, D. C., Bird, B. L., & Masek, B. J. Assessment issues in behavioral medicine. *Behavioral Assessment*, 1980, *2*, 1–18.
Sackett, D. L., & Haynes, R. B. (Eds.). *Compliance with therapeutic regimens*. Baltimore: Johns Hopkins University Press, 1976.
Sackett, D. L., & Snow, J. C. The magnitude of compliance and noncompliance. In R. B. Haynes, D. W. Taylor, & D. L. Sackett (Eds.), *Compliance in health care*. Baltimore: Johns Hopkins University Press, 1979.
Sackett, D. L., Haynes, R. B., Gibson, E. S., Hackett, B. C., Taylor, D. W., Roberts, R. S., & Johnson, A. L. Randomised clinical trial of strategies for improving medication compliance in primary hypertension. *Lancet*, 1975, *1*, 1205–1207.
Shapera, R. M., Hable, K. A., & Matsen, J. M. Erythromycin therapy twice daily for streptococcal pharyngitis: Controlled comparison with erythromycin or penicillin phenoxymethyl four times daily or penicillin G benzathine. *Journal of the American Medical Association*, 1973, *226*, 531–537.
Sharpe, T. R., & Mikeal, R. L. Patient compliance with antibiotic regimens. *American Journal of Hospital Pharmacy*, 1974, *31*, 479–484.
Sherwin, A. L., Robb, J. P., & Lechter, M. Improved control of epilepsy by monitoring plasma ethosuximide. *Archives of Neurology*, 1973, *28*, 178–181.
Simonds, J. F. Study of diabetic attitudes in children and parents. In *Seventh allied health postgraduate course in diabetes*. Kansas City, Mo.: American Diabetes Association, 1975.
Simonds, J. F. Emotions and compliance in diabetic children. *Psychosomatics*, 1979, *20*, 544–551.
Soutter, R. B., & Kennedy, M. C. Patient compliance assessment in drug trials: Usage and methods. *Australian and New Zealand Journal of Medicine*, 1974, *4*, 360–364.
Stansfeld, J. M., & Webb, J. K. G. Plea for longer treatment of chronic pyelonephritis in children. *British Medical Journal*, 1954, *1*, 616.
Stuart, R. B., & Davis, B. *Slim chance in a fat world: Behavioral control of obesity*. Champaign, Ill.: Research Press, 1972.
Stunkard, A. J. New therapy for the eating disorders: Behavior modification of obesity and anorexia nervosa. *Archives of General Psychiatry*, 1972, *26*, 391–398.
Weintraub, M., Au, W. Y. W., & Lasagna, L. Compliance as a determinant of serum digoxin concentration. *Journal of the American Medical Association*, 1973, *224*, 481–485.
Wilson, C. J., Potter, D. E., & Holliday, M. A. Treatment of uremic child. In R. W. Wiaters (Ed.), *Body fluids in pediatrics*. Boston: Little, Brown, 1973.

Index